MANUFACTURING AUTOMATION

THE IRWIN SERIES IN PRODUCTION OPERATIONS MANAGEMENT

Aquilano, Chase, and Davis,
Fundamentals of Operations Management,
Second Edition

Chase and Aquilano,
Production and Operations Management: Manufacturing and Services,
Seventh Edition

Cohen and Apte,
Manufacturing Automation,
First Edition

Hill,
Manufacturing Strategy: Text & Cases,
Second Edition

Chu, Hottenstein, and Greenlaw,
PROSIM III for Windows,
Third Edition

Lambert and Stock,
Strategic Logistics Management,
Third Edition

Leenders and Fearon,
Purchasing and Materials Management,
Third Edition

Lotfi and Pegels,
Decision Support Systems for Operations & Management Science,
Third Edition

Melnyk and Denzler,
Operations Management: A Value-Driven Approach,
First Edition

Nahmias,
Production and Operations Analysis,
Third Edition

Niebel,
Motion and Time Study,
Ninth Edition

Sanderson and Uzumeri,
Managing Product Families,
First Edition

Schonberger and Knod,
Operations Management: Customer-Focused Principles,
Sixth Edition

Stevenson,
Production/Operations Management,
Fifth Edition

Vollmann, Berry, and Whybark,
Manufacturing Planning & Control Systems,
Fourth Edition

Zimmerman,
ABC Operations Software,
Beta Edition

THE IRWIN SERIES IN STATISTICS

Aczel,
Statistics: Concepts and Applications,
First Edition

Aczel,
Complete Business Statistics,
Third Edition

Bluman,
Elementary Statistics: A Step-by-Step Approach,
Second Edition

Bowerman and O'Connell,
Applied Statistics: Improving Business Processes,
First Edition

Bryant and Smith,
Practical Data Analysis: Case Studies in Business Statistics, Volumes I and II,
First Edition

Delurgio,
Forecasting: Principles and Applications,
First Edition

Doane, Mathieson, and Tracy,
Visual Statistics Part I,
First Edition

Duncan,
Quality Control & Industrial Statistics,
Fifth Edition

Cooper and Emory,
Business Research Methods,
Fifth Edition

Gitlow, Oppenheim, and Oppenheim,
Quality Management: Tools and Methods for Improvements,
Second Edition

Hall,
Computerized Business Statistics,
Fourth Edition

Hanke and Reitsch,
Understanding Business Statistics,
Second Edition

Lind and Mason,
Basic Statistics for Business and Economics,
Second Edition

Mason and Lind,
Statistical Techniques in Business and Economics,
Ninth Edition

Neter, Kutner, Nachtsheim, and Wasserman,
Applied Linear Statistical Models,
Fourth Edition and
Applied Linear Regression Models,
Third Edition

Shin,
The Irwin Statistical Software Series: Minitab, SAS, SPSS Guides, *Second Editions;* **Statgraphics,**
First Edition

Siegel,
Practical Business Statistics,
Third Edition

Webster,
Applied Statistics for Business and Economics,
Second Edition

Wilson and Keating,
Business Forecasting,
Second Edition

THE IRWIN SERIES IN QUANTITATIVE METHODS AND MANAGEMENT SCIENCE

Bodily, Carraway, Frey, Pfeifer,
Quantitative Business Analysis Casebook,
First Edition and
Quantitative Business Analysis: Text and Cases,
First Edition

Bonini, Hausman, and Bierman,
Quantitative Analysis for Management,
Ninth Edition

Hesse,
Managerial Spreadsheet Modeling and Analysis,
First Edition

Stevenson,
Introduction to Management Science,
Second Edition

Turban and Meredith,
Fundamentals of Management Science,
Sixth Edition

F I R S T E D I T I O N

Manufacturing Automation

Morris A. Cohen
The Wharton School
University of Pennsylvania

Uday M. Apte
Edwin L. Cox School of Business
Southern Methodist University

IRWIN

Chicago • Bogotá • Boston • Buenos Aires • Caracas
London • Madrid • Mexico City • Sydney • Toronto

Library of Congress Cataloging-in-Publication Data

Cohen, Morris (Morris A.)
Manufacturing automation/Morris Cohen, Uday Apte.
p. cm.
Includes index.
ISBN 0-256-14606-3
1. Manufacturing processes-Automation. 2. Production management-Data processing. 3. Technological innovations. I. Apte, Uday. II. Title.
TS183.C63 1997
670.42'7—dc21 96-29709

McGraw-Hill
A Division of the McGraw-Hill Companies

MANUFACTURING AUTOMATION

1 2 3 4 5 6 7 8 9 0 DOC/ DOC 9 0 9 8 7

ISBN 0-256-14606-3

Publisher: *Tom Casson*
Executive editor: *Richard T. Hercher, Jr.*
Editorial coordinator: *Alisa Watson*
Marketing manager: *Colleen J. Suljic*
Project supervisor: *Jim Labeots*
Production supervisor: *Karen Thigpen*
Designer: *Matthew Baldwin*
Cover designer: *Stuart Patterson*
Prepress buyer: *Charlene R. Perez*
Compositor: *Interactive Composition Corporation*
Typeface: *10/12 Times Roman*
Printer: *R. R. Donnelley & Sons Company*

http://www.mhcollege.com

D E D I C A T I O N

To my beloved wife Hedy for her wise counsel, understanding, and support for this particular project, as well as for all of the other ventures which bring us together,

and

To my father, Charles Cohen and to the memory of my mother Ethel Cohen, who have always encouraged me to seek learning and to pursue knowledge.

Morris A. Cohen
Bala Cynwyd, Pennsylvalnia
October 1996

To my wife, Aruna, for her love and support through the years, and to my father, Marutirao, and to the memory of my mother, Indira, for inspiration through their own writings and love of books.

Uday M. Apte
Dallas, Texas
October 1996

P R E F A C E

Manufacturing technology is a body of knowledge which affects the design, production, and delivery of goods and services. It is embodied in the equipment, processes, procedures, and methodologies used throughout the firm as it generates value for its customers. Changes to these technologies, based on automation and computerization, is accelerating and these changes have had a profound impact on company organization, the definition of work itself, and the expectations of consumers for product performance, value, and quality. This technology also has fueled the creation and demise of different industries and has altered the competitive structure of the survivors. Finally, the introduction of automation and computerization technologies has affected the productivity of all segments of the economy and thereby has influenced the overall prosperity of the national economy.

In spite of these influences, most companies still struggle with the management of technology. How much should a manager know about the underlying technology before he or she can make an informed decision concerning investment or deployment? How should cost, quality, and service be measured and managed in the new environment? How should existing business functions and processes be reengineered with the aid of the new technologies? This book is directed towards providing some answers to these questions.

This book was developed to support a course which provides students with an understanding, from a *managerial* perspective, of the emerging automation and information technologies which are used in production. The emphasis is to develop an appreciation of what these emerging technologies are and how they have been (or should be) used to provide firms with competitive advantage. The focus throughout is the operations function and, hence, the complete range of value-adding activities are scrutinized. Thus, we look both inside the factory walls and beyond, throughout the supply chain for examples of the application of technological innovation. We also consider cases drawn from the service sector and believe that the subject matter of this book is equally relevant for operations management in this sector as well. Our aim throughout is to balance discussion of current and emerging technologies with discussion of key managerial issues.

This book is an outgrowth of a course on Operations Technology which has been taught for 10 years at the Wharton School. Versions of the course have also been taught at Stanford University and at Southern Methodist University. This book can be used to support courses directed towards business administration, engineering systems, and/or industrial engineering students at either the advanced undergraduate or masters levels. It has evolved from the recognition that the current level of understanding of managing technology among business and engineering students is superficial and that the relevance of this topic is high, regardless of the ultimate career path of the student.

We have found that students with some prior knowledge of operations management, operations research, and computer science are adequately prepared for the material. In some cases, students without this background, but with a strong interest in the subject of technology have taken the course and have done well. The instructor can supplement the book with a wide variety of ancillary and supporting materials (some of which is included in the instructor's guide), to tailor the course and the use of the book to their specific requirements.

This book supports a flexible approach to teaching the subject, since we recognize that subject material concerning technology and its use in operations, has a short shelf life. Consequently, the book emphasizes basic principles concerning either the function of different classes of technology or management decisions in an enviromment where such technologies are in place. Wherever possible, we have introduced concept frameworks to support understanding of these principles.

SPECIAL FEATURES OF THE BOOK

- Close integration with managerial cases which provide in-depth insight into specific technologies, i.e., how they work, their use in operations, the challenges faced in their implementation, their impact (actual and potential), how they were justified, etc.
- Design of a web page which includes downloadable course materials (readings, lectures, software), Internet supported case discussion interactions, hot links to relevant company pages, and other academic web sites.
- Opportunities to use multimedia instruction materials such as videos, Internet sources, and interactive cases.
- Opportunities to use various software tools (e.g., simulation models of a factory, schedule optimization models, CAD systems, etc.) which can provide students with hands-on experience.
- A comprehensive set of supplementary readings which can be adjusted in response to changes in the underlying technologies and the relevant managerial issues.
- A comprehensive instructor's manual which includes detailed overheads of lectures and instructions for case discussions.
- Concept-building problems at the end of each chapter.
- An extensive bibliography of current titles drawn from both the business and academic literature concerning the management issues covered in this book.
- A comprehensive glossary of technical terms commonly encountered in the field of manufacturing automation.

ACKNOWLEDGMENTS

This book has benefited from the interchange of ideas among a wide range of colleagues. We acknowledge, in particular, the input of Ramachandran Jaikumar, of the Harvard Business School, who was one of the first to develop a managerial course in the area of factory automation more than 10 years ago. Suggestions have also been received and incorporated, along the way, from Michael Harrison, of Stanford University, and numerous colleagues at the Wharton School. We especially note the contribution of materials from Karen Donohue and Paul Kleindorfer. We would like to thank David Upton, from the Harvard Business School, who has freely shared his new case materials, articles, and ideas concerning the interface between technology and operations management with us.

Finally, we thank those individuals who helped by reviewing the final draft of this first edition: Joseph Blackburn (Vanderbilt University, Nashville, TN), Nicholas A. Glaskowsky Jr. (University of Miami, Coral Gables, FL), John McCreery (Ohio State University, Columbus, OH), Jeff Rummel (Duke University, Durham, NC), and John Usher (Mississippi State University, Mississippi State, MS). Their suggestions and comments have immensely helped us improve the book.

We would also like to acknowledge the efforts and patience of our editors at Irwin, Dick Hercher and Alisa Watson. They never faltered (throughout the many delays and diversions which were experienced) in the production of the manuscript.

The contributions of Marge Weiler, who acted as administrative assistant on this project must be especially acknowledged. She has provided us with support and valued suggestions throughout the process of writing and revising the material. Her good humor and insight into the human condition was especially important to us.

CONTENTS

MANUFACTURING AUTOMATION

INTRODUCTION

The processes associated with designing, producing, and distributing products, which we refer to as the operations function, are undergoing profound change. Many firms that were dominant in their markets for generations have gone out of business. Competition has increasingly become global. New products, based on developments in the basic and applied sciences, appear at a rapid rate. These new products compete on the basis of their quality and performance, as well as on cost. As a result, the survivors in the year 2000 and beyond must modify the practices and policies used to manufacture and distribute their products.

Against this backdrop of change, we see that there have been fundamental developments in the technology used to manufacture goods (process technologies) and to design the goods themselves (product technologies). These developments began with the Industrial Revolution, but, for the most part, have been occurring over the past 30–40 years. Information processing, for example, has become cheap, fast and ubiquitous. This has led to computerization of many of the basic manufacturing tasks (design, process analysis, scheduling, material control, process control, etc.). New raw materials also have been developed (e.g., composites, ceramics, plastics and alloys). Intelligence (whereby systems can learn and make decisions) has been introduced in process control and management systems. Similarly, electronic displays and data processors have been incorporated in the design of many products (e.g., fuel injection, voice recognition). Finally, the level of communication between and among processing units is reaching unprecedented levels.

Many observers feel that the development of automated manufacturing is just now reaching a level of implementation and accessibility which can lead to fundamental change. In particular, the competitive structure of industries is being altered. How then should firms respond as such change comes about? It is our view that *integration,* of diverse component technologies among business functions and even across corporate boundaries, is the key to making effective use of these emerging technologies. In this book, we shall explore the status of the component technologies for automating manufacturing and related operations functions, as well as issues related to managing their successful implementation.

The following questions have become critical as firms try to come to terms with the new realities of manufacturing technology:

1. What competitive strategies should a firm adopt to take advantage of the opportunities afforded by the development of the new process and product technologies?
2. How will enhanced process capabilities, material attributes and product performance, all enabled by technology, influence a firm's choice of new markets to pursue and new products to develop?
3. How should relationships with suppliers, customers, and competitors be modified to take advantage of the enhanced linkages enabled by the communication capabilities associated with information intensive operations technologies and the rapidly emerging world wide web of electronic commerce?
4. How should plant designs and equipment (shop floor) layouts be adjusted?
5. What is the new role of labor (both direct and indirect) in the automated manufacturing environment?
6. Are new organizational structures within the firm and among firms required?
7. Have existing accounting and performance measurement procedures become obsolete? If yes, how should they be changed?
8. What will be the impact of these technologies on society in terms of employment and profitability at the national and international levels?

Our purpose in this book is to address both the managerial and technical aspects of the questions raised above. We do so at a level of technical detail which is appropriate for an informed managerial perspective. Thus we will develop a framework for understanding the managerial implications of these technologies. We then will introduce a set of concepts to support their evaluation and where possible, specific methods and tools to assess their potential impact. Our approach does not mean avoidance of the technical details. Indeed, it is impossible to consider the management of the operations function without delving into the intricacies of construction and utilization of product and process technologies. At the same time, we aim to discuss technologies in a manner which will make this text accessible to a broad spectrum of readers in all functional areas of business.

The key issues to be addressed in this book are:

- **Selection:** There are a variety of competing technologies which affect the design of the product, the manufacturing process, and the host of supporting activities required to manufacture and sell the product. What are the choices? What are the criteria for selection? How should alternatives be evaluated and chosen?
- **Justification:** Manufacturing technologies require major investments of capital, attention, and enthusiasm. Such investments must ultimately prove to be

worthwhile in terms of their costs and benefits. Most firms require justification prior to investment as well as evaluations during and after implementation. How should these analyses be carried out? Are financial criteria relevant or have they become a barrier to successful use of the technologies? What other performance metrics are affected by these technologies?

- **Implementation:** New and innovative organizational procedures are required to use the emerging technologies successfully. Implementation takes place in the context of the firm's infrastructure of manufacturing resources for its existing product lines. Most failures of the new technologies can be traced to poor planning and execution of implementation.
- **Technology Strategy Consistency:** The range of technologies relevant to operations is diverse. Many of the technologies can be thought of as building blocks. While some can stand on their own, it is clear that the full benefit of individual technologies is enhanced if adoption and use are coordinated as a part of a coherent strategy for manufacturing technology management. This strategy must be made to be consistent with the firm's overall competitive strategy and its collection of functional strategies (e.g., marketing, finance, human resources).
- **Restructuring:** The realization of competitive advantage based on the impact of manufacturing automation will only occur if there is significant interaction among the various business functions of the firm. In order to automate manufacturing processes, firms must enhance communication, share common databases and achieve a high level of coordination in managerial decision making. As the costs and risks of technology investments increase, it is becoming increasingly clear that even the largest global firms will need to form joint ventures with other companies to share the burden. In this context, the adoption of automated (computerized) technologies for manufacturing and other value adding operations activities can act as a catalyst for fundamental restructuring of internal organizational structures and for forming cooperative relationships with other firms.

Outline of the Book

The fundamental purpose of this book is to address the problems raised and opportunities made available by modern manufacturing automation technologies. We will address these issues from a managerial perspective in order to develop the level of understanding of manufacturing technologies that is required for making informed managerial decisions. Therefore, the capabilities and performance characteristics of each technology will be considered in a sufficient level of detail so as to answer questions such as:

1. What technologies are currently available for automating operations (which includes manufacturing, distribution, supply, product design, engineering)?

2. What technologies will be available in the future and what capabilities will they afford?
3. How do the automated manufacturing technologies actually work (in terms of their use of information, their control systems, the role of operators, etc.)?
4. What are the costs and benefits of these technologies? Can they be evaluated?
5. How can individual technologies be integrated? What does it take to make automated manufacturing technology work successfully?

The book includes a collection of readings and lists of cases appropriate for an advanced undergraduate or graduate level course dealing with the management of manufacturing automation. The material is divided into eight chapters. The arrangement of this material includes key management processes, for example, product design, process design, production scheduling, distribution, and performance evaluation.

The managerial issues and our perspective on the evolution of manufacturing automation technology are presented in chapter one. This chapter also deals with the development of manufacturing strategy as a coherent link with the broader corporate business strategy. The second and third chapters deal with the technologies and concepts useful in product and process design, namely, computer-aided design and engineering, group technology, computer-aided process planning, and so forth.

The fourth chapter presents an overview of the key factory floor automation technologies such as computer numerical control machines, programmable logic controllers, robotics, flexible manufacturing systems, etc. The fifth chapter explains the rudiments of the infrastructure technologies that provide the foundation for using manufacturing automation technologies. Examples of these technologies include: control hardware and software, communications networks, communications standards and protocols, vision and nonvision sensors, factory floor data collection systems, automated material handling, and storage systems.

The sixth chapter discusses concepts and methodologies available for production planning and scheduling in an automated environment. In this chapter, we provide an understanding of such topics as capacity planning, aggregate planning, manufacturing resource planning, shop floor control, detailed scheduling, and simulation. The seventh chapter deals with the issue of interorganizational integration. Specifically, we discuss electronic data interchange (EDI), which has emerged as standard for the exchange of information with suppliers, dealers, and customers, and its impact on industry structure.

The eighth, and final, chapter explores the managerial issues raised in our analysis of technology. It includes discussions of the human resource issues associated with the introduction of manufacturing automation, justification, selection and acquisition of technology, implementation of technology and methodology for performance measurement, and cost control of manufacturing automation.

C H A P T E R

1 THE EVOLUTION OF MANUFACTURING TECHNOLOGY

1.1 Evolution of Manufacturing Technology

This chapter begins with a review of some recent history which highlights the impact of past technology developments on operations practices and performance. These lessons from history are instructive, since they indicate that the development of fundamental product and process technologies have led to major realignment of markets, the invention of new competitive organizational structures, and the development of new management paradigms. It is our contention that, today, the economy is in the midst of a new period of industrial restructuring.

Before we start our review of the evolution of manufacturing technology over the last two centuries, it would be useful to state our main hypothesis about the pattern of this evolution. We identify three stages of manufacturing technology: *Craftsman* (pre-1870), *Mass Manufacturing* (1870–1945), and *Automation Technology* (1945–present). The names of the manufacturing stages suggest the main characteristics of each stage. The associated time spans are, of course, approximate. We suggest that by understanding the characteristics of these stages we can identify a pattern of improvement along the fundamental *output* performance dimensions of manufacturing processes: product cost (or productivity), product variety, and volume or capacity. These improvements are meant to be cumulative in the sense that, as one moves from an earlier to a later stage, the advantages of the earlier stage are maintained while some improvements of the later stage are added. The manufacturing technology stages and main characteristics of each are briefly stated in Table 1.1. A stylized representation of the evolution of improvements is depicted in Figure 1.1. It suggests that progress involves improvement in all three performance dimensions.

It is important to note that assessment of the performance of a manufacturing system goes beyond the output performance dimensions noted above. The capabilities of any production system, with or without the use of automation technology, should be evaluated along multiple dimensions. Competition based on enhanced product quality, system flexibility, and cycle time reduction has become intense. Our interest in this book is to understand the linkage between such external (visible to the customer) met-

TABLE 1.1 Stages of Manufacturing Technology

Stage	*Characteristics*
1. *Craftsman* (Pre-1870)	High degree of manufacturing flexibility, product variety, and potential for quality were some of the main advantages. Limitations included low volume and high cost.
2. *Mass Manufacturing* (1870–1945)	
A. Scale Strategy	Parts standardization and mechanization of parts fabrication allowed for significant increases in volume. Production costs decreased, but manufacturing flexibility and product variety suffered.
B. Variety Strategy	Product variety and manufacturing flexibility improved while the advantages of cost and scale were maintained.
3. *Automation Technology* (1945–Present)	Automation of individual technologies allowed lowering of cost and improvement in quality. The degree of flexibility and product variety is maintained and in some cases improved. The main limiting factor is the difficulty of achieving coordination between individual automation technologies.

FIGURE 1.1

Evolution of manufacturing technology improvement

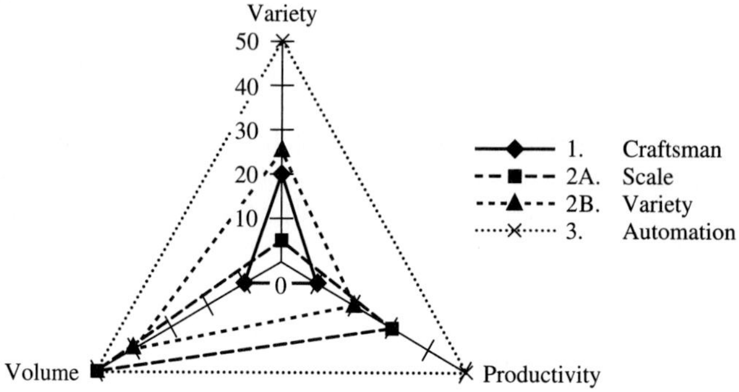

rics and those internal performance capabilities that are affected by the adoption of technology. The following seven *internal* performance attributes are relevant since many managers believe that these attributes are especially sensitive to technology:

- *Speed:* The rate at which output can be produced or material handled. The higher the speed, the less time it takes to complete the process, thus adding value for customers.
- *Power:* The ability to move and/or machine heavy, hard, raw material. Higher power potentially improves the efficiency of material transformation processes by reducing the resources required—manpower, time, and so forth.

- *Accuracy:* The degree to which product specifications can be adhered to. Accuracy requires low deviation from target levels concerning dimensions, physical properties, and so forth. Meeting specifications means higher quality; thus machines with higher accuracy can manufacture higher quality products.
- *Repeatability:* The ability to perform the same task in exactly the same manner over and over again. With high repeatability, the variability of output can be reduced, which means higher quality products.
- *Flexibility:* The ability to respond to changing market needs by having the capacity to produce a variety of products, using a variety of processes and materials. Flexibility allows a manufacturing enterprise to better deal with uncertain environments. As discussed below, the term *flexibility* can be understood and defined in many alternate ways.
- *Judgment:* The ability to recognize the state of the manufacturing system (e.g., operating condition is normal, machine tool is broken, material handling system has a malfunction, etc.) and to make an appropriate decision for generating a suitable action plan. Sound judgment is critical in dealing with contingencies that are inevitable in any manufacturing system. This ability is crucial in approaching the sometimes stated goal of factory automation—that is, to operate in a "lights out" manner without human intervention.
- *Learning:* The ability to learn from past experiences as they relate to outputs, processes, and inputs of the manufacturing system. Learning ability is particularly important in today's competitive world where making continuous improvement is the key to long-term survival and success.

Figure 1.2 indicates how the capabilities of automated machines compare with those of human beings. We can see that automated systems are generally superior in the areas of power, speed, accuracy, and repeatability, whereas humans are superior in the

FIGURE 1.2

Comparison of capabilities: Humans vs. automated machines

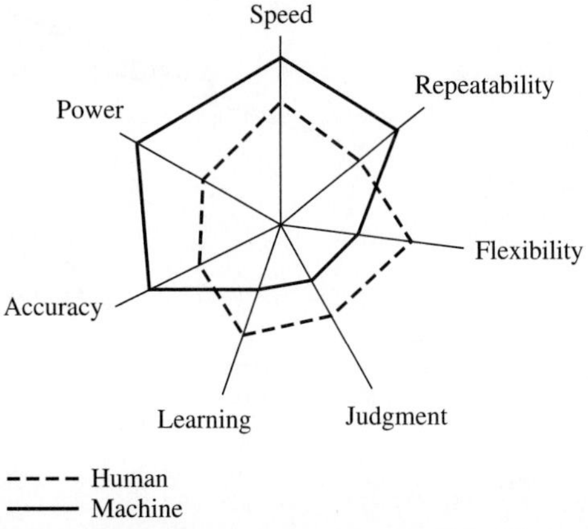

areas of flexibility, judgment, and learning. We will return to the issue of flexibility and its measurement later in this chapter. We also note that the most recent capability term which has gained currency is *manufacturing agility.* This term in many ways is an aggregate of all of the internal and external performance metrics that we have mentioned so far.

Ayres (1991) has collected some data that further illustrates the historical progression of *product complexity* (as measured by the number of parts) and *precision* (as measured by dimensional tolerances). Figure 1.3 illustrates the long-term trend in complexity. A Colt musket (ca. 1800) required about 50 parts, all of which were made from the same armory. An early sewing machine (ca. 1860) required less than 150 parts. Estimates of the number of parts used in an early automobile (ca. 1900) range from 1,500 to 2,000, while today's automobiles require more than 30,000 parts (only 10 to 15 percent of which are made by the nameplate manufacturer). A trend with comparable change has also occurred in the level of *precision/repeatability* that can be attained in parts fabrication. Ayres (1991) notes that in 1800 the limit was on the order of 100 microns. By the turn of the century, state-of-the-art equipment could achieve precision levels of 10 microns. Ten-micron precision in machining became attainable by the 1920s, and today we have surpassed the submicron level (0.01). Comparable tolerances are achieved in modern surface mount and integrated circuit applications.

Our review of historical developments suggests that a principal driving force behind the progress of manufacturing automation technology has been to develop pre-

FIGURE 1.3

Trends in product complexity

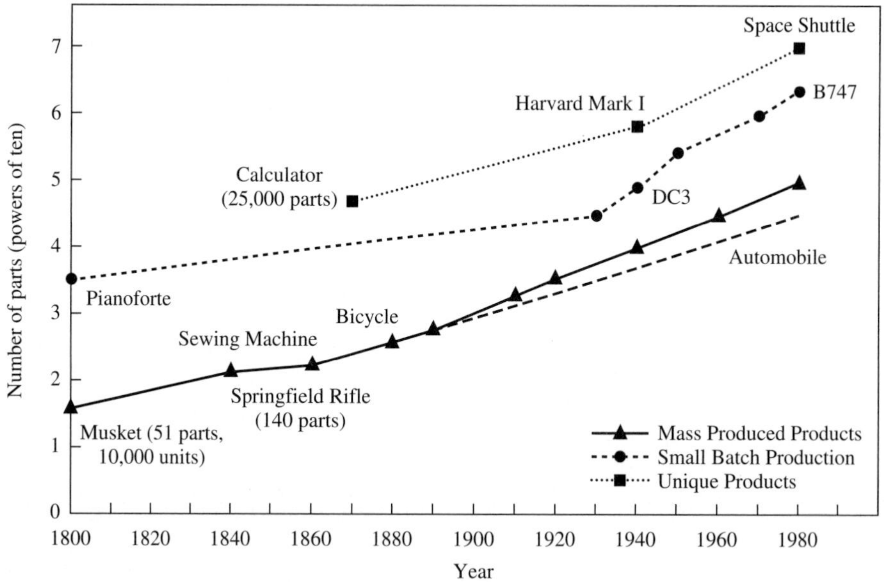

Source: Derived from "Computer Integrated Manufacturing, vol. I: Revolution in Progress", R. Ayres, 1991, Chapman & Hall, London.

cisely those capabilities where (automated) machines fall short of human workers. As we shall see throughout this book, significant progress has been made in terms of flexibility, albeit in a narrowly defined way. Yet, even today, manufacturing systems are far behind the abilities of humans in the attributes of learning and judement. Until such a time that automated system capabilities surpass all human capabilities (which may never occur), firms must rely on combining automation technologies and humans in an effective manner as they configure their manufacturing systems. Consequently, the role of human resources is still critical to the success of manufacturing systems. We review this role of human resources in Chapter 8.

Stage 1: Craftsman (Pre-1870)

A main feature of the early stage of manufacturing technology was the dominance of individual craftsmen. The production and delivery of goods and services were accomplished by a vast array of small, disconnected craftsman/owner-operated enterprises. These enterprises operated in well-defined geographic areas, dealt with single product lines, and served in well-defined market niches. The flow of goods among these enterprises was completely determined by market transactions. The craftsman was often assisted by one or two apprentices or journeymen who were treated as a part of the family. A distinguishing characteristic of this stage was that all stages of production and distribution—from product/process design, parts manufacturing and assembly, to packaging and distribution of final products—were controlled by a single craftsman.

A major advantage of craftsman-based manufacturing technology was its high level of manufacturing flexibility and product variety. A skilled craftsman using only simple tools was capable of producing a vast array of products and designs involving a number of different skills. While product quality depended on the level of craftsman skill, the potential for excellent quality was always present. The main limitations were that the production volume was necessarily low and the production costs were relatively high. As a result, manufactured goods were luxuries that were not affordable to the majority of the population.

Stage 2: Mass Manufacturing and Distribution (1870–1945)

American industry went through a revolutionary upheaval in the period 1870 to 1917. In one generation, the structure of industry changed in a fundamental manner. Through extensive vertical and horizontal integration, large production/distribution firms evolved. These new, integrated firms emerged almost overnight and operated on a national (and sometimes global) scale. They consisted of many operating units which spanned multiple stages of the production/distribution value-added chain. Raw material, intermediate good, and finished product flows were coordinated by a hierarchy of salaried managers. Out of necessity, these new corporations developed management practices (such as advertising, cost accounting, and logistics) to support the new, integrated organizational structure. As a result, the role of competitive market transactions in matching supply and demand was replaced, to some extent, by internal transactions within vertically integrated firms operating in oligopolistic markets. As Chandler (1977) puts it, we went from an economy where the "invisible hand" of the marketplace was replaced by the "visible hand" of management decision making.

Chandler suggests that these events were triggered by technology developments and not simply by the random arrival of gifted, avaricious entrepreneurs (i.e. the "Robber Baron" theory). From an economic perspective, the shift in industrial organization can be explained by the notion of *transaction costs.*[1] Consider a situation where a firm has a choice between market procurement (buy) or internal production (make) for each of its inputs. If the cost of acquisition within the firm (i.e., the internal production plus transaction cost) is less than the cost of purchasing the input in the marketplace, internal coordination will be used in place of market coordination. This shift in costs was enabled by technology developments. In particular, Chandler highlights the following changes:

1. Development of a *national communication/transportation system.* By 1870, national telegraph and the transcontinental railroad network were in place. Thus, for the first time in history, material and information flow on a continental scale was feasible.
2. Invention of *high-speed, continuous process manufacturing equipment.* This equipment supported production of standardized products on an unprecedented scale. For example, 15 Bonsack machines could produce enough cigarettes to supply the national market of 1880. Users of this equipment had to develop markets to absorb their vast output and had to design new products to take advantage of the capabilities of this equipment.
3. Application of *product support technologies*. For example, the invention of the refrigerated rail car made it possible to transport perishable products (like beef) from the Midwest to the eastern market. Rather than ship "beef on the hoof" to markets in the East, it was now possible to utilize large-scale, efficient meat-packing plants at the rail terminal cities which were closer to the sources of the raw material.

   Another example of the role of product support was the Singer company which invented a sewing machine for popular use during this period. The sewing machine was a highly sophisticated product (for its day), and its widespread adoption led to a breakdown of traditional retailing and distribution channels by independent wholesalers and retailers. Indeed, support for the sale and servicing of this product necessitated the development of a vertically integrated network of company-owned retail outlets. These outlets explained the product, taught consumers how to use it, repaired it, and provided owners with new ideas for its application (similar to personal computer dealers who emerged with the introduction of low-cost computers for home use).

[1] *See Markets and Hierarchies,* by O. E. Williamson (1975), for a discussion of transaction cost theory and how it affects firm structure.

Chandler observed that three classes of new organizational structures were established to take advantage of these technological developments. These structures were:

- *Mass distributors* who received material inputs from many producers and resold these products to many consumers through networks of warehouses and retail facilities,
- *Mass producers* who developed large scale, vertically integrated manufacturing capabilities, and
- *Integrated* mass producers/distributors.

It is interesting to note how the technological developments of this period affected operational performance capabilities. *Economies of speed* were achieved by sustaining a high volume of material flow. This, in turn, led to a high level of cash flow that was used to internally finance the firm's activities. Thus, first movers could take advantage of this cash flow to lower prices, thereby increasing market share, and to build large integrated organizations without surrendering equity. *Economies of scale* were also supported by the technology developments noted above. Large-scale production and distribution were also feasible. This led to lower unit costs which again supported the restructuring process described above.

More recently, Kogut (1992) examined the organizational capabilities that were required to validate Chandler's hypothesis that technology enabled competitive success through first movers' exploitation of scale and scope economies. His analysis of empirical data suggests that the level of organizational capability that existed in the United States during the relevant period was greater than that of other industrialized countries. As a result, a greater number of large-scale firms developed in this country.

Scale Strategy

Following 1917, we saw continued evolution of operations strategies. Again, technology was the key. The most significant event was the development of mass production by Henry Ford (actually introduced in 1913). The roots of mass production can be traced to the development of the Armory manufacturing system (ca. 1850) which was used to produce rifles for the U.S. Army (see Nevins and Whitney [1989]).[2] The Armory system introduced the following key concepts for manufacturing:

- **Interchangeability of parts:** Any part produced in a batch should be acceptable for assembly without requiring custom fitting or adjustment. This requires standardization of product/part design and close conformance to specifications in parts fabrication. Interchangeability thus reduces production costs and improves product serviceability, quality, and performance.
- **Mechanization of parts fabrication and product assembly:** While interchangeability itself does not necessarily lower costs, mechanization clearly does. Mechanization, of course, is not feasible without interchangeability. Mechanization can also lead to dramatic increases in productivity and capacity.

[2]Under the Armory system, suppliers were required to assemble working rifles from randomly selected parts. It took many years for competing vendors to reach this benchmark.

These concepts were demonstrated by early 19th century pioneers such as Eli Whitney and Marc Brunel (see *The Economist* [1994]). One of the first companies to exploit what came to be known as the "American" system of production was Beretta, the northern Italian arms producer, (see Jaikumar [1989]). Indeed, as a result of this change, Beretta was able to reduce its employment level from about 13 workers per machine to 3, with a 50 percent rework rate for three separate models.

Thus the Armory system enables production of high-volume, standardized products. It is important to note that it involves the transfer of worker production skills to machines and, as such, is an early case of automation in manufacturing. Higher levels of skills, however, are now required for product design, machine construction, and maintenance.

It was Henry Ford who exploited parts interchangeability and mechanization on a scale which had never been seen before. His mass production line for the Model T automobile revolutionized manufacturing. It is characterized by the following attributes:

- High volume (ultimately up to 2 million units per year).
- No "fitting" during final assembly (i.e., perfect interchangeability).
- A flow shop layout where all products follow the same path and sequence of operations.
- A moving assembly line.

It is interesting to note that the mass production line is an early example of *just-in-time (JIT)* material flow. Indeed, Toyota and other Japanese developers of JIT trace their roots back to Henry Ford. In Ford's system, it took 10 days for iron ore to be transformed into a car.

The productivity improvements of mass production were dramatic (e.g., magnetos that took 13 minutes to produce in the old system were produced in 5 minutes on the line, and a chassis that used to take 750 minutes to manufacture now took 93 minutes). Beretta adopted the Ford system in 1918 and realized a 3:1 productivity increase. They also saw rework reduce to 25 percent (from 50 percent) and attained the capability to produce 10 separate models. Such changes in mass production can lead to very low costs for a given standardized product design. It is still the only system which can accommodate production volumes of complex products in the order of hundreds of thousands of units per year.

The division of labor and the repetitive nature of mass production led to resistance on the part of factory floor workers. The use of machine-paced standards to control work flow places pressure on workers to maintain a high pace of operation. The labor-management problems of the automobile industry are still with us today and, in part, can be traced to worker resistance to the "dehumanizing" aspect of mass production.

Mass production lines are also characterized by specialized, dedicated equipment. In today's terminology, this is referred to as *hard automation.* Such lines are inflexible since any model change requires reconfiguration of the line, retooling, and other structural changes. As Henry Ford was fond of saying, "You can have any color you want as long as it is black."

The main features of mass production were: (1) low-cost, large-volume production of standard products and (2) low level of manufacturing flexibility and product variety.

Variety Strategy

A competitive response to Ford's low-cost, uniform product concept was the *model variety strategy* of Alfred Sloan of General Motors. During the 1920s, the American automobile industry had significant excess production capacity. In order to promote the sales of its products, GM developed a strategy of market segmentation whereby a separate product is targeted for each market segment. This strategy also involved annual model changeovers and a wide range of product designs.

The manufacturing system used by GM to support its strategy of variety and segmentation included use of generic machines with interchangeable workheads. As a result, different products could be produced by the same set of production equipment. In this way, GM completed retooling for its annual model changeover in three weeks. In contrast, when Ford introduced its first model change in 1926, the disruption, investment costs, and production delay almost ruined the company.

These developments in the automobile industry illustrate a fundamental trade-off between *flexibility and efficiency* in manufacturing systems. It is interesting to note that this trade-off is still crucial for understanding the costs and benefits of the new automation technologies which we are concerned with in this book.

Implementation of GM's system led to the development of cost accounting procedures for manufacturing. Overhead allocation, based on direct labor content, was used to compute the full cost of producing a unit. These practices became standard for manufacturing in the United States and abroad (and have only recently been challenged by activity based methods; see Chapter 8). Other managerial innovations of this era included:

- *Divisional* organization structure,
- *Return on investment* as a basis for justification (driven primarily by direct labor reduction associated with mechanization), and
- *Time standards* to optimize manufacturing performance (through assembly line balancing and worker productivity measurement).

These innovations illustrate again how a firm must adapt its management procedures and performance evaluation methods if it is to be a successful user of a new operations technology.

Stage 3: Automation Technology (1945–Present)

During the postwar period, especially during the period 1960–1990, American firms have been preoccupied with organizational and financial restructuring strategies: the "conglomerate mania" of the 60s, the "merger and acquisition mania" of the 70s and 80s, and the "right sizing" movement of the 90s. It is interesting to note that the role of operations policy in general, and manufacturing technology in particular, to enhance a firm's competitiveness has only recently been rediscovered by American firms. As we enter the latter half of the 90s, we see that use of information technology is accelerating and its role as enabler of corporate reengineering has been enhanced (see Davenport [1993]).

In contrast, the historical development of industry in Japan after World War II exhibits a strong and earlier linkage between manufacturing technology, production process selection, business strategy, and competitive performance. At each stage in this industrial development process the appropriate manufacturing technology was adopted. As noted by Stalk and Hauk (1990), the following sequence of manufacturing strategies was adopted in Japan:

1. 1945–1960: *Low Labor Cost Strategy.* Labor-intensive industries, such as textiles, steel, and shipbuilding were targeted to lead the reconstruction of the Japanese economy. They exploited the then inexpensive pool of labor.
2. 1960–1965: *Scale-Based Strategy.* Capital-intensive industries dominated. The investment in capital led to increased labor productivity.
3. 1965–1970: *Focused Factories.* Product lines were narrowed and high-volume markets for the surviving products were targeted (e.g., ball bearings).
4. 1970–present: *Flexible Factories.* A strategy based on low volume and high variety at world competitive prices.

During the postwar period, the economies of other industrialized nations in Europe and the Pacific Rim also followed paths of development that placed emphasis on operations and manufacturing technology.

In spite of these regional differences, the pace of technological development in manufacturing has accelerated throughout the world. With the development of computers and communications technology over the last three decades, the use of *automation technology* in all phases of product life cycle has flourished. Examples of these automation technologies include:

- **Product and process design technologies:** Computer-aided design and engineering, group technology, computer-aided process planning.
- **Factory floor automation technologies:** Computer numerical control machines, robotics, flexible manufacturing systems, and so forth.
- **Planning and control technologies/software:** Inventory control systems, manufacturing resource planning systems, shop floor control, scheduling systems.
- **Infrastructure technologies:** Communications standards and protocols, automated materials handling, sensors.

The term *computer integrated manufacturing (CIM)* has come to be associated with many of these developments. We will consider the various aspects of CIM in detail in subsequent chapters. Figure 1.4 illustrates the components of the CIM "wheel," in terms of the technology classification noted above. The three wheels at the perimeter of the CIM wheel refer to the first three automation technology groups. These technologies are integrated through the center hub of infrastructure technologies. See Table 1.2 for definitions of individual automation technologies.

FIGURE 1.4

The CIM wheel

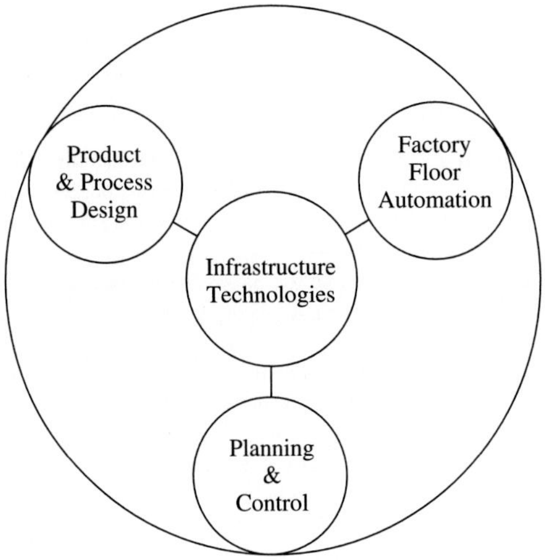

The automation technologies noted in Table 1.2 have the potential to impart a variety of benefits. Automation of individual processes, used in various phases of the product life cycle, can lower costs and improve quality while maintaining the volume characteristics of mass manufacturing. The level of manufacturing flexibility can also be improved, where *flexibility* is defined in general as "the ability to change or react with little penalty in time, effort, cost or performance" (Upton [1994]). This definition can be particularized to operations contexts by referring to various production system capabilities, for example, the ability to switch from one product to another without cost or significant down time, the ability to produce in lot sizes as small as one, the ability to respond to customer orders with reduced lead times. Indeed, the notion of flexibility has been classified extensively (see Browne, et al. [1984], Cohen and Zysman [1987], Gerwin [1985], Noori [1990], Swamidass [1985]). Based on these studies, we can generalize the notion of flexibility in terms of three areas:

- *Product Flexibility:* ability to deal with changes in product mix or range, product design, innovation, product volume.
- *Process Flexibility:* ability to deal with changes in machine, routing, raw material.
- *Infrastructure Flexibility:* ability of the organization itself to adapt to changes in its structure and deal with changes in automation technology.

Table 1.3 illustrates a framework for linking these dimensions of flexibility to the major sources of uncertainty in the manufacturing environment. One can view the flexibilities noted here as a basis for the capability for a manufacturing system to respond in a contingent fashion to these uncertainties.

TABLE 1.2 Automation Technologies

Technology	Description
Computer-Aided Design (CAD):	Computer software systems for assisting product designers. These systems (1) support design and layout of products and components, (2) display and manipulate product images, (3) create drafts of completed designs, (4) develop specifications for manufacturing the product.
Computer-Aided Engineering (CAE):	Assists design engineers in selecting components and materials for products and in performing engineering analysis, that is, mathematical modeling and analysis of performance of designed products.
Group Technology (GT):	A system for classification and coding of parts and products based on geometric, engineering, manufacturing, material, and other attributes. GT software systems include a coding scheme as well as software to support classification and retrieval of parts data (designs, routings). GT also refers to plant layout of equipment into cells, which are characterized by equipment dedicated to a collection of products.
Computer-Aided Process Planning (CAPP):	Use of a computer to develop a detailed process plan for production of a part or assembly.
Automated Materials Handling (AMH):	The automated movement of materials, parts, or assemblies through the use of powered equipment, conveyers, or robots.
Vision Sensors:	Devices used for optical, noncontact sensing to automatically receive and interpret an image of a real scene, to obtain information, and/or to control machines or processes.
Programmable Logic Controllers (PLC):	A programmable device (computer) which is used to control the operating functions (e.g., on/off, timing, sequencing, motion) of individual pieces of equipment and processes.
Computer Numerical Control (CNC) Machines:	Machines in which a computer provides instructions for controlling the position and speed of machine tools through the use of numbers and symbols.
Robot:	A programmable device with a multifunction manipulator that can move parts or tools from one point to another. Functionally, today's robot resembles a human arm.
Local Area Network (LAN):	A communication system that supports high speed data flows in any office, building, factory, campus, and so forth. It connects input and output devices under a variety of passively linked topologies.
Flexible Manufacturing System (FMS):	A system that incorporates many individual automation technologies and displays flexibility in producing a variety of parts. The FMS typically consists of CNC machines and AMH systems operating under the control of a central computer.
Material Requirements Planning (MRP) System:	A system for creating a time-phased plan for production, or procurement, of required quantities of materials, parts, or subassemblies. The MRP system uses the master production schedule of products, the bill of materials, and the current inventory levels of various materials, parts, and subassemblies for creating this plan.
Manufacturing Resource Requirements (MRPII) System:	An integrated, hierarchical approach to link business planning decisions to manufacturing and nonmanufacturing functions. Coordination and information flow is supported to link manufacturing related decisions such as material sourcing, aggregate production scheduling, factory capacity load planning, and distribution and order fulfillment to nonmanufacturing functions (e.g., accounting, sales).

TABLE 1.3 Flexibility Classification and Advantages

Environmental Uncertainty	*Type of Flexibility*	*Potential Strategic Advantage*
Product volume demanded	PROCESS	Respond to market sales variations
Delivery time required by customer Delivery of input materials	PROCESS	Respond to short lead times required by customers and unreliability of suppliers
Machine downtime	PROCESS	Delivery and dependability
Degree of market segmentation	PRODUCT	Demand for product diversity and customization
Length of product life cycles	PRODUCT	Respond to innovation and competitor actions
Customer specifications required	PRODUCT	Competition based on product value
Raw material quality	PRODUCT	Product quality and consistency
End-market information made available	INFRASTRUCTURE	Responsive to true demand signal
Appropriate organization design and incentives	INFRASTRUCTURE	Supports learning and adoption of technological change

Upton (1994) notes that pinning down the notion of flexibility is difficult and that efforts to attain it often are not successful. He argues that achieved flexibility is, in fact, highly situation specific. Accordingly, Upton suggests a framework in which management begins by defining the particular *dimension* or attribute of the system for which flexibility is required (e.g., product design specifications, factory product mix). The next issue is the *time horizon* over which changes will occur (minutes, days, weeks, years). Finally, it is necessary to specify the specific *elements* of flexibility (as noted in Table 1.3), that are actually required.

It has been argued that the set of manufacturing automation technologies introduced in this chapter can be combined in various ways to raise the level of a firm's flexibility. Upton (1995) reports on an extensive analysis of flexibility in the paper industry which challenges the premise that investment in automation leads to enhanced flexibility. Indeed, his data showed that paper plants with higher investment in CIM have less range (breadth of product characteristics producible) and longer changeover times. Moreover, these two aspects of flexibility appear to be uncorrelated. Therefore, the attainment of flexibility will require a careful blend of humans and computers on the factory floor.

The actual diffusion of technologies associated with operations automation is quite uneven. Some surveys indicated that Japanese manufacturers set the pace in the use of robots and flexible manufacturing systems in the 70s and 80s. On the other hand, manufacturers in Europe and the United States have made greater use of more ambitious, fully integrated "factories of the future" (see the *Economist* [1987]). More recently, the development of real time systems that reach beyond the factory to the customer and to suppliers has occurred in leading companies (see *The Wall Street Journal* [1990]). Globalization has also increased the rate of diffusion of certain automation technologies in developing countries.

FIGURE 1.5

Technology Penetration (% Using)

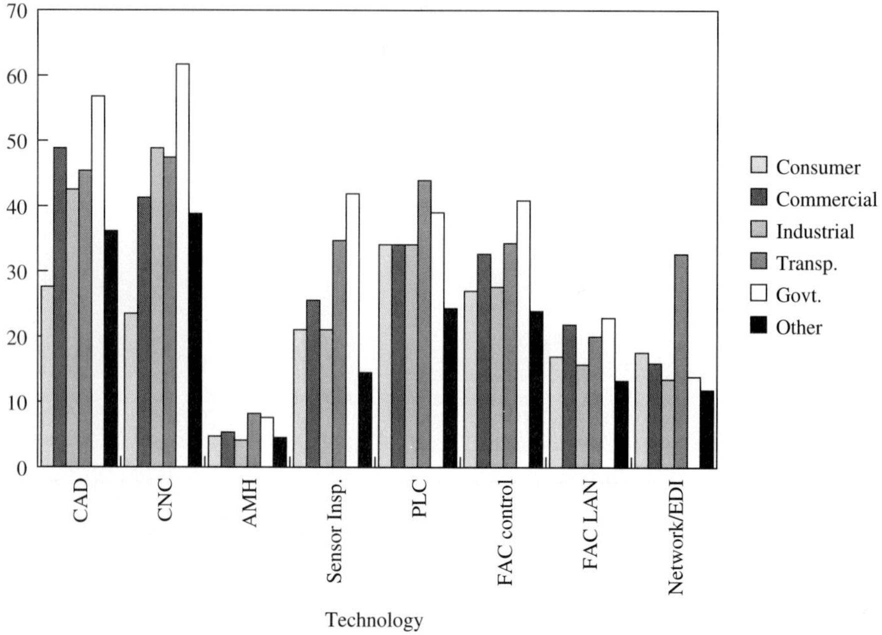

Source: U.S. Bureau of the Census, Current Industrial Reports, Manufacturing Technology (1988)

Figure 1.5 indicates the penetration of various CIM technologies in major industries in the United States in 1988. Not surprisingly, the most pervasive technologies are CAD and CNC. Variation by market for most products suggests that the government and the transportation industries are leaders in the use of "high-tech" products. Based on these data, it is also clear that there is a considerable distance to go before we can declare that automation technologies have become dominant.

1.2 Today's Challenges

What, then, are the problems, organizational obstacles, and opportunities associated with adoption of automation technologies in manufacturing, and how should they be addressed? As noted in the introduction, management is faced with the need to answer some fundamental questions in order to develop an effective manufacturing automation strategy and program.

It is our thesis that *coordination* and *integration* are critical for successful adoption of automation technology. Observers such as Hayes and Jaikumar (1988) have noted that linkage of the individual technologies often has been delayed because there are no universally accepted communications standards. Without the ability to connect and coordinate the use of technologies throughout the manufacturing process, benefits will be limited. In Chapter 4, we examine such infrastructure technologies in more detail.

Justification is also an important issue. Many observers have suggested that inappropriate cost accounting and use of inadequate financial analysis tools have limited the introduction of automation technology (see Cooper and Kaplan [1991]). We will consider this factor in Chapter 8, which focuses on the challenges in quantifying the costs and the benefits of automation.

Planning for implementation is also crucial. Many firms have found it difficult to move from prototype to full-scale production. This, in part, has been a result of poor predictions of resource needs and bottlenecks. The recognition that organizational changes are needed to support automation has been slow, and the inevitable resistance to such changes has occurred. As Kogut (1992) noted in his study of U.S. firms at the turn of the century, it was such organizational developments which determined who would be a success or failure in using the new technologies of the day.

The suggested implementation strategies for CIM come in many flavors, but they all require at least three basic steps: (1) *simplify,* (2) *automate,* and (3) *integrate* (or, alternatively, wait for the competition to *obliterate* you). By *simplification,* we means the preconditions needed for successful adoption of an automation strategy. Thus, waste in the process must be removed. As noted in Suzaki (1987), Toyota has identified the "seven wastes" of manufacturing. They include:

1. *Waste from overproduction:* this leads to excessive use of manufacturing resources.
2. *Waste of waiting time:* it has been estimated that only 5 percent of total time spent in a factory results in added value.
3. *Transportation waste:* through double handling and increased travel distance.
4. *Processing waste:* unnecessary production steps can be eliminated.
5. *Inventory waste:* in addition to idling resources, inventory works to hide production inefficiencies and quality problems.
6. *Waste of motion:* layout and material flow patterns can significantly affect throughput.
7. *Waste of product defects:* which increases cost (through rework) and can decrease sales due to customer dissatisfaction.

Strategies for the elimination of waste (prior to the introduction of automated technology) include adoption of *JIT* and *TQM (total quality management).* JIT leads to the development of manufacturing environments whose operational performance approaches that of continuous flow systems (e.g., an oil refinery).[3] Critical steps here include the reduction of production batch setup time, product/process redesign, and use of Kanban ("pull") inventory controls.

TQM adopts a comprehensive system viewpoint of the firm. At each stage in the value-added chain, performance goals are set, measurements are made and actions for improvement are adopted. A market-driven approach, or customer orientation, is used to link the performance of each stage to support competitive requirements.

---

[3]Continuous flow processes are distinguished from discrete processes (batch flow or job shops) in terms of the manner by which work moves through the facility. In a continuous flow plant material moves from process step to process step without interruption.

It is important to avoid using manufacturing automation as a substitute for good manufacturing practice. The procedures noted above for simplification are an essential prerequisite for the successful implementation of technological solutions. They, in their own right, do much to improve a firm's competitive situation. By linking simplification with the subsequent steps of automation and integration, firms will be able to derive additional competitive advantage. These latter two steps are the major concern of this book.

In the *automate* step, firms install the appropriate technology(ies) drawn from the CIM wheel. Clearly, an automation plan is a requirement. This requires knowledge of what technologies are available and how they can contribute to improved performance.

Finally, *integration* refers to the connection of *islands of automation* through material and data links. It is based on an enhanced level of coordination in making manufacturing control decisions. It also requires a high degree of functional interaction. Such integration will be enabled through installation of information and material handling networks (i.e., the "hub" of the wheel) and the adoption of more comprehensive control architecture. In addition, firms must develop new organization structures (such as teams) and incentive schemes for the automated manufacturing environment.

One successful model for CIM planning is the hierarchical *pyramid* approach used by automation vendors such as Allen Bradley (see Figure 1.6). At the bottom level, the firm must select individual *machines* and *processes.* The next highest level combines these pieces of equipment into automated *workstations.* Work stations, in turn, can be combined to form functionally oriented *cells.* Such cells require a high degree of coordination among multiple stations. Managerial issues concerning production scheduling and factory information management are prominent at the next level, which consists of groups of cells organized into *divisions or departments.* Integrated software systems are typically used to manage the flow of material and work at the division level. Finally, at

FIGURE 1.6

The pyramid—a hierarchical system for manufacturing automation

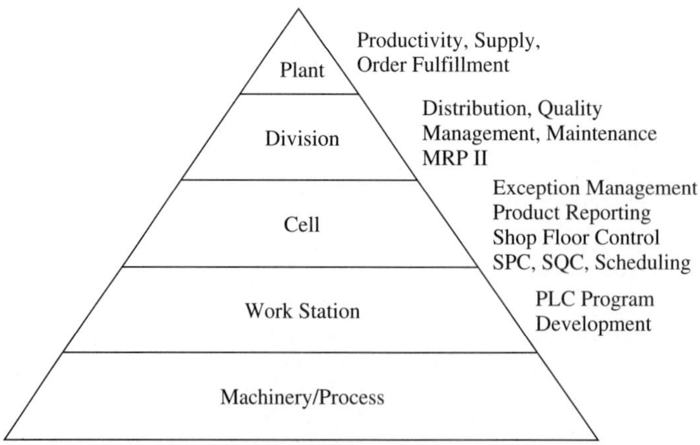

the peak of the pyramid, we have a *factory level view* of the automated enterprise. At this level, management is concerned with strategy and planning. (See Avishai's 1990 interview with Tracey O'Rourke of Allen Bradley and the Allen Bradley case by Goldstein and Klein [1987], for a description of the application of this structure to the automation of a switching plant).

Another useful framework for understanding successful technology strategy in manufacturing is the *learning laboratory* concept introduced by Leonard-Barton (1992). Figure 1.7 illustrates the four major knowledge creation activities in a factory which acts as a learning laboratory (i.e., integration of internal knowledge, integration of external knowledge, independent problem solving, and continuous innovation and experimentation). Leonard-Barton documents how these four activities have been used, with brilliant success, by Chaparral Steel, one of the leading mini-mill operators in the U.S. steel industry. The reader is referred to the discussion of the case in Leonard-Barton and Preuss (1992), for a description of the application of this framework.

FIGURE 1.7

Knowledge creation and control activities in a learning laboratory

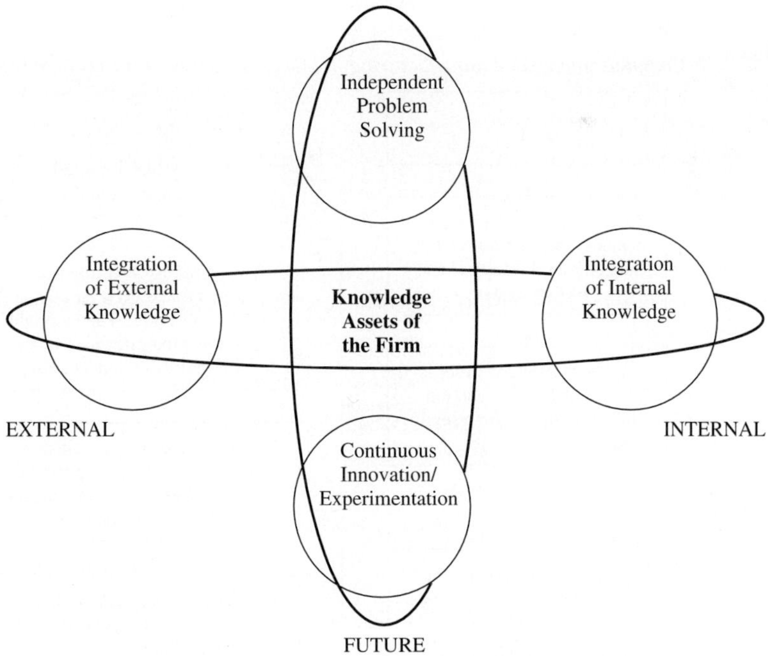

Source: Dorothy Leonard-Barton, "The Factory as a Learning Laboratory," *Sloan Management Review*, Vol. 34, no. 1, Fall 1992. Reprinted by permission of publisher.

Recently, Hammer and others (see Davenport [1993] and Hammer and Champy [1993]) have popularized a new approach for improving firm performance. This approach, known as *reengineering,* seeks to make major improvements in metrics such as cost, quality, and responsiveness, and is based on fundamental restructuring of the key processes that a firm uses to add value to its outputs. This radical restructuring is enabled by advances in information technology (IT) and is thus closely related to the themes of this book. The emphasis in reengineering on process analysis, organizational change, customer focus, and use of technology to enhance communication, coordination, and efficiency is shared by our view of technology management.

It is also interesting to note how firms are using automation technologies to respond to the increased competitive pressure for product variety. Panasonic Bicycle (see Fisher [1993]), for example, has developed a strategy of *mass customization* that allows them to build custom designed bicycles, to order, through the use of automated design, processing and communication technologies.

One emerging vision is that of a highly automated factory that is a part of a flexible, globally integrated organization. The firm of the future will design, produce, and deliver products rapidly in response to changing markets, competitive situations, and technology developments. These products, moreover, will be produced with zero defects, will use minimal inventory levels in the production/distribution process, and will be deliv-

FIGURE 1.8

Computer integrated manufacturing (CIM) from customer order to delivery

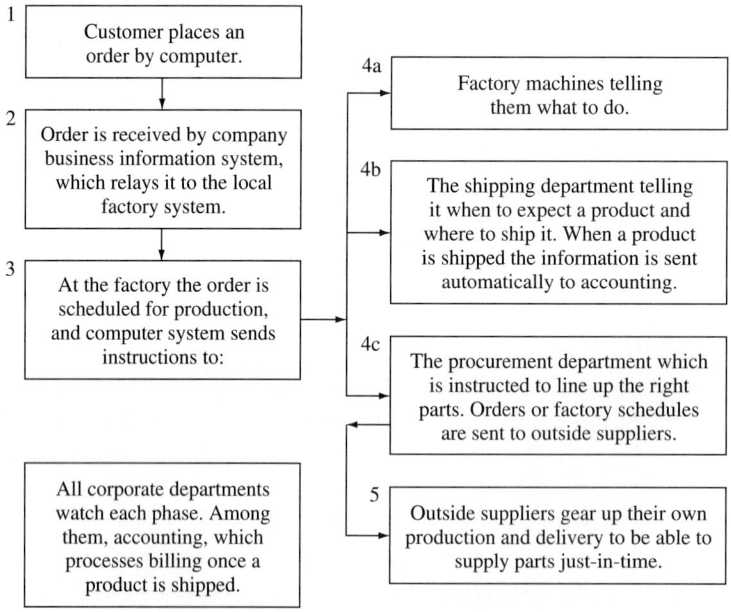

ered within highly compressed production lead times. There is some evidence to suggest that this future, in fact, is emerging. The integrating role of information and automated technology in this scenario is essential. Figure 1.8 illustrates how information provides the connectivity between supply chain processes to make this scenario a reality.

The recently coined term *agile manufacturing* reflects this approach. The vision is of a virtual enterprise formed to respond to a new product concept. Designs are executed; parts are fabricated; product is assembled, marketed, and distributed; and customer needs are met. This process is carried out by a coalition of possibly separate entities who have access to the best technology and capabilities. Alliances can form quickly, leverage off partner skills, and meet customer needs because processes are precise and efficient (i.e., automated) and all stages can be linked electronically.

Having reviewed the patterns of evolution in manufacturing technology, it is natural for a manager to ask: What does this all mean for me? What trends in manufacturing technologies may I expect? How can I use these predictions in planning for my manufacturing strategy?

It is our hypothesis that the key trend in operations over the next couple of decades will be movement towards the attainment of greater levels of flexibility, with retention of all of the cost, volume, variety, and quality advantages of current production technologies. Moreover, the role of automated, information-intensive technology in reaching this goal is central. Firms must therefore develop manufacturing strategies that focus on a policy of integrating flexible automation into their current processes in an appropriate and cost-effective manner. The principal management challenge is to understand which technologies to choose, and how to use them successfully.

Study Questions

1. What are the principal performance attributes of manufacturing systems that can be affected by the introduction of automation technology? How does automation affect each of these attributes?
2. What are the historical stages of development of manufacturing technology (over the past 150 years)? How do these stages compare in terms of the dimensions of volume, variety, and productivity? Give an example for each historical stage.
3. What are the major technology components of *computer integrative manufacturing (CIM)?*
4. Define *flexibility* for manufacturing systems. How does automation impact flexibility? Why is flexibility important in today's business environment?
5. What are the four major knowledge creation activities in a factory?

References

Avishai, B. (1989), "A CEO's Common Sense of CIM: An Interview with J. Tracy O'Rourke," *Harvard Business Review,* January–February, pp. 110–117.

Ayres, R. U. (1991), *Computer Integrated Manufacturing,* Revolution in Progress Series, Vol. 1, Chapman Hall, New York.

Browne, J., K. Rathmill, S. P. Sethi, and K. E. Stecke (1984), "Classification of Flexible Manufacturing Systems," *The FMS Magazine,* Vol. 2(2), pp. 114–117.

Bylinski, G. (1994), "The Digital Factory," *Fortune,* November 14, pp. 92–110.

Chandler, A. D., Jr. (1977), *The Visible Hand: The Managerial Revolution in American Business,* Belknap Press of Harvard University Press, Cambridge, MA.

Cohen, S. and J. Zysman (1987), *Manufacturing Matters: The Myth of Post-Industrial Economy,* Basic Books, New York.

Cooper, R. and Kaplan, R. S. (1991), *The Design of Cost Management Systems,* Prentice-Hall, Englewood Cliffs, NJ.

Davenport, T. H. (1993), *Process Innovation: Reengineering Work through Information Technology,* Harvard Business School Press, Boston, MA.

The Economist (1987), "Factory of the Future," May 30.

The Economist (1994), "A Survey of Manufacturing Technology," March 5.

Fisher, M. L. (1993), "National Bicycle Industrial Co.," Case, The Wharton School, University of Pennsylvania.

Gerwin, D. (1985), "Organizational Implications of CAM," *International Journal of Management Science,* Vol. 13(5), pp. 443–451.

Goldstein, S. and J. Klein (1987), "Allen Bradley (A)," Harvard Business School Press, Boston, Case 9-687-073.

Hammer, M. and J. Champey (1993), *Reengineering the Corporation,* Harper Business Books, New York.

Hayes, R. and R. Jaikumar (1988), "Manufacturing's Crisis: New Technologies, Obsolete Organizations, *Harvard Business Review,* Vol. 66, no. 5, Sept/Oct, pp. 77–85.

Jaikumar, R. (1989), "From Filing and Filling to Flexible Manufacturing: A Study in the Evolution of Process Control," in J. P. Ranta, ed., *Trends and Impacts of CIM,* Working Paper 89-1, IIASA, Laxenburg, Austria.

Kogut, B. (1992), "National Organizing Principles of Work and the Erstwhile Dominance of the American Multinational Corporation," *Industrial and Corporate Change,* Vol. 1(2), Oxford University Press, pp. 285–325.

Leonard-Barton, D. (1992), "The Factory as a Learning Laboratory," *Sloan Management Review,* Fall, pp. 23–38.

Leonard-Barton, D. and G. Preuss (1992), "Chaparral Steel: Rapid Product and Process Development," Harvard Business School Press, Boston, Case 9-692-018.

Nevins, J. and D. Whitney (1987), *Concurrent Design of Products and Processes,* McGraw-Hill, New York.

Noori, H. (1990), Chapter 3, *Managing the Dynamics of New Technology,* Prentice-Hall, Englewood Cliffs, NJ, pp. 69–97.

Stalk, G. Jr. and T. M. Hauk (1990), "Competing Against Time," *Research and Technology Management* Vol. 33, no. 2, Mar/Apr, pp. 19–24.

Suzaki, K. (1987), *The New Manufacturing Challenge: Techniques for Continuous Improvement,* Free Press, New York.

Swamidass, P. (1985), "Manufacturing Flexibility: Strategic Issues," *Discussion Paper No. 305,* Graduate School of Business, Indiana University.

Upton, D. (1994), "The Management of Manufacturing Flexibility," California Management Review, Berkeley Vol. 36, no. 2, pp. 72--89.

Upton, D. (1995), "What Really Makes Factories Flexible," *Harvard Business Review,* July–August, pp. 74–84.

Yoder, S. K. (1990) *The Wall Street Journal* "Putting It All Together" June.

Williamson, O. E. (1975), "Markets and Hierarchies: Analysis and Antitrust Implications," Free Press, New York.

Reading 1.1
Manufacturing Technology

The Economist

On the Cutting Edge

The machines that dominate the factory floor thrill the senses. Their tools whirl under the workers' sharp eyes; some are being taught what to do, while others, having learnt, go through the motions with inhuman speed and precision. Mechanisms as big as houses arch protectively over metal they are cutting to an accuracy of a few ten-thousandths of an inch. Smells from lubricants and the cutting fluids that splash across the tools hang in the air, as though the great machines marked their territory with scent like crouching animals. Work as shifting and complex as the aromas flows through the whole building.

The Ingersoll Milling Machine company is one of the last machine-tool businesses left in Rockford, Illinois, a town that once proclaimed itself the machine-tool capital of the world. Its machines produce more machines; they in turn produce machines for consumers to buy and use. The floor of its main works looks just as it should, a factory that would be recognisable as such throughout the world. In its research laboratory across the road, though, there is something quite different.

The octahedral hexapod does not look like any machine that you have seen before. It is a bit like an illustration in a geometry book, a bit like a sculpture, a bit like a time machine. Steel cylinders are joined in a framework with eight open, triangular faces: a top and a base held apart by six triangular sides, half of them pointing up and half pointing down. The three topmost corners are shoulders; from each of them, two arms reach towards a platform that they hold in the centre of the framework. On the platform is a light motor and a machine tool capable of cutting, drilling or boring. When the six arms expand and contract—they do not bend—the platform moves; it is like the pointer on a ouija board, picking up messages from another world. Its cutting edge can be positioned anywhere within a metre or so, pointing in any direction.

The hexapod is exceptionally strong and rigid. It can be installed anywhere, with no special foundations. It can be made cheaply. At heart, though, it is there to do the same thing as the hulking machines next door, and the key technology is the same: computer control. For making complex machines, machines controlled by computers are now essential; they are becoming more and more common in all kinds of manufacturing. The computer has taken over the tool. That has a profound effect on the way jobs are done and on the ways in which firms compete, an effect profound enough for the computer's arrival to be heralded as a new industrial revolution. But there is more to come.

A Tale of Two Worlds. New machines permit revolutions, but they do not start them. The changes going on in manufacturing are not all about machines; but their rate and nature are influenced by the extraordinary range of new manufacturing technologies now available. That is what the revolutionaries are looking at. However, they do not all see the same thing. There is no old Indian proverb about seven blind men trying to make sense of one barricade, but there should be. The revolutionaries do not know what they will get, nor do they agree on what they want. But they do agree that the hour for revolution has come.

Manufacturing companies face a bewildering range of technologies. The ones that matter will be those that let firms play most fruitfully with information, predicts Oliver Morton.

It has been brought about to a large extent by trade. Manufacturing industries in rich industrial countries are facing more and more competition from the rest of the world. Rockford's workshops were overtaken by Japan's; Japan now sees the industrialising Asian countries moving into its machine-tool markets. Manufacturing has entered an age of global competition that is forcing it to change. Low wages elsewhere are forcing companies to seek higher productivity, often through automation. And the basis of competition has changed, too. In the 1980s, in response to Japanese excellence, higher quality shot to the top of corporate wish-lists across the world. Managers put huge efforts into achieving it.

At the same time, another change became apparent. Markets are fragmenting, both in size and in time. Product cycles have grown shorter, making total production runs smaller and reducing the time that any manufacturing system designed for a single product could be useful. And consumers are asking for products that fit their needs as if designed for them alone.The age of mass consumption may be drawing to a close; if it does, the age of mass production must end too.

Many companies are responding by seeking to produce exactly what the customer wants, paying as little regard as possible to the question of whether anyone else wants it and whether the firm has ever made it before. Mass production in lots of one: that is the aim. Automated machines that can turn their handlessness to many tasks are seen as the answer, offering perfect, personalised products cheaply in double-quick time. This is the sort of equipment that a bespoke machine-tool company such as Ingersoll hopes to make its living from, now that the commodity machine-tool business has gone.

It is not an impossible aim; some companies are already doing it. Not all the revolutionaries will reap such benefits, though. A highly automated system can make a limited range of shoddy goods, and that sad eventuality is quite likely to be visited on companies that fail to grasp the nature of the revolution. It is not, in its essence, about quality, though in the right circumstances computerised machining can provide quality better and more dependable than even the most skilled human. Nor is it a revolution about flexibility, though computerised machines can go from making sprockets to poltroons and back in the time it takes to click on a screen with a mouse. Nor is it about time, though today's machines work quickly and ceaselessly. It is a revolution about information.

The cutting edge of a computer-controlled machine is the point where information is made flesh, without human intervention. The two worlds that meet are very different. The immaterial world of information allows perfection that dull earth cannot match; pure forms, absolute certainties. Matter is awkward, cussed, unpredictable, flawed. But matter is solid, workable, graspable; information is fluid, hard to grip or control. It is easy to share, potentially boundless; matter is limited, which is its strength as well as its weakness.

There has been, until recently, only one place where the two worlds could be seen to meet: in humanity. People live in both worlds, creatures of thought and deed, and have always used their thought to make their deeds better. They have found ways to make the material world carry information around, not just explicitly, as writing or recordings, but also implicitly. A well-designed tool embodies information, even if it has never been near a computer. Now, though, the information can flow free of mind or matter, liberated in the perfect realm that exists within and between computers. This freedom offers great potential; knowledge can be shared more freely, can be multiplied endlessly, and can be changed ceaselessly, without cost. The computer lets manufacturers try out new ideas and play with hypotheses in ways they never could before.

The challenge is to make this new realm of thought compatible with the old world of deed when thought and deed are not combined in the one body. Sometimes this means limiting the awkwardness of the material world. A computer-controlled cutting tool is useless unless exactly the right part is held in exactly the right place; a robot cannot pick up a slightly misplaced component. The machines are too much part of the information world to have the thoughtless dexterity that the human mind picks up through years of stopping its body running in to things. At other times, it is the world of information that must be restrained, its freedom and perfection limited; computer programs must be made to behave more like the objects in the material world that they refer to. This survey will look at how such approaches can be used in the places where the worlds still have much to offer each other that has not yet been realised; in design, in modeling, in planning.

The technological details differ from industry to industry. The tools used for making Tupperware and printing newspapers are not at all similar. Technical advances are introducing new products, new materials to make them from and new tools to apply to the materials—something that would not be possible without the new information technologies, which is what they have in common. Semiconductor manufacturing, for example, uses a suite of

Working elsewhere

Manufacturing employment % of total labour force ranked by change

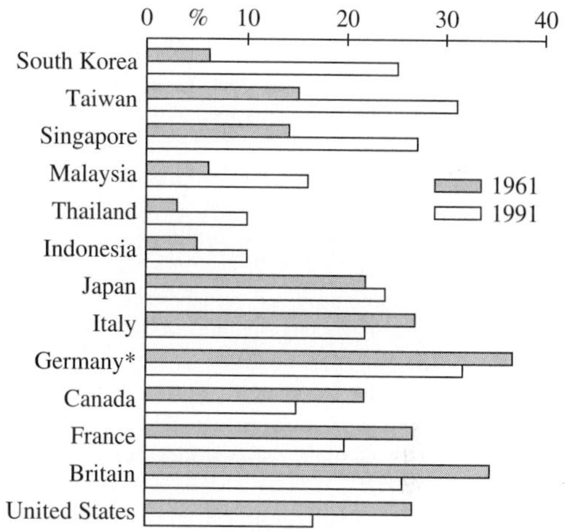

**Western*

Sources: ILO; Taiwanese national statistics

technologies largely developed within the past 30 years to build products undreamt of before that time; it is the basis of computing, but would also be impossible without it. Increasingly, all the technologies of processing, creating and assembly, from die-casting to spinning cotton to printing chips, are co-ordinated by computers.

The Feeling of Feedback. That "by" means "by means of"; it does not denote agency. Tools, no matter how sophisticated and no matter whether made of steel or information, need users. Companies that think they can fix problems with nothing but a bit of technology—or even a lot of it—are sorely mistaken. General Motors invested billions in fruitless automation in the 1980s. When technology changes, the work changes. And when work changes, the company should change, too.

Work has been changed through technology since long before the spinning jenny. In the information age, though, the changes made when incorporating a new technology have become strangely reflexive. When companies were people, plant and product, the information infrastructure lay mostly in the organisation of people. Companies arranged themselves in functional groups and management hierarchies designed to get the appropriate informatin to the right level for decision-making, and to transmit the resulting commands.

Now that information has been freed from the confines of material manifestation in ledgers and files, and can flow along preset channels without human intervention, much of that structure needs to be looked at afresh. Manufacturing companies that were structured to transfer and control information can now be rebuilt around the production process. The information technology used in that process can help the rebuilding. It is information technology that forges the link between engineering the product and re-engineering the company.

Think back to the octahedral hexapod. It looks weird, but its purpose is the same as that of the huge arched milling machines in the factory next door. The difference is that those machines had ancestors that were controlled entirely by men. Although they are now controlled by men indirectly, for the most part, their origin is still evident in their structure. In this, they are like living things. The information that gives them form is embedded within them, like a creature's genes; this embodied history imposes constraints on the forms that life can take.

The hexapod, on the other hand, has no ancestors. Its designers tried to work out how they would build a computer-operated milling machine if there had never been any other sort. The idea that informed their design did not come from previous machines. It was not inherited, like a gene; it was invented. The octahedron is a platonic solid straight out of the realm of ideal forms—information made flesh, again.

The future of manufacturing will be played out in a similar way—through contrasting fits of evolution and invention. At the moment manufacturing looks relatively unchanged, as do many machines; a deceptive surface above depths that have changed quite profoundly. New technologies that tie shop floor to shopping mall more tightly than even before could change that. Some companies will reach their new shape by evolution; some will reach it from a *tabula rasa,* unconstrained by history. Their structure will reflect the strengths that come from incorporating information technology into the design of a business from the beginning, all the way from the corporate strategy of the board to the cutting edge of the tool.

The Arsenals of Progress

To see historical constraints at their most concrete, visit the factory of Fabbrica d'Armi Pietro Beretta in Gardone, northern Italy; thousands of feet tall, they hem it in. The factory is not particularly large, but it clogs the narrow valley in which it sits. The mountain torrent that runs down from the hills and the main road that runs up to them both pass through the middle of the labyrinthine sheds and halls; it is not convenient. But because of the iron ore in the hillsides, gun barrels have been made there since the Middle Ages, and despite the fact that the ore now comes from far afield, they still are.

The Beretta family has been in the business since 1526 or so, and claims to be the world's oldest industrial dynasty; it is now a highly automated, flexible manufacturer, a clever user of up-to-date technology. Its barrels, still the company's pride and joy, are hammered from cold steel by a machine that can be programmed for shotguns or rifles of all sorts. In its evolution from craft guilds to global company, it has witnessed revolution as dramatic as any its guns have been fired in.

The greatest drama came in America at the turn of the 19th century, when the shape of modern manufacturing was laid down. In New Haven, Eli Whitney put into practice his idea that if gun parts were machined accurately enough, guns could be assembled from the first parts that came to hand, rather than from parts tailored to the individual gun, as was previously the custom. The city engineer of New York, Marc Brunel, saw how to break down the job of making wooden blocks used in the rigging of men-of-war into a series of specific shapings. He took the idea to England and commissioned 43 specialised machines for the tasks from Henry Maudslay; with them, ten men could do the work of 100.

Lessons from the past about men and machines

The system and since

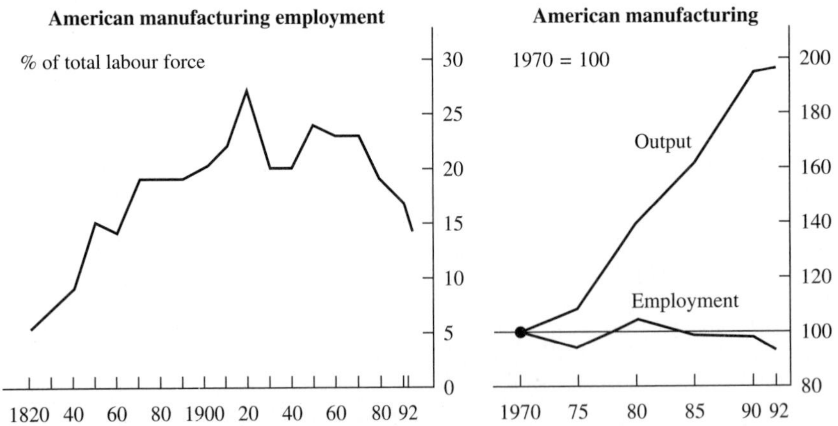

Source: U.S. Bureau of Labour Statistics; U.S. Department of Commerce

Whitney's concept of interchangeable parts was the basis of the "American system" of manufacture, which soon spread from the arms business to farm machinery, and then to almost all mechanical production. It required ever more precise machining to work; and ever more precise machining became available, thanks to Henry Maudslay's lathes, among other things. Brunel's notion of cutting up work, and creating specialised tools for each of the atomistic functions this division produced, eventually led to 20th-century mass production, brought to its peak in the second world war, then Ford's production line delivered a B-17 bomber every 36 minutes—and was still criticised for being slack.

Only new types of companies capable of large investments and strict control over workers could reap the benefits of these technologies. Beretta was more or less pushed into this; Napoleon outlawed the guild system at the end of the 18th century, forcing the forge owners to become modern capitalists. Before then, the gunmakers' crafts passed from father to son, except on the numerous occasions when the son was shot first; the citizens of Gardone showed an alarming propensity for using their wares to settle personal and business disputes.

In the new companies of the 19th and 20th centuries, the relation between men, machines, and work was transformed. The change was not a matter of more machinery; in 18th-century Gardone, and throughout Europe, the age of the machine was already at hand. Machines were helping men make things better; and men were beginning to think of themselves and their world as mechanical. Clocks had taken time from the skies and put it in houses and pockets, while Newton had turned the skies themselves to clockwork; Descartes had done something similar to men. The pieces of a man were like the pieces of a machine. The image of the machine pervaded the enlightenment, from its philosophy to its pornography.

The science of mechanics, the study of force and matter, was in the ascendant. Everything seemed soluble: Laplace, France's leading physicist, talked of a supreme intellect that could understand the entire cosmos, and predict its movements indefinitely, as a way of celebrating the powers of the mechanistic approach. Not surprisingly, this mechanism was reflected in an intense interest in the study of machines. When he was not touring Italy picking out art for his friend Napoleon to plunder, Gaspard Monge, a colleague of Laplace's, developed the mathematics of descriptive geometry, applying them to the tasks of describing machinery. His work is the basis of modern technical drawing, and made possible the blossoming of the American system. It allowed people who had never seen a given part to make it with tolerances that were close enough to what was required. Previously, people learnt what a part should look like by seeing one.

Soon they were to learn what a part was like by being one. In the 19th century, it was the factories that were like machines: their occupants became more like components. Monge's pupils developed mathematical descriptions of work, transforming what had hitherto been a human activity into a mathematically calculable quantity in the science of machines, a quantity defined as the product of force and distance. As work was quantified it became a commodity. People were becoming not just parts, but interchangeable parts. The trend reached its first peak at the beginning of the 20th century with Frederick Taylor and his precise measurement of the capabilities and tolerances needed for the smallest sub-divisions of labour: the scientific management of the production line, in which man was a source only of work and individual error.

Fruit of the Loom. The most spectacular working wonder of the early 19th century was Jacquard's loom. Developed in 1804, it could be programmed to produce different designs according to which punched cards were fed into it. It chewed up information and turned its abstractions into realised patterns. To men like Laplace and Monge, it was a marvel. The state awarded Jacquard a pension; the workers rioted, fearing for their jobs. When Charles Babbage tried to build the world's first computer in England, a mechanical embodiment of Laplace's ideas, he used Jacquard's cards for the programming; in a sense, all subsequent programmable machines are the fruit of the loom.

It was the technology of the Jacquard loom, with its ability to turn abstraction into physical patterns, that inspired early forays into the automation of machine tools after the second world war. It was not the only technology available. It was possible, for example, to record the movements of an expert machinist and then play them back, rather than conceive of the movements as mathematical perfections, encode them on to a storage medium (a numbing task before cheap computing power), and get a machine to realise them. But it was numerical control, not record/playback, that won the backing of America's military-industrial complex. It allowed them to do things they could not trust to machinists, making finely worked pieces of material for new high-performance aircraft, for instance. It also allowed them, they believed, to replace skills on the shop floor with programming skills. The technology of Jacquard's loom was always conceived of as a technology to free capital from labour.

The numerically-controlled (NC) tool transmuted itself, reasonably painlessly, into the computer-numerically-controlled tool (though there are hangovers; the people who work with them still call CNC programs tapes). At the same time a new sort of automation started to emerge: industrial robotics. Here computer control took on a new importance. It was not just mimicking human control of a tool; it was mimicking, poorly, a human. While there were tools before automatic control, there were no robots. They are pointless without some inhuman source of information to animate them.

However, robots have not displaced men and women (robots often do assembly work traditionally left to women) entirely, despite the fact that their advent gave rise to yet another wave of speculation about the workerless factory. They have a role in manufacturing, and have been used well in Japan, where they have helped bring about large increases in productivity. The Japanese have understood that, if work is designed properly for robots, they

The Rise of the Robots

Industrial robots in use

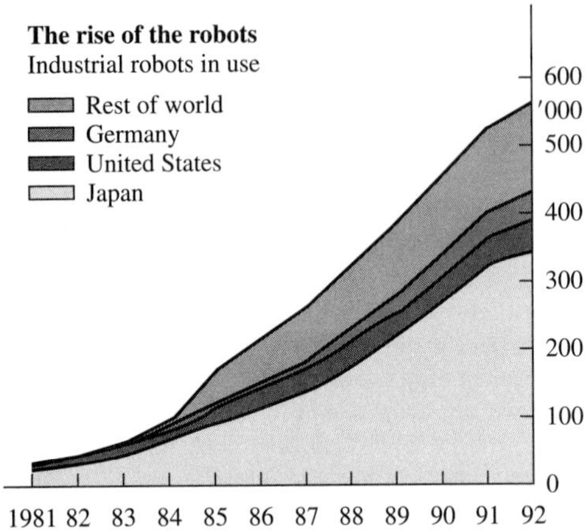

Source: United Nations

will do it well—but they are not able to replace people at jobs that have evolved to need a human's innate ability to fit the world of ideas and intentions to that of deeds and objects.

That does not mean robots cannot do things that people do now. At the Fraunhofer Gesellschaft's research institute in Stuttgart, robots do things previously unthinkable. They wire up lights, attach jacks to the end of printer cables, fit rubber padding around car doors, string tennis rackets. The robots can do this not because they are clever, but because the people behind them are. The cleverness lies in the tricks that let robots do their tasks in ways quite unlike those of people. The aim is to save jobs, not lose them—at least, not locally. With robots, hopes Jens-Gunter Neugebauer of the Fraunhofer, German factories can become productive enough to keep such work as tennis-racket stringing from moving to cheap-labour countries. There may be fewer jobs in the factories, but the factories are more likely to stay in Germany, and new service jobs will be created to support them.

Beretta uses only a few robots. However it saw the possibilities of NC reasonably early, and invested at a time when the payback was fairly slow. It went as far as buying an NC machine-tool maker to understand the technology better. It reorganised its production to take advantage of the increases in accuracy, reproducibility and flexibility that the machines offered. The reward came in the 1980s, when it was able to offer better guns at lower prices and win a sidearms contract for the American military. Eli Whitney would have been impressed.

Beretta has captured the benefits of past industrial revolutions. It can deliver huge numbers—the American military contract was for roughly half a million. Its accurate tools and its corporate attitude provide extremely high quality. It offers a wide range of guns, with each specific model available in a bewildering number of versions, in part because of computerised flexibility. But there is still progress to make, not in machining, but in handling information. Beretta has only recently started to exploit computer-aided design, for example. It is time to assess the benefits this could bring.

The Mind's Eye

In the biggest factory in the world, the first Boeing 777 is being put together. In April this year, it will roll out of the doors and, fairly soon thereafter, lift itself into the air over Seattle. Boeing and its customers hope that this twin-engined long-haul airliner will be a remarkable machine in many ways. In some ways, it already is. Normally the company would build mock-ups of a new aeroplane to make sure that its millions of parts fit together. The workers have not assembled any trial 777s, though; the first produced, the one now being assembled, will be a functional flying machine.

But a mock-up exists elsewhere: in trillions of bytes of information stored in the company's mainframe computers. The development of the 777 has been the biggest single trial of compueraided design (CAD). Using the computer's representations of the aeroplane's parts, engineers and their programs checked that the whole thing would fit together before any pieces were made. Their checking appears, so far, to have been pretty good. When the wings of the first 777 were attached to its waist, the port wing tip was out of position by a thousandth of an inch; the starboard wing was postioned as accurately as the gauges could measure.

When computer graphics can give verisimilitude to dinosaurs, turn actors into androids and make even the most tedious news broadcast a kaleidoscope of shifting imagery, it is easy to underestimate the power of the simple images CAD programs put on to screens. They are not spectacular. But CAD makes it possible to transmit designs directly to machines that will follow them perfectly; it gives the designer new tools; and it allows companies to change the way they organise design.

Automating design, the most important step in a product's life

Good design is the key to manufacturing. It is the difference between a product that does a great job reliably, is easily fixed if damaged and is made cheaply and quickly; and a pile of junk. As the diagram on the next page shows, half the life-cycle cost of a product is fixed by the original design; 80% may be set by the time that production actually begins. The designer needs all the help he can get.

Changing the Parameters. The classical advantages of automation are time and accuracy; both are provided by CAD. Technical drawing is a time-consuming task. If CAD screens were nothing more than high-tech drawing boards, they would still be useful; studies show that they can cut the time it takes to draw a design by as much as 90%. Perfect copies are made at the touch of a button, changes incorporated quickly and easily.

A design made with CAD can be transmitted perfectly from one place to another, if they both use the same system, and passed

FRONT-LOADED

Impact of early design decisions on life-cycle costs

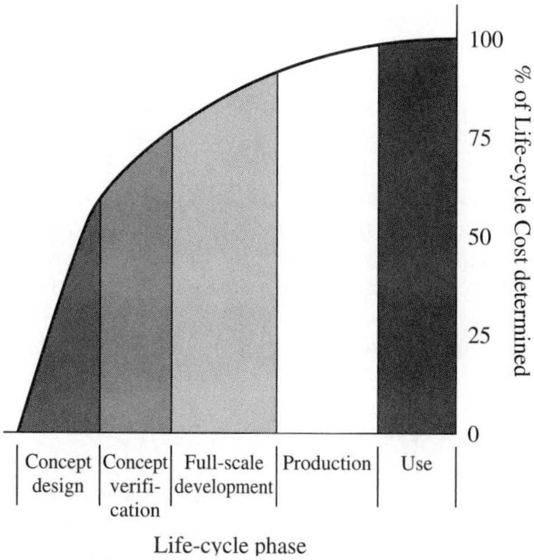

Source: Draper Laboratory, Cambridge

without much fuss into CNC tool systems. So subcontractors know exactly what is wanted; and if they make it accurately, as their CNC tools should, the parts will fit perfectly. The impact of CAD's ability to provide "digital preassembly" are felt in big things, like slotting in the 777's wings, and in lots of small things too.

Aeroplanes are built in slices, but there are many systems—the air supply, the heating, the plumbing and the electronics, for example—that permeate the whole. (The data pathways are, in the case of the 777, particularly impressive, which is another aspect of the information age—where would an airliner be without 100 channels of television in the back of every seat?) Normally the conduits and tubing required for these things are added during assembly. Now, though, the slices of aeroplane come with all their stanchions, tubing and whatnot already attached, and it all lines up as it did on screen. When new floor space was added to Boeing's Everett plant for the 777, it came with power outlets for hand tools, normally needed to trim and shape the parts during assembly. They have hardly been used.

Such gains in precision are seen in all sorts of automation. Take spot welding. Designers used to put in more spots than were needed, because the space between spots varies when the welders are people. Robots get it spot on, and so, on top of any other advantages gained, there is a 10% decrease in the amount of work.

The advantages of CAD mean that large complex engineering tasks now depend on it. If big companies work with CAD, subcontractors need to be able to receive designs as CAD data and feed them into their computer-controlled machines. This can be a problem. There are many different CAD systems available, and transmitting data from one to another and into manufacturing systems is not as easy as it should be. According to Mitchell Fleischer at the Industrial Technology Institute in Ann Arbor, Michigan, as many as 25% of designs transmitted to small tool and die shops as CAD files have to be printed out as blueprints and re-entered into the computer because translations between programs are so poor. In these circumstances, lead times and costs go up, not down, with the installation of CAD. To simplify things, many companies impose their choice of CAD on subcontractors. That means some subcontractors end up with many systems. Mr. Fleischer knowns of one with 17.

There is a standard for CAD transmission, IGES, but different systems can approach it in different ways, which causes problems. STEP, a new standard, should be less vulnerable to that. And not only is it better than IGES, it also offers far more. IGES is a standard for technical drawings, and describes things in terms only of geometry. Its creators have designed STEP to go beyond a vocabulary of figures to provide a way of describing parts in terms of what they are to a designer, and to include all sorts of other data about the product. In this it reflects the change in the CAD industry which has turned CAD systems from computerised drafting boards to the seeds of something far grander.

There are still a lot of computerised drafting boards around, offering simple CAD capabilities to people with personal computers. But at the high end, advanced systems are increasingly capable of working with features, not just with geometry. A designer can point at one of a part's faces, tell the machine that he wants a hole in it—and the hole will appear. He does not have to specify its geometrical relations to all the faces concerned, or even draw it. If he wants a knob, or a flange, then a knob or flange will appear. By using such natural concepts, the tool gets closer to the designer's thoughts.

Parametric Technology of Waltham, Massachusetts, has led the way in a particularly fruitful application of this technology with their ProEngineer software. If a feature of the design has to be changed for some reason, the program will make the change, and then make further changes to keep things consistent. The impacts of the change ripple through, and the design remains a consistent whole. The system works by remembering the history of the design on the screen; when the change is made, the program goes back to the point in the part's history where the feature being changed first appeared, makes the change, and then runs forward through history, amending everything that has to do with the changed feature.

This technology has been the basis of Parametric's impressive growth. Other companies have taken it up, although some in the field distrust it and are using other techniques to achieve the same end. The new systems provide the designer with one of the greatest advantages of the world of information; the ability to ask "what if" and look at alternatives without actually trying them out. All that is used is processing power, the costs of which are increasingly negligible. They also provide flexibility. Make a late change on a parametric system, and as soon as the numbers have been crunched, it is as if the change had been made early on. Look again at the above diagram. The curve dictates the cost of making changes; the flatter the curve, the higher the cost. Modern CAD tools can reduce the costs of late changes and straighten the curve.

The ideal, though, is to avoid late changes by getting everything right early on. The effort to increase the quality of products and, most of all, to reduce the time to market has led to "concurrent design." Traditionally, design and production are separated by a metaphorical wall. The finished design is thrown over it, at which point the designers go on to the next thing and the production people realise they cannot make the current one. The idea of concurrent design is that designers should work closely with everyone else who will at some point have a stake in the product; what used to be done at length should now be done in parallel, with manufacturing people having their say all through the design process, and building up the manufacturing capability early on.

Until the 19th century, there was no need for concurrent design; everything happened in one mind. The designer would build the product, or at least be a master of the building craft. The increasing complexity of products is one reason that stopped; but information played a role, too. Companies were cut into divisions, to ease the flow of information; and the advances in drawing started by Gaspard Monge made designs portable.

The aim of a good concurrent design team is to recreate a master craftsman's mind and see all aspects of the process at once. For that to happen, the divisions that give structure to the company have to be overcome. That is why concurrent design requires the time and effort of management. Boeing put people into the 777 design teams according to the section of the aircraft they were working on, rather than their functional division within the company. Manufacturing and maintenance were represented all along—as were customers. British Airways, for example, spotted a way to get four more seats on by rejigging the aft toilets and galleys: a 1% increase in capacity for no extra cost.

Mimicking a mind, though, is still a hard thing for a team to do, even if the structure is right. It has to know what it is thinking about, to be able to communicate its thoughts clearly and unambiguously, to know when a change has been made and what its implications are. That is what CAD systems provide. They allow designers to see the effects of suggested changes in ways that were previously impossible except for those with exceptional mental abilities; they give everyone the spatial awareness and rigour that only freaks of nature had before. And they make these acutely perceived visions portable. As well as a design tool, CAD is a communications technology, tying all the knowledge in the team together. It is the team's common imagination; the mind's eye looking into the world of information.

Living in the Material World

The thoughts that float on CAD screens may be fathers to the deeds of manufacturing; but a lot of pregnancy and labour separates the two. During this, a concurrent-design team needs more than visualisation. Parts are not just physical objects—they are problems for manufacturing engineers, they are fulfilments of specifications, they are objects of consumer desire. The design process has to take all this on board too, and for that it needs a model of the product that contains all sorts of information beyond its shape.

Steve Gutierrez, who works on Hewlett Packard's advanced CAD software, talks about the product model as a wheel. Its hub is a database that holds all the necessary information about the product. Its rim is the CAD representation of the product. The spokes are other types of software, which advise on ways of cutting sheet metal to make the parts, or look at how the product will deform if you hit it, or allow the electronics to be sorted out. The spokes all do their job in the context of the CAD model, and depend on it; in turn, they give it solidity, and connect it to the rest of the world. The different parts of the model can come from different suppliers, as long as they can communicate. The STEP standard should let them do so.

How much development can be done in a computer; and how much needs to be done there?

Frank-Lothar Krause, of the Production Technology Centre in Berlin, offers a goal for such a product model in terms of the front-loaded diagram on page 31, which showed how much cost depends on design. You can add a second set of curves that show how much is known about the product (see diagram on next page). Today, lots of life-cycle cost is defined when hardly anything is known about the product; CAD improves that a bit. Detailed product modelling might improve it a lot, pushing the knowledge curve above the cost curve. Mr. Krause suggests that knowing how to make product models could soon be as important as making the product, and might be the ability companies most need to develop. George Kuper, head of Ann Arbor's Industrial Technology Institute, sees the transformation of data into information as a process equal in importance to the transformation of raw materials into a product.

Fumihiko Kimura of Tokyo University offers a vision of what such models could be: "virtual manufacturing" systems so realistic that they can simulate everything that matters about both a product and the processes that produce it. Imagine a simulation of a car that can be assembled in the computer by a simulation of the production plant, that safety engineers can crash in the computer and customers can test-drive in it. Researchers around the world are developing virtual-reality systems that will let designers walk around their products and computer-generated dummies try them out for comfort. In America, the army is now using flight simulators to provide feedback for helicopter designers. The pilots tell them where the controls should be best placed.

Flight simulators and virtual reality are enticingly sexy but they are not enough; there is more to virtual manufacturing than visualisation. It needs programs that can model all sorts of aspects of the real world. Some of this modelling cannot yet be done; some of the modelling that can be done is hugely expensive in computing time. Computers deal easily with the mathematical perfection of geometry. That is what made simple CAD possible in the dark ages of two decades ago. The physical world, though,

FOREWARNED

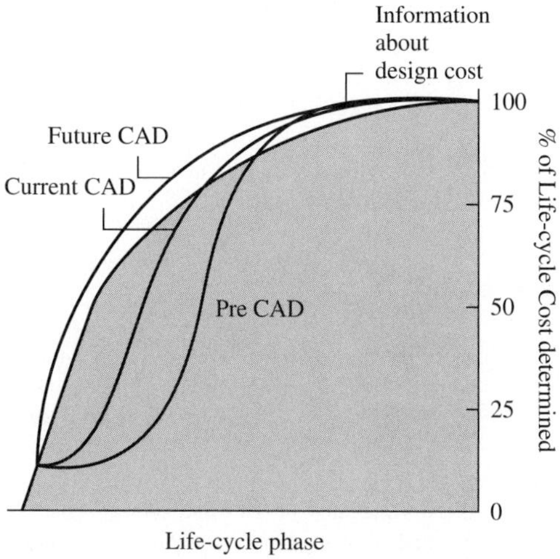

Source: Goering

is very different. Working out how a product will bend in a crash, or how heat will flow through it, is hard. The best computers can do is to cut up a model into tiny little bits, and use rules derived from physics and chemistry to predict the effects that stress, heat or whatever will have on them, and how the changes in each will affect its neighbours. This soaks up ever more computer power as the analysis becomes more detailed.

Such computerised analysis is widely used to test designs, especially in products that are expensive, produced in small numbers or pushing the boundaries of the possible. However, it is often unwieldy and time-consuming, and it is less versatile than other information technologies. Tools that analyse the material world this way seem to take on much of its intractability. They describe how a part will behave, but they do not help to make it behave better. They are not good for asking "what if" questions, because each answer takes too long.

Get Real. The alternative to using a computer to model the physical processes of the real world is to get direct access to them with a prototype. This is perhaps the fastest changing area of manufacturing technology. There are new technologies around that will build almost any part a CAD system can dream up in hours. They are, as yet, limited in the materials they can use: but they are getting better all the time, and they already offer capabilities that appear magical, opening new doors between the material world and the realm of information.

Most of these rapid prototyping technologies work by building a three-dimensional structure out of a pile of two-dimensional cross-sections. Selective laser sintering, for example, works with layers of fine powder. A computer system takes a CAD representation and cuts it into thin slices. Then a laser beam draws the outline of one of those slices on the powder, and its heat makes the particles in the powder join together: it sinters them. The outline completed, the laser scans back and forth within it, sticking together all the powder so that it forms a solid slice. A new layer of powder is spread over the top, the outline of the next slice is drawn, and the process continues.

Laser sintering can produce models in wax or nylon. If the powder particles are thinly coated specks of metal, it can make metal parts; the coating is removed after the part is made, leaving a foamy metal that is then infused with liquid copper to fill up the spaces. Similar techniques can make things layer by layer using liquid plastics that a laser can solidify, or using layers of paper that a laser can cut away, or using a technology more like photocopying to produce each layer. There are also ways of building a model drop by drop, and others, descended from ink-jet printers, that bind together powders with finely-controlled sprays.

All depend on CAD; you could not bring a new design from the world of information to the world of matter unless there were ways to put the information in the right form. But the idea of turning CAD into matter has been treated as a new challenge, and solved in several different ways. The new technologies have been invented, not evolved.

According to a report produced by Quo-tec, a British consultancy, on the basis of a trip to America by European experts, there are about 500 such machines now in use. All but one of the companies that build them are American, as are most of the users—only the laser-in-a-bath-of-plastics method, the oldest, is widely used outside America. They are based there because rapid prototyping is a computer-based innovation—at which America excels—rather than a new variation on old production machines, which are equally advanced in Europe and Japan.

The systems are, as yet, far from perfect. They are quite expensive, though prices may fall quickly, especially for the machines based on ink-jet printing technologies. Perhaps the biggest drawback is the narrow range of materials available. The systems make CAD data real in geometrical terms, but they often do not produce real parts, just models in the wrong material. That limits the scope for physical testing. But models are still useful. If a product involves mechanical linkages—trains of gears and cogs, for example—physical models can show that the pieces will all move together the right way more convincingly than a computer rendering can. And prototypes plonked on desks can make things seem real to a non-technical mind—a finance director's, perhaps—in ways a CAD file cannot.

Perhaps virtual manufacturing within computers will one day be so good, and so widely used, that there will be no need for any parts of the process to move from the information world into the material one. But that day is not one to seek for its own sake. The reason for doing as much as possible in the world of information is that changes are made more easily when intractable matter is kept out of the way. If movement between the world of information and that of matter becomes easier, and physical artefacts become part of the process without slowing it down, fine. If it is easier to build a model of a wing and put it in a wind tunnel than to simulate the air-flow in a supercomputer, then developers will do it. The world of information is only preferable when it offers real advantages.

Even if everything is done in a virtual environment, though, parts must at some point be made for production. Devotees of rapid prototyping technologies are already aiming at that point; to show that they are not limited to approximate models and stop gaps, they are rechristening their craft free-form fabrication (FFF). For some mould-making, an FFF model is as good as anything used today, and available far more quickly. A metal FFF part can work as a tool on an existing machine. It may not be as tough as a traditional tool, but it does not need to be if it can be reproduced perfectly. Ceramics can be made with FFF printing technologies, in which powders are glued together and then cast; and ceramics are increasingly sought after.

Though virtual products and FFF may look like different solutions to the problems of rapid design, they could be complementary solutions to the problems of rapid manufacture. That, at least, is what Arlan Andrews of America's Sandia National Laboratories hopes. Sandia has developed a multi-user virtual environment for design, in which people can put together and test any contraption they can think of. Any problems with analysis are alleviated by the fact that Sandia has some extremely powerful supercomputers. But the system does not need supercomputers; and anyway, supercomputers are becoming cheaper to buy and, thanks to high-speed data links, easier to gain access to.

On the other side of the corridor, Sandia is working with FFF systems, both to fill in the detail that for the present eludes the most super of computers, and to see how much final manufacturing can eventually be passed on direct to FFF systems. For some materials, machining may soon be a technology completely superseded. The raw material will be transformed into a finished part, with no chips or tailings left over. Such processes may never be quite as fast as mass production, but they have the much-sought-after attribute of flexibility built in from the very beginning. They are far more flexible than today's computer-controlled machine tools could ever be. And they need no factory, no supply chain; just the raw material, the design and, perhaps, an operator. Some types of product—personalised porcelain, perhaps—may go all the way from design to department store in the world of information.

Putting It All Together

Companies want products designed well and quickly; information technology can make that easier, if the commitment to design is there. They also want a wide range of products, and they want them produced at exactly the right time. Traditional automation—production lines with tools and fixtures built into them—cannot do this. It offers high efficiency only with constant output and a small range of products, so it makes sense only for large product runs. The advent of programmable machines was meant to change that by providing flexibility.

Gather computer-controlled machine tools together, give them robots to fetch and carry, and put computers in charge of the whole thing: you now have a flexible manufacturing system, and can use the abbreviation FMS to prove it. That abbreviation is dangerous. It suggests that flexibility is a single quality, that can be bought in the form of machines and programming. In fact, flexibility covers a multitude of virtues and some sins as well; and those virtues reside in the whole company, not just in the machines on the shop floor. The good news is that once flexibility is understood, computers may be able to help more than sorry past experience gives any reason to believe.

Flexibility does not come naturally to factories or computers; it has to be planned.

Daniel Whitney of the Massachusetts Institute of Technology identifies a number of different types of flexibility. There is flexibility about inputs, which lets the same thing be done even when the parts needed are presented in unusual ways. There is flexibility within production, which allows many things to be made in different batch sizes at the same time in the same system. There is process flexibility, which allows entirely new things to be made quickly and easily. And there is flexibility as a way of having a second chance if you make a wrong decision. Many companies, he suspects, really want the last of those: but it may not be on offer. As for the first, it may not be needed in a well-run system. But flexibility in production and process can be achieved, as Dr Whitney's case study of Nippondenso, Japan's biggest car-component company, demonstrates.

The robots and programmable machines used in FMS reflect a belief in flexibility as part of the process of manufacture. But there is more to it than that. Flexibility can be part of the product, designed in from the start. With the the proper design you can limit the need for flexibility in the manufacturing process and, perhaps more important, define the bounds of that flexibility in advance. The aim is to be no more flexible than necessary, given the variety of products being produced. Nippondenso achieves this largely by getting the flexibility implicit in the design realised only in the last stage of manufacture, assembly. Its assembly is highly automated, but the work involved is not complex: the robots can afford to be simple because the intelligence is in the products.

Nippondenso puts it there in several ways. One is by minimising the number of parts. A second is modularity and standardisation. If the difference between two radiators is their length, they can have a large number of components in common. With proper design, many different products can be made from lists of parts that differ only slightly. A third is the principle of partial assembly. If a half-assembled object hangs together, there is no need for special jigs and fixtures to hold it in place. Fewer jigs: more flexibility. This, too, is an accomplishment of good design.

If the same parts are used for different products, and the different products require little by way of different handling, then flexibility can be achieved without losing efficiency. The manufacturing facility is arranged to maximise such advantages. Many of the robots and other devices are developed in-house, as part of the same process as the design of the products. The robots are not designed to be particularly dexterous or smart; the work is designed not to need such skills. Developments of product and process feed off each other continuously.

Heading for Heterarchy. Nippondenso deliberately minimises the role of flexibility in fabrication—the making of individual parts, which is what flexible manufacturing systems are meant to excel at. Much of this fabrication is done by subcontractors. Dr Whitney points to a tendency for Japanese companies to buy parts but make their own systems, while American companies tend to do the reverse. But the fact that flexibility can be achieved without flexible fabrication does not explain why flexible fabrication, the supposed goal of FMS, is hard to achieve.

It is not impossible, given proper design; there are well conceived flexible manufacturing systems. But there are also many that are not quite what was wanted, and the growth in FMS sales has not been what early proponents hoped for. David Upton of the Harvard Business School puts the blame on the bit of the system that is meant to provide flexibility: the computer. It is the computer that makes an FMS into a system; without it, an FMS is just a bunch of tools and robots. It is the computer that has animated the dream of going beyond FMS and integrating whole factories and companies to achieve a new kind of manufacturing. The tools and robots on the shop floor would be just part of some grand immaterial structure that spread beyond the bounds of their world: the machines in the ghost.

Ghosts are undoubtedly fabulous things; but they have their drawbacks. They tend to stay in one spot, be overburdened with guilty knowledge, harp on about the past and never change, just fade away. Not, in short, very flexible. Traditional computer systems share many of these flaws. The fact that information (like ectoplasm) can be very fluid does not mean that the systems in which it runs allow flexibility. Quite the reverse: they are often excessively rigid. A system running an FMS has to be complex in order to keep track of everything, to recognise different failures and to deal with them. If the range of products the FMS is making changes, this software often has to be changed too. If the FMS itself is changed—one type of mill or robot replaced by another, say—there must be more changes. The software to deal with this is complex and extremely hard to amend; in fact it is far more difficult to change than the hardware, which can just be bought and bolted in place.

These rigidities are not unique to flexible manufacturing systems. Dr Upton has found it in other places where computers have been put in charge. Take the fine-paper industry: there is a strong correlation between flexibility, in the sense of the ability to move from one product to another, and the advent of computer-integrated manufacturing. But it is negative. The flexible plants are those where people are incessantly striving for flexibility, not those into which flexibility has been programmed.

The problem is that manufacturing is extraordinarily complicated; and the capacity computers have for discovering perfect solutions in the world of information cannot, as yet, deal with such complexity. Most companies do not really know how they do what they are doing; their solutions have evolved unconsciously. A good manufacturing system is self-regulating: staff who can see what is going on continuously tweak and titivate, keeping it productively poised on the edge of chaos. But a centralised, hierarchical computer system of the sort that most companies have cannot capture this continuous functional jiggling. Lacking the ability to evolve, a system of tight controls must be invented for it, controls that constrain the system's ability to adapt. Even so, chaos sometimes breaks through. Dr Upton found that computer-integrated paper plants were often more likely to have catastrophic failures.

People cannot provide the different products that producers now want at the speeds that they could manage for mass production when helped by the machinery of a production line. So Dr Upton offers a new view of automation, one that he calls heterarchy rather than hierarchy. A part comes into a factory with a chip sitting on top of it. The chip knows what needs to be done to the part, and broadcasts the first step in the process. The machines in the factory ask themselves whether they can comply; those which can, ask themselves how much work they have to do and, on the basis of that, bid for the right to machine the part. The chip listens and makes it choice. When everything has been done, the part leaves the floor but the chip goes back to the beginning, to negotiate for another part. The chips learn which machines to trust.

The charm of this idea is that there is no hierarchy. Each chip needs to know only about one thing—a product or a machine—and its needs and capabilities. Failures in the system are always local. Efficiency comes about through the invisible hand of the market. Changes are easy. Add new products and the machines will still make them, if they can, with no new programming. Add new machines and the products that need them will use them.

In the real world, where jigs and fixtures, among other things, are needed, such systems may be pipedreams. But they are useful thought experiments—and a useful metaphor for a form of computer programming that can provide similar advantages. Object-oriented computing began as a way of modelling the real world. The "objects" within it are packages of data wrapped in programming that applies to that data. An object in a factory control system might represent an automatic vehicle. The data would say what it was and what it carried; the programming would allow it to be sent hither and thither by other objects in the system.

Imagine an object-oriented solution to controlling an FMS. Instead of trying to control the state of all parts of the system centrally it would have different objects within it representing the products, the robots and the tools, which knew how they should relate to each other, and which could follow rules like those in Dr Upton's market-based factory. Change the system, and you give the computer new objects. You do not need to rewrite the program from scratch. There is no need for continuous invention; evolution is powerful in a world where information is told to behave as if it were embodied. Such programs are proving highly useful as ways of organising the factors of production, parts, labour, capital and, as important as all three, time.

The Uses of Time

In the modern world, everyone can know the time exactly and carry it with them anywhere. In the pre-industrial world, only precise observations of the heavens could provide accuracy, and time was encased in all-but immovable clocks. The difference was both brought about and necessitated by industrialisation. Greenwich mean time was taken from London to the provinces by clocks on steam engines, the better to synchronise the rhythms of the nation's work. Time's unmeasured flow became controlled, paid for, subdivided. Factory whistles punctuated it, Frederick Taylor and his stopwatches measured it exactly, punching the clock gave it value.

Fitting processes together, pulling companies apart

At present, the industrial obsession with time is focused on two issues; time to market, and being just-in-time. Time to market depends largely on design; just-in-time depends largely on working practices. In a company convinced of the value of being just-in-time, orders trigger production, rather than triggering a trip to a warehouse that is being continuously filled by the production process. It is a large part of what is meant by "leanness" in manufacturing, and much in favour. It has not, to date, had much to do with technology: it has been about management and the design of work.

The idea is to avoid tying up capital in work in progress. If the production system is thought of as a machine, work in progress corresponds to the tolerances between the parts. It is tolerance in time and money, not space: the inventory between two parts of the process represents the degree of uncertainty over the time they take, and over the speed at which the whole system may have to work. Computer-controlled machining and CAD have allowed engineers to make pieces with tolerances finer than ever before as a matter of routine. Can analogous technologies do the same for time tolerances?

According to 12 Technologies, formerly known as Intellection, they can. The Dallas-based company starts from an analysis of the weakness of the just-in-time approach. In this, the message to up the pace or slow it down travels upstream from the customer through to the component supplier like a tidal bore running up a river. If the process is simple, that is easy enough. However, if the process is complicated and many-branched—that is, flexible enough to turn out many different products—things get trickier. There are more likely to be bottlenecks in the process that distort its speed. And if the changes in the market are extremely quick, quicker than the time it takes for information to travel step by step up the manufacturing stream, then the system loses stability like a pendulum swung beyond its accustomed range.

12's solution looks simple; it is a program that plans the path of work through a manufacturing plant, saying what parts will be needed where and which machines will do what. Such programs have been around for a long time—there is a large installed base of software dealing with material-resource planning (MRP). But Rhythm, 12's product, is smarter and better at details. An object-oriented approach allows it to solve scheduling problems by looking at the starting condition and the desired outome, and bringing the two together. Traditional programs are more idealised, starting with assumptions such as an infinite supply of components and slowly running through long loops of calculations before coming to an answer.

The result is that Rhythm can produce schedules in minutes, not days. Work can be planned by the hour rather than the week. That means the tolerances are tighter; if the system gives good results by the hour, there is no need to keep a week's worth of production around. Another advantage of Rhythm is that it can be used to schedule things machine by machine. It, and better software like it which will surely follow, can take a floor full of machines and arrange the flow of work through them so that they become "virtual cells", linking machines far apart into the sort of tight-knit unit seen in an FMS. They will not quite match an FMS in throughput, perhaps; but they can be reconfigured quickly and easily. Factory layout moves from the material world into the world of information.

Rhythm's speed allows it to reroute work more or less on the fly. Earl Mott, who has put Rhythm into the heart of Black & Decker's manufacturing division, enthuses about its ability to answer hypothetical questions. It can tell you what is gained and lost by giving an order high priority. And according to Sanjiv Sidhu, president of 12, it can do this for a whole company, not just for a factory. It can tell you whether shipping parts over from the Singapore plant is an easier way to fulfil an urgent order than building them on the spot. It allows you to commit to a delivery date with a high degree of confidence. Delivering on time is a problem for a lot of companies. Many do not even meet the date they commit to, let alone the date that the customer really wants. Improving that performance would provide a competitive edge. Mr Sidhu notes that many of his company's customers are firms that were quick to tackle quality, a crucial issue of the 1980s. They now see timeliness taking its place as the ground for competition. Quality is no longer a stake to be raised. It is the ante necessary to enter the game. Time is where the stakes are being laid.

With Rhythm, a company can go beyond simply having a flexible facility; it can know just how flexible the system is through simulations, and it can know what making use of that flexibility to prefer one job over another will cost. Remember Dr Upton's notion of a factory floor as a marketplace, where machines bid for parts, and the work ends up efficiently allocated. Rhythm provides similar capabilities; but rather than setting up a surrogate market, it allows the scheduler to let the real market in to the factory, and set priorities accordingly.

There are other technologies that can offer similar insights into how factories actually work—a whole new category of software that can be seen as fitting between traditional MRP and what actually happens, called manufacturing execution software, is forecast to grow by 20% a year through the 1990s by Frost & Sullivan, a consultancy. And there are technologies to allow computers to know what is really going on. Computers work on the idea that what should be happening is happening. Once something goes wrong, this approach can amplify the problem and widen the divide between idea and reality. Stand upstream of a robot on a production line and push the pieces it is working on out of kilter; it is fun if you enjoy watching futility. Now factories are turning towards barcodes and other sorts of labelling to keep track of reality. If the software and the people running it are flexible enough to cope with the data, this does a lot of good.

To see such technologies working together, look at Motorola's Fusion production facility. Fusion is the company's name for a family of pagers, designed to be easily customised, and the processes of design, manufacture and marketing that surround it.A customer phones in a requirement—a pager, say, of a particular size and colour (it matches his tie) that plays a particular tune. The Fusion computer system turns the specifications into a set of parts, and nips into a virtual manufacturing world to reassemble them and make sure they add up to a product. Then it releases the design as a series of routing instructions to the manufacturing facility. As the piece goes through the facility, it is continuously tracked. Nothing is made until it is ordered; mass production in lots of one.

The advantages go beyond single facilities, though. Hewlett-Packard has realised that, like many companies, it is not yet delivering to its customers at the speed they would like. A company-wide object-oriented database, showing exactly where everything is, is part of the solution; so are changes in management and in attitude. Their new technology is part of a general corporate re-engineering; the changes involved make sense only in the context of one another.

The Tools Turn In. Such re-engineering is all the rage, and technology is a good tool for it. What sort of companies might eventually emerge from the makeovers that the new tools will offer?

One of the most influential visions of future manufacturing in the past few years goes by the name of "agile manufacturing". It is outlined in the "21st century manufacturing enterprise strategy", produced by Lehigh University's Iacocca Institute for America's defence department. Agile manufacturing requires companies to have more than just the fast, lean and pro-active virtues of the 1980s. Like "new men", they have to be butch but caring, ever responsive to the needs of others. They also have to co-operate. Agility resides in the provision of a product, rather than in any single company. The visionaries like to talk of virtual enterprises, ad hoc coalitions put together around a promising new idea. An idea is born, a design made, a design verified, parts built, a product assembled, marketed, shipped: a need is fulfilled. Each step could be done by someone different; the last step is the product of the whole.

There are various preconditions for success in such a venture. One is clear communication. Here the new standard for product data, STEP, will undoutedly be useful, especially since it can carry quite a lot of data about the product on top of its geometrical form. Standards have their drawbacks; they always lag the leading technologies. The history-based approach of Parametric's CAD software, for example, which allows the far-flung repercussions of changes to be felt, cannot be communicated through STEP: when a model is moved through the STEP standard, it loses this aspect of its history. But interoperability will, in the end, matter more than pure performance, and assuring that systems in different companies work together will definitely require standards.

The precondition that matters most, though, is predictability. The essence of agility is sensitivity to time. The different companies involved have to know their capabilities exactly, and the time they take exactly. This is what new factory management technologies make possible. When a virtual enterprise is assembling itself, it has to know precisely the dimensions of its parts, not in breadth, length and depth, but in terms of such things as process time and quality. At present, few companies can accurately measure themselves in many of these dimensions.

There are hints of these virtual worlds already to be seen. The American military is encouraging them, as it encouraged NC machining. A programme called CALS will require every new piece of equipment to come with a form of electronic product model, and in so doing begin to tie the supply chain into a sort of virtual enterprise. And there are less technologically advanced models available elsewhere. One is found south of Beretta's home in Gardone: the fabric mills of Prato. In the 18th century, the old guilds were unified into vertically integrated mills. In the 1970s the mills subdivided themselves back into small units. They now provide a marketplace of capacity co-ordinated by a few brokers and served by a united retailing operation. The industry, once moribund, is now flourishing.

Dr Upton and his colleague Ramchandran Jaikumar have pointed out that there is a growing potential for this to happen elsewhere, in other industries. Manufacturing plants that can work from CAD files are spreading. The size of many manufacturers is shrinking. With a modern multi-purpose CNC tool, a man in a garage can bid for work that would have needed a job shop in the past. Versatile scheduling can give a few such machines great capabilities. The information networks needed to join such places together are almost in place, offering high data-rates to any user in an industrial country that needs them. And the software for good product modelling is being written. Without product models, such schemes are doomed.

It may be that another great change is on the way. Once parts became interchangeable, people soon followed suit. Now the factories themselves may go the same way. The parts will evolve, their histories guiding and limiting them. The new whole, the new machines of production, remains to be invented.

The Multiplication of Labour

Frederick Taylor searched for the one best way to do the job with his measures and stopwatches. But there may not be only one best way. In a flexible world, by definition, there are many ways, and many trade-offs between them. Which is best depends on what you value. The same is true of possible responses to new technology. Agility, virtual enterprises and the like provide a range of choices all of which may be correct for some, but not for others.

In his book "Die Fraktale Fabrik" (Springer-Verlag), Hans-Jürgen Warnecke, head of the Fraunhofer Institute in Stuttgart, uses an image from geometry. Not the perfect Euclidean geometry of CAD drawings and platonic forms, but the new geometry of fractals, endlessly complex shapes generated from simple mathematics that organise themselves in unpredictable but almost regular ways; their distinctive forms repeat on different scales, geometric echoes fading to the infinitely small. The image is one of worker empowerment through technology, with the local organisation of teams on the shop floor echoing the organisation of the whole company. This repeated complexity is the working out of a single corporate aim.

Fractals are computer-generated; before the advent of cheap number-crunching, such patterns remained hidden in dull equations, the outcome of which could never be seen. A fractal factory depends on computers, but also on people, and the new possibilities created within jobs that have sometimes been dull. A man working with a tool used to look inwards, managing the tool. Now that the tool takes care of the job on its own, he can look outwards, at the flow of work to and from the tool, at how the tool could be better used, at maintenance requirements. Taylorised work required men to ignore everything except what they did; computerised work makes it possible for the nature of work—its flow, its quality, its design—to be reclaimed as an area of workers' collective expertise.

Possible, but not necessary. Technology does not force; it permits. The future in some factories may be one of a newly sophisticated Taylorism, with work controlled by networks beyond the reach of the workers. John Alic, of the American Congress's Office of Technology Assessment, points out that new technologies can make work more opaque, harder for a worker to understand: alienating. However, there is research showing that in workplaces where the shop floor is given local control over new technology, the technology is better used than when it is controlled elsewhere. People offer solutions, as well as problems; they often turn out to be better at organising themselves than bosses expect. Decentralised decision-making seems an obvious corollary of decentralised information.

Perhaps machines will do this themselves one day. Engels believed that the human mind had evolved to take full advantage of the human hand; more recently, many researchers have preferred the idea that the mind is an organ evolved to allow complex social interations. Computers will face similar challenges. The fact that solutions can always be invented for computers by humans who have already evolved consciousness may mean that computer consciousness never develops. If it does, though, it might well be in response to the tasks of the factory. Manufacturing systems, unlike other places where computers are used, offer a place to act, not just data to process; two worlds to live in at once, their border endless in its complexity.

Frederick Taylor saw no fractals. His was a mechanical sociology, a theory of the factory as machine in which the flow of information and work was embodied in material products. Workers' intelligence was largely wasted. Now it needs to be captured. Information is becoming disembodied. For the moment, at least, bridging the gaps this opens up between ideas and deeds requires people. A post-Taylorist view of manufacturing needs explicitly to understand the non-mechanical aspects of the process provided by the people within: the ability to respond and adapt, to form coalitions, to maintain a complex internal balance.

Machines cannot yet come to life. But factories and manufacturing can. The image of the machine has been projected on to factories since the 19th century heyday of mechanisation. Now biology is the dominant science. It is in the ability of living things to evolve and to create, to act and to process information that new metaphors for manufacturing will be found. A machine is more than the sum of its parts; but a living creature is far, far more.

Reading 1.2
The Factory as a Learning Laboratory
Sloan Management Review

Dorothy Leonard-Barton
Harvard Business School

What is the next production frontier? The author argues that it is operating factories as "learning laboratories." These are complex organizational ecosystems that integrate problem solving, internal knowledge, innovation and experimentation, and external information. Chaparral Steel is the model example, and the experiences of its managers and employees are used throughout the article to show how the learning laboratory works.

Just after 2:00 A.M. on 26 March 1991, the night crew at the Chaparral Steel minimill in Midlothian, Texas, cast the first production run of "near netshape" steel beams in the United States. General Manager Lou Colatriano grinned with tired satisfaction and headed home for his first break in forty-four hours. From a sketch on a paper napkin to the first red-hot slab of metal that emerged from a patented mold to streak down the new mill line, the elapsed time was twenty-seven months. That included design, mold development, modeling, pilot runs, and construction, and it represented expertise from five companies on three continents. "One of our core competencies," explained CEO Gordon Forward, "is the rapid realization of new technology into products. We are a learning organization."

Every manufacturing company in the United States would like to be able to make that statement. Yet a steel mill seems an unlikely place to look for lessons on the quick commercial realization of inventions. In fact, so does any factory, as innovation is generally associated with research laboratories and development organizations. Moreover, we usually assume that the pressure to get product out the door conflicts with learning. But as speed to market becomes an increasingly important criterion of competitive success, we need to rethink our concept of what a factory is. Factories *can* be learning laboratories.

For decades, U.S. factories were passive service organizations to the rest of the company, churning out products designed without benefit of manufacturing input and burying product and process defects under mountains of inventory. The past decade has seen a transformation of many of these operations into low-inventory organizations dedicated to total quality and to active participation in new product development. What is the next frontier for production? In this article I suggest that it is running operations as learning laboratories.

Source: Reprinted from "The Factory as a Learning Laboratory," by Dorothy Leonard-Barton, *Sloan Management Review,* Fall 1992, pp. 23-38, by permission of publisher.

What Is a Learning Laboratory?

A learning laboratory is an organization dedicated to knowledge creation, collection, and control. Contribution to knowledge is a key criterion for all activities, albeit not the only one. In a learning laboratory, tremendous amounts of knowledge and skill are embedded in physical equipment and processes and embodied in people. More important, however, are the nontechnical aspects, the managerial practices and underlying values that constantly renew and support the knowledge bases.

In this article I put Chaparral Steel under the microscope as an example of a highly successful learning laboratory; its leadership has put tremendous effort into creating a consistent learning system. Many of Chaparral's practices are found piecemeal (often experimentally) in other U.S. organizations. As my references suggest, scholars studying best practices among Japanese manufacturers have made strikingly parallel observations, and Chaparral's policies are consistent with prescriptions by organizational learning theorists. Whether you are managing a fabrication shop or claims processing in an insurance office, Chaparral's management system offers a potentially useful model.

Taking an Organic System View

Chaparral is the tenth largest U.S. steel producer. Its high-quality standards have been rewarded by the market, and it consistently sets records for productivity, compared to both U.S. and Asian competitors (see the sidebar, "The Evidence on Chaparral Steel" on page 40). Clearly, whatever it is doing works.

A close look at the company reveals an organic learning system so tightly integrated that Forward says he can tour competitors through the plant, show them almost "*everything,* and we will be giving away *nothing* because they can't take it home with them." His confidence derives from the fact that the learning laboratory cannot be constructed piecemeal. It is comprehensible only as an organic whole; close scrutiny is required to appreciate its delicacy. To complicate matters, such a corporate ecosystem is in continuous flux, constantly regenerating itself. Even if a competitor identifies important elements of the system, emulation will require time. By then, Chaparral managers trust they will have moved on to the next innovation. The Chaparral system has evolved in response to a turbulent competitive environment, as Forward observes: "We have to go like hell all the time. If the price of what we sell goes up too high, . . . all of a sudden lots of folks will be jumping in. And they can get

The Evidence on Chaparral Steel

Chaparral has been setting records in steel production almost since its inception in 1975. In October 1982, it set a world record by producing 67,666 tons in one month—then the highest monthly tonnage produced from a single electric furnace/continuous casting combination. By 1984, the company was listed in *Fortune*'s May 28 cover story on the ten best-managed factories on the United States.

Chaparral's pace of improved performance continues unabated. Compared to a Japanese average output of 600 tons per worker-year and a U.S. output of 350 tons, Chaparral put out 1,100 tons of steel per worker-year in 1989. Its 1990 productivity figures of 1.5 worker-hours per rolled ton of steel compared to an overall U.S. average of 5.3 (including integrated steel mills), a Japanese average of 5.6, and a German average of 5.7. Chaparral is currently the largest supplier of steel rod for the oil industry and the largest supplier of rod for mobile home frames in the United States.

The company has won significant international recognition for quality as well. In spring 1989, Chaparral was awarded the right to use the Japanese Industrial Standard certification on its general structural steel products. It was the first U.S. steel company outside Japan given that privilage. Out of approximately fourteen U.S. companies producing hot-rolled carbon bars and shapes, the compamy claims to be the only one certified by the AIME (American Institute of Mining, Metallurgical, and Petroleum Engineers) for nuclear applications. Similarly, it is one of two companies, out of fifteen, whose steel is certified by the American Builders of Ships and that therefore do not have to seek certification for each individual job.

Chaparral has a reputatuion for quality at the low-cost end of its product line as well. A customer's study comparing scrap rates of Chaparral steel to that of the only other company from which the customer purchased, revealed a difference ranging from four to twenty-five times for the much more expensive alloys.

into business in eighteen months or so. . . . We constantly chip away the ground we stand on. We have to keep out front all the time."[1]

A learning laboratory does not occur spontaneously but is designed, created, and maintained through constant managerial attention to communicating the underlying values, checking the management systems' smallest details for consistency, and adapting any inharmonious elements. Thus managers designing a learning laboratory need to adopt holistic, systems thinking. They will have to acknowledge the practical utility and bottom-line impact of corporate values. They may have to confront chronic underestimation of the interdependence between incentive and education systems and corporate strategy. Moreover, such systems thinking must permeate the organization's every level. Everyone in the firm must appreciate the self-reinforcing nature of knowledge-creating activities. Only by comprehending the whole system can one understand why, when a fragment of the learning laboratory is pulled out to be examined (a particular project, a specific learning activity), it comes out vinelike, trailing roots back to deeply held values and widely observed management practices. It is this intense interconnectedness that makes such systems difficult to imitate and fragile—but effective.[2]

Learning requires creation and control of both external and internal knowledge for both current and future operations. Therefore, four distinguishing activities are critical to a learning laboratory: (1) problem solving (in current operations); (2) internal knowledge integration (across functions and projects); (3) innovation and experimentation (to build for the future); and (4) integration of external information flows (see Figure 1).[3] Each activity is the operational expression of an underlying value and is strongly supported by a compatible managerial system of procedures and incentives. Thus each activity, value, and managerial system functions as an internally consistent subsystem. Although I will describe each subsystem separately, the four are mutually aligned and interrelated; that is, values and managerial systems underlying one subsystem also support the other three. An in-depth look at the four subsystems at Chaparral suggests some of the principles that distinguish a learning laboratory.[4]

Before exploring these subsystems, however, we need a brief explanation of Chaparral's near net-shape project, an innovative foray directly into the traditional territory of "Big Steel" that illustrates the company's knowledge creation and control subsystems.

The Near Net-Shape Project

One way to push equipment performance and ensure learning is to set goals for each project considerably beyond current production capabilities.[5] Chaparral managers set a very ambitious goal for the near net-shape project: to produce large (eighteen- and twenty-four-inch wide) structural steel I-beams for the same per-pound cost as the simple rounding reinforcing bars ("rebars"), the company's first product. "We knew what it took to make rebar," explains General Manager Duff Hunt. "The challenge was to produce a more difficult product for the same cost . . . . We decided on the twenty-four-inch because we wanted to explore the most technically challenging product. Becuase there may be some idiosyncracies about the process that we need to learn, we also chose a second size . . . based on the amount of North American consumption." Reaching this cost objective (half of Big Steel's) required drastically reducing the energy costs (roughly 25 percent of total) of rolling the steel into the required end shape. The nearer to the

FIGURE 1

Knowledge Creation and Control Activities in a Learning Laboratory

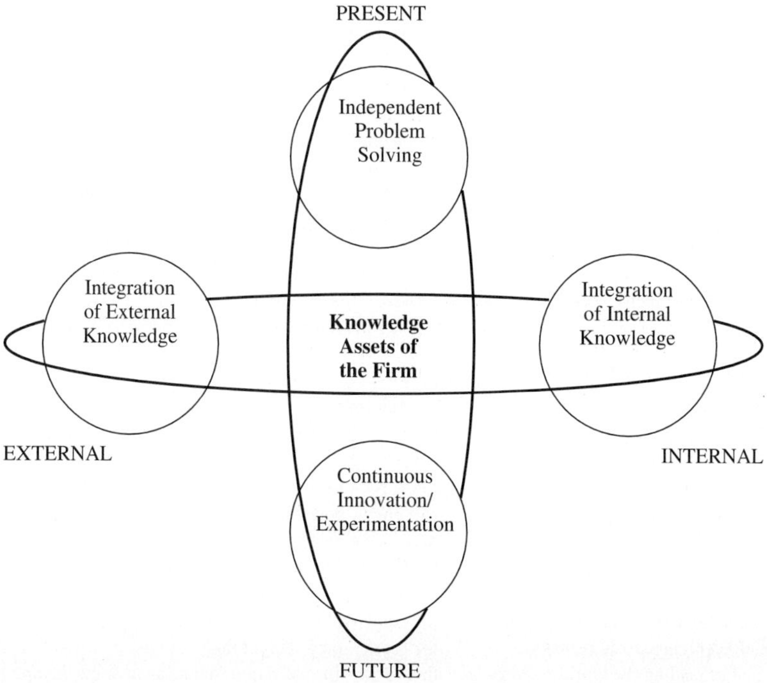

final ("net") shape the molten steel could be cast, the less rolling required.

The only known processes for casting steel to its near net shape were far too capital and labor intensive for Chaparral, which views large investments in either capital or labor as a threat to future flexibility. The steel-making molds and processes developed for the near netshape project therefore embody knowledge beyond anything available on the market and, in fact, beyond anything the leading vendors of steel-making molds thought possible. The approximately four-foot copper alloy tube through which the molten steel passes to emerge in nearly I-beam shape is deceptively simple in appearance. However, the mold is patented, which suggests that the combined expertise of Chaparral designers and Italian and German mold fabricators endowed it with cutting-edge knowledge.

> **Empowered individuals have the self-confidence, freedom, and motivation to continuously solve problems.**

The mold is contoured so the hot metal won't bind as it shrinks going through, and the fabrication methods produce an absolutely smooth finish on the sides, to lessen chances of rending the thin skin surrounding the molten steel as it emerges. It is difficult to imagine how Chaparral could have succeeded in this knowledge-intensive project had the company not been run as a learning laboratory.

The Learning Laboratory System

Some of the activities, values, and managerial practices described below characterize any high-quality organization. Others are very unusual—at least in the United States. Since they operate as a system, we have to look at the whole operation to understand why emulating just some parts may never produce true learning laboratories.

Subsystem One: Owning the Problem and Solving It. The first critical learning subsystem is the triad of (1) the independent problem solving required for continuous improvement of current processes, (2) egalitarianism as an underlying value, and (3) shared rewards as the reinforcing incentives system (see Figure 2). Empowered individuals, who command respect in the organization and who feel ownership in the system, have the self-confidence, freedom, and motivation to continuously solve problems. The principle involved here turns the old production saw of "if it ain't broke, don't fix it" on its head and maintains: "if it ain't being fixed continuously, it's broke."

- **Activity: Indpendent Problem Solving.** Learning starts with empowered individuals who can identify and solve problems independently because they have a clear sense of operating objectives. The often-articulated vision at

FIGURE 2

Subsystem One: Owning the Problem and Solving It

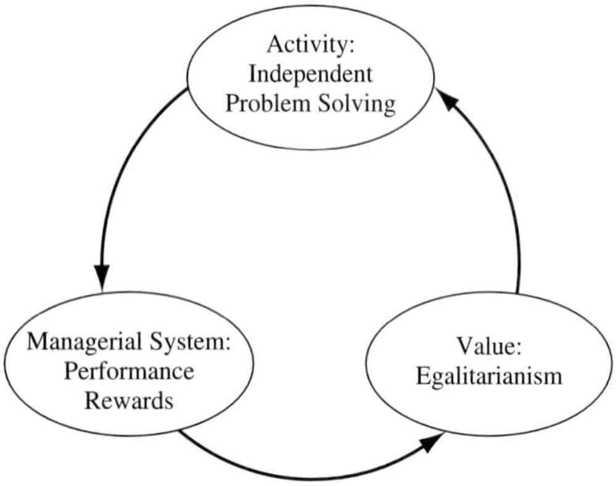

> Chaparral is to lead the world in the low-cost, safe production of high-quality steel. This resembles the mission of many other companies that are not learning organizations. For a learning lab, the test of a vision is the extent to which is permeates the whole organization, guiding every micro and macro decision. Unless the vision can be directly translated into operational principles, that is, into guidelines for running the factory lines, it may have little effect on actual shop-floor behavior. Chaparral's goal of leading the world requires innovation beyond the current cutting edge of production techniques. Maintaining a cost advantage requires constant improvements in productivity. The vision dictates that those improvements cannot come at the expense of quality or employee safety. Therefore the goal for every hour, the criterion for every person's activity, is crystal clear: make ever more steel—increasingly better than anyone else. "We never stray far from the market," says Hunt.

In a learning environment, progress has to be everyone's business—not just that of a few specialists. At Chaparral, who owns a production problem and responsibility for its solution? An incident during the first few weeks of operating the near net-shape caster, when cooling hoses were bursting, provides some insight. "When something like that comes up, and there seems no immediate solution," explains a senior operator, "you go see what the problem is. You don't say, 'That's not my area,' or 'I don't know that much about it.' You just show up." In this case, a group of operators, a welder, some foremen, and a buyer spontaneously scattered to seek solutions. "Everybody telephoned some person they thought might know how to fix the problem—vendors, experts—and within three to four hours we were getting calls back," says the senior operator. "Service people were showing up, and we worked the problem out. If it had been just one guy, probably a foreman, and everyone walked out, . . . it would have taken him ten times longer to find a solution."

Because the performance goal is very clear, supervisors do not need to micromanage the line, and the organization runs lean. For instance, two months after the first run of the near net-shape casting, the pulpit controls operator is carefully checking the timing on the line with a stopwatch. The red-hot beams pass through the rolling mill stand once, then stop, reverse, and go through again. Meanwhile, the flow of steel behind the beam being rolled is diverted. Every second of unnecessary diversion costs money because the diverted steel will have to be reheated to be rolled. Therefore the operator wants to achieve split-second timing. Asked who suggested he perform this function (which is often given to a process engineer elsewhere), he is surprised at the question: "No one." He considers it obvious that improvement is always a part of his job: "We're still learning, but we will get it. You just take it upon youself to roll as much steel as you can whether you are the pulpit operator, the roller, the furnace operator. . . . We want to do the best we possibly can. Everybody here has the attitude to pick up the pace."

One of the greatest advantages of this attitude, as a maintenance foreman points out, is that "ideas come from just about everybody. The operators working on the equipment have a lot of input because they see the exact problems when they happen." Moreover, potential improvements are immediately enacted with no wait for management approval or standardization of "best practices." If it works, it is the *de facto* standard. If it improves performance, everyone will imitate it. "Whoever can come up with an idea on how to fix it, from the millwrights or myself right on up to the top,. . . does it right then," explains a foreman. At Chaparral, there is no formal requirement, as at some Japanese companies, for a certain number of improvement suggestions from each employee. Everyone is involved in some process improvement projects, and, as the foreman explains, "We are all out here to make it run. Probably 90 percent of the problems never even make it to the morning meetings [held among everyone on the shift to discuss problems]. They are fixed in the field."[6]

The downside to this intrapreneurial attitude is that although managers set the goals, no one has the authority to tell another employee *how* to accomplish a task. Process engineers and supervisors who know a better procedure often have difficulty convincing operators on the line—much less their peers. "You can't tell them how to do it," Administration Vice President Dennis Beach admits somewhat ruefully, "and they don't do it the way you would." The engineers concur. They are not called upon enough, in their opinions. However, the benefits from general ownership of all problems are that it is not possible to "pass the buck" and no one expects a steady-state manufacturing process—ever.

- **Value Egalitarianism and Respect for the Individual.** A learning environment is premised on egalitarianism, the assumption that all individuals have potential to contribute to the joint enterprise (if they are willing to develop competence). Forward has observed, "We figure that if we could tap the egos of everyone in the company, we could move mountains."[7] Outward symbols of values are important because they convey meaningful messages to all employees. At Chaparral there are no assigned parking places, no different colored hard hats or uniforms reflecting title or position, and the company dining room is a local diner. More unusual, a scant two levels separate the CEO from operators in the rolling mill; a visitor is surprised when an operator stops Foward on a walk through the plant to discuss a new product's problems. A millwright notes: "If you have tact, you can tell anybody from Mr. Forward on down exactly what is on your mind. There is no problem in expressing your opinion here." Repeated requests to interview a particularly knowledgeable project manager are politely rebuffed with a shrug and an explanation from a vice president: "Sorry, he just doesn't want to. I can't *make* him." However, respect for the individual does not mean equality of responsibility, lack of discipline, or even consensual decision making. Chaparral managers believe that a supervisor should be a leader, trained to make good decisions—including hiring and firing.

- **Managerial System: Performance Rewards.** Positive thinking and slogans alone cannot create genuine employee investment in innovation and in identifying and solving problems.[8] Chaparral has taken unusual steps to ensure that peformance and incentive systems back up management's belief in egalitarianism. In 1986–1987, when employment leveled off, and management confronted potential stagnation, the pay structure was overhauled to reward accumulation of skills as well as performance. Even more radical (especially for a steel mill) was the switch from hourly wages to salary for everyone. There are no time clocks at Chaparral. Forward explains, "When I am ill, I get a day off. Why shouldn't everyone else?" He is fond of saying that the management system was designed for the 97 percent who are "conscientious people who want to put in a full day's work." The 3 percent who abused the system were let go. Beach summarizes the philosophy: "We manage by adultery. We treat everyone like an adult." Moreoever, the pain and pleasure of work are spread around. For instance, in contrast to many factories, operators are not assigned to shifts according to seniority. Everyone rotates into night shift, a practice that fosters knowledge accumulation twenty-four hours a day and guards against a possible disproportionate accumulation of skills and knowledge in the most favored hours. The reward for seniority is thus not greater comfort, but more challenges. How do senior employees feel about that? "During the summer in Texas," one foreman rationalizes with a grin, "you'd *rather* work at night."

Steel-making is hot, dirty, demanding work; the pace of innovation and the constant pressure to produce more, better, and more safely in this organization exacerbates the brutal conditions imposed by the technology. What keeps the workforce from burning out? One factor is the incentive systems that have evolved.[9]

Bonus schemes are linked to company profits—for everyone. "We think janitors and secretaries are important in our success, too, and they should share in the rewards," observes Forward. An operator comments, "The more money the company makes, the more money I make. The profit-sharing system creates built-in pride." Further, 93 percent of the employees are stockholders and together own 3 percent of the stock. In 1988, each employee received one share for every year worked at the company; 62 percent buy additional shares every month through payroll deductions. Although the monetary implications are small, this policy is consistent with the rest of the rewards structure, and some employees find it symbolically important. A furnace controls operator comments, "I feel like this company partly belongs to me. Owning part of the company makes you care. I take better care not to waste anything because I feel like I am paying for it."

These could be just words, of course. Scholars studying the relationship between such incentive systems and productivity have not reached consensus about the benefits of employee shareholding (partly because shareholding agreements vary widely). However, there is strong support for the notion that profit sharing generally does raise productivity.[10] At Chaparral, the dual incentives of employee profit sharing and shareholding seem a natural complement to the rest of the learning laboratory system.

Subsystem Two: Garnering and Integrating Knowledge. The second subsystem revolves around knowledge accrual (see Figure 3). In the learning laboratory, knowledge is highly and visibly valued. Management invests in educating the whole person, not just the technical side, and knowledge flows freely across boundaries. The principle is this: every day, in every project, add to the knowledge resources.

- **Activity: Integrating Internal Knowledge.** In a learning laboratory, one would expect to see visible embodiment of knowledge creation and control in highly innovative physical systems. Chaparral boasts of such cutting-edge equipment as an automobile shredder that they believe is the fastest and most efficient in the world, a horizontal (instead of vertical) caster, and some of the most advanced digital furnace controls anywhere. Because of the constant push to improve production, Chaparral managers have to design what they need, rather than purchase the best available equipment off-the-shelf. Why design in-house? "To keep the knowledge here," a mill manager explains. Moreover, managers assume that the performance of *any* purchased equipment can be improved. Some improvements are novel enough to be patented. Rolling mill equipment that its vendor believed was limited to eight-inch slabs is turning out fourteen-inch, and the vendor has tried to buy back the redesign. The two electric arc furnaces, designed originally to melt annual rates of 250,000 and 500,000 tons of scrap metal, respectively, now produce over 600,000 and 1 million tons, respectively.

Chaparral management emphasizes homogenizing the level of knowledge throughout.

FIGURE 3

Subsystem Two: Garnering and Integrating Knowledge

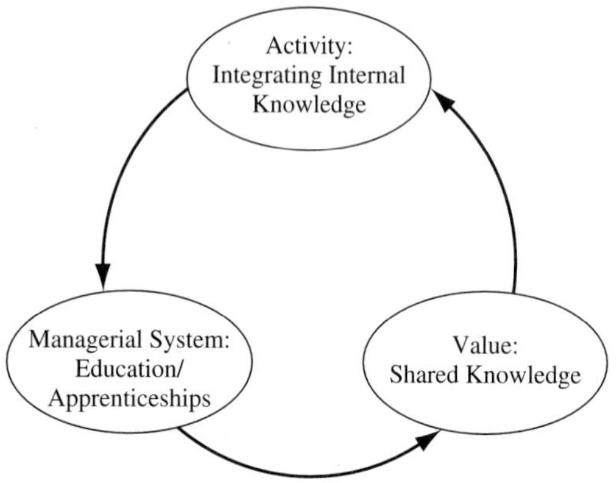

The physical processes following the molding step are also knowledge intensive. Only one other steel mill in the world does the hot-link rolling employed in the near net-shape casting project, whereby hot cast steel is sent directly into the rolling mills. Using this technique was a big risk because no one knew exactly what the properties of the thinner cast steel would be. The hot link meant less subsequent rolling than usual and hence less "working" of the steel to obtain the desired crystalline structure. The risk paid off when the combination near net-shape casting and hot link turned out to produce steel with exceptionally good metallurgical characteristics. Chaparral processes seem clear outgrowths of skill, yet how does such expertise and knowledge accumulate?

Just as continuous processing has great advantages for manufacturing over most batch processing, so the unimpeded flow of information aids learning more than fragmented, batch-processed information.[11] Most organizations are physically structured to emphasize vertical (hierarchical) and horizontal (functional) boundaries. In contrast, Chaparral management emphasizes homogenizing the level of knowledge throughout; few pockets of information are isolated by position, function, or working shift.[12]

Information flow at Chaparral is obviously aided by its size, deliberately held to under a thousand employees. An individual garnering knowledge on a trip, at a conference, or from an experiment can readily transmit it, as all employees are located in the same place and know each other. The company was also designed to facilitate knowledge flows by encouraging as many accidental meetings as possible. The plant layout accommodates the hands-on style of management favored at Chaparral; even Forward's office is just steps away from the furnaces and mills. The locker room is located here also, so that at least once a day employees cycle through the one-story headquarters building.[13] Consequently, meetings are as likely to be held in the halls as in the conference rooms. Since many decisions go unrecorded and memos are anathema, it is important that people see each other frequently.

Hierarchical boundaries are minimal. This is a do-it-yourself company with no acknowledged staff positions and only a few positions that seem stafflike, such as personnel. There are fifty graduate engineers and technicians, all with line duties. In fact, everyone has line responsibilities, most of them tied directly to steel production, and decision making is pushed down to the lowest possible supervisory level, "where the knowledge is." Lead operators are selected for their knowledge-transmitting as well as knowledge-creating skills, because much knowledge flows horizontally among peers. Work is structured with the objective of disseminating knowledge. For instance, in commissioning the new mill that receives the near net-shape product (that is, ramping it up to problem-free production), only two teams of operators are being trained. Each team works a twelve-hour shift (with paid overtime). After the initial eight weeks of this grueling schedule, these operators will be dispersed among the rest of the crews to diffuse the knowledge they have created and assimilated about the new process's idiosyncrasies.

Chaparral has also proven that traditional horizontal boundaries can be redrawn and expectations altered. The quality control department is responsible for reacting to quality problems identified by operators on the line—not by separate inspectors. Production workers do 40 percent of maintenance tasks. A maintenance foreman notes that "at Chaparral, we get involved with the whole process. We are not just tied to one area."

Although the company has a marketing department, everyone is considered a salesperson. Every employee from CEO to receptionist has a business card to use with customers. Security guards do data entry while on night duty and are trained paramedics as well. Such multifunctional experience is encouraged not only to make the organization more flexible but because management believes it discourages territorial possessiveness over information.

No research and development (R&D) department exists separate from production. Forward maintains that "everybody is in research and development. The plant is our laboratory." This statement has implications beyond the obvious that there is no research function. Some of the problem solving and experimentation that are done in the midst of production involve research. However, interfacing with the possessors of the latest scientific knowledge is not restricted to an elite group of specialists. Knowledge accumulation in a learning lab cannot be the responsibility of a few special people. Forward describes the large, separate research centers in some companies as "lovely, really nice. But the first time I went into one of them I thought I was entering Forest Lawn (Cemetery). After you spend some time there, you realize you *are* in Forest Lawn. Not because there are no good ideas there, but because the good ideas are dying there all the time."[14]

At Chaparral, even operators participate in R&D activities. When the world's leading supplier was constructing the patented mold for the near net-shape project, the three-person team that shuttled back and forth to the site in Germany, learning about the fabrication process and serving as a source of information about the intended production process, included an operator. As described later in Subsystem Four, Chaparral employees at all levels are constantly tapping into the latest, most current knowledge banks around the world.

- **Value: Shared Knowledge.** Since performance drives everything in this company, and individual incentives are tied to performance, employees seem engaged in a marathon relay race, where winning as a company team takes precedence over individual ownership of ideas, and knowledge is liberally shared. There are acknowledged experts, such as the director of operations who is an "equipment whiz," and the mold expert who was largely responsible for the near-net shape design, and the organization does not lack for large egos. However, Vice President of Operations Dave Fournie says that "you don't *have* to have credit for particular ideas to be thought good at your job. Lots of innovations take more than one good idea. They go through a gestation period, and lots of people figure out how to make sense of it. The point is to focus on the good of the *whole.* That's why we don't have suggestion boxes, where you hide ideas so someone else won't steal them."

Chaparral employees are often unable to identify the source of production innovation. Production Manager Paul Wilson explains, "It is hard to say who fathers an idea. It doesn't make any difference. Everyone share in the pride of doing, and if the experiment fails, everyone shares in the failure. In other places, a few poeple do a lot of innovating. Here a lot of people do little bits that add up." A millwright makes similar observations: "No one is looked down upon. If I am supposed to know more than somebody else but that other person catches [an oversight], well that's fine—he just bailed me out. I am more than likely going to help him out a different time."

- **Managerial System: Apprenticeships and Education.** An organization that values knowledge must provide mechanisms for continuous learning. Chaparral management has sent some employees to school to obtain advanced degrees, but it also invests heavily in an unusual formal apprenticeship program for everyone in the plant, which is developed with the Bureau of Apprenticeship and Training in the U.S. Department of Labor. (Most apprenticeships are run by unions.) As Forward notes, "Expertise must be in the hands of the people that make the product."

The roughly three-and-a-half-year program allows apprentices to progress to the level of senior operator/craftsman by successfully completing 7,280 hours of on-the-job training and designated formal schooling. The foremen of individual crews schedule the on-the-job training and evaluate the candidate's systematic progression through various tasks in the factory. For example, 2,200 hours in steel-pouring operations is one qualification in the ladle metallurgy apprenticeship. In addition, all apprentices complete study programs in safety, operating processes across the entire company, mathematics as related to operations, metallurgy, and basic mechanical maintenance "in an effort to give all operators a solid basic understanding of the equipment used and the operating processes." Scheduling this study, which requires four hours of unpaid work a week, is left up to the individuals, who have five options: home study with manuals and videos; in-plant study in instructor-monitored study rooms; formal classes; personalized tutoring by instructors; and study in work areas as operations permit. (Operators may also get credit for prior experience by proving knowledge on the job.) Courses include basic engineering knowledge such as conduction in liquids and gases, as well as very specific skills such as ladder logic programming for understanding programmable controllers and lubricant storage and handling. Perhaps more surprising are such sessions as "working with other people" under "troubleshooting skills."

The most unusual aspect is the instructors. Selected foremen rotate in from the factory floor to teach. "It creates a lot of credibility for the education program on the factory floor," explains one instructor. "What's more," he adds wryly, "I have to live with what I teach. So I'd better do a good job."

Chaparral also invests in outside courses of all types, tending to train its own people rather than hiring those with a particular skill. A maintenance foreman attended a special engineering course because "no one knew anything about vibration analysis so I was asked to go study it." External education includes nontechnical courses as well. In few factories would a furnace controls operator attend a Dale Carnegie-type course to help him enhance the interpersonal skills critical to smooth teamwork. Line managers are expected to be technical experts, constantly up-to-date with cutting-edge technology—whether that be some aspect of metallurgy or human resources. Therefore every manager attends conferences and cultivates a professional network of information sources. The manager of the scrap shredding operations, for instance, has just attended a conference on recycling—in the future, Chaparral's expertise in recycling may be an important corporate capability.[15]

FIGURE 4

Subsystem Three: Challenging the Status Quo

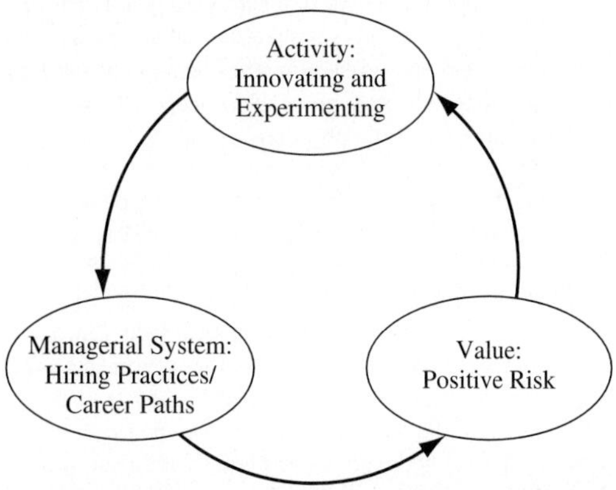

Subsystem Three: Challenging the Status Quo. The third subsystem in a learning laboratory involves constantly pushing knowledge frontiers (see Figure 4). The company must select employees for their desire to challenge their own and others' thinking. They must see risk as positive, because it comes with the experimentation critical to innovation. The company must select suppliers for their superior capabilities—and for their willingness to be pushed beyond the bounds of their current knowledge. The principle involved is this: always reach beyond your grasp.

- **Activity: Continuous Experimenation.** Learning requires constant pushing beyond the known, and Chaparral employees are skilled experimenters. A visitor was surprised to find that extensive overhead slides explaing the formal Taguchi experimental designs that were guiding the development of the horizontal caster had been prepared for the board of directors. This extremely technical presentation was to help directors understand the methodical knowledge-creation process, enable them to identify critical decision points, and thereby better equip them to evaluate the risk. Another example of knowledge-creating experimentation more often found in research laboratories than factories is the one-sixth scale model of the near net-shape caster, which uses water to approximate the flow of steel. Standing alongside the rolling mill, close to a water source, this model allowed testing of several types of baffles and lengths of mold. Resulting observations guided much of the eventual mold design decisions.

Innumerable large and small experiments are more of the "cut-and-try" variety. "We aren't always as systematic about our learning as we could be," Forward admits. Many creative simulations are conducted right on the production line. The advantage is that the more closely the experimental environment approximates the final production environment, the more immediately relevant is the information generated. The disadvantage is the obvious potential to disrupt production—which is why most factories prefer to isolate experimentation. In one of many projects leading to the near net-shape casting, a prototype of metal splashboards was first constructed out of plywood. By continuously soaking the wood in water, the crews were able to keep it from being consumed by molten steel just long enough to prove the concept. "We were the local hardware store's favorite plywood customers for a while," one employee recalled. Similarly, during the design of the near net-shape caster, several prototype molds of almost pure copper were used to determine if casting a dog-bone-shaped slab was possible. The soft molds held together just long enough to show this new casting process's feasibility.[16]

The operating rule is this: if you have an idea, try it. Line managers authorize tens of thousands of dollars for experiments without higher authority. Hunt explains, "We use products to do research. We can close the feedback loop between researchers and users by using new methods and new materials within our own facility." Wilson agrees: "In other companies, the word is—don't rock the boat. Here we rock the hell out of the boat. We don't know the factory's limits. We want it to change, to evolve."

Not every suggestion is instantly accepted. When a maintenance operator at a conference spied new digital furnace pulpit controls, he then had to convince his supervisor to invest. For almost two years, the supervisor remained uninterested in visiting the Mexican vendor and a site using the controls. Yet the operator persisted. Finally, his supervisor understood the potential of these state-of-the-art controls and agreed to their installation and customization to Chaparral operators' specifications.

During the development of the near net-shape beams, Chaparral frequently disconcerted development partners by its innovation speed. One innovation followed so hard on the heels of the previous one that they almost overlapped.

If everyone experiments, learns, and innovates, then neither success nor failure can be heavily personalized.

When the German mold developers finished a prototype on schedule and called to ask how they should ship it, they were astonished to be told, "Don't ship it at all; cut it up and make it look like this." The designers had already learned enough from the prototyping process to improve the design. A new mold would be required. What surprised partners was that Chaparral did not appear to begrudge the "wasted" $40,000 per mold, since the knowledge engendered by each prototype enabled the next step in innovation.

- **Value: Positive Risk.** In a research laboratory, risk is accepted as the norm, since the cutting edge is always fraught with uncertainty. In contrast, risk is usually anathema in a production environment. Managers of a learning factory must tolerate, even welcome, a certain amount of risk as a concomitant of knowledge acquisition. Chaparral managers avoid riskless projects because a "sure thing" holds no promise of competitive advantage—no opportunity to outlearn competitors. Says Forward, "We look at risk differently from other people. We always ask what is the risk of doing *nothing*. We don't bet the company, but if we're not taking some calculated risks, if we stop growing, we may die."

This positive attitude toward risk permeates the company. If everyone experiments, learns, and innovates, then neither success nor failure can be heavily personalized. If individuals are singled out for praise, then they have an incentive to protect ideas as intellectual property rather than seek embellishment from friendly critics and codevelopers. If individuals are singled out for blame, then the risk of failure may overwhelm the impulse to innovate. What happens when you try something and it doesn't work? What is the penalty? "Everybody makes mistakes," a Chaparral foreman responds. "You don't have to cover up a mistake here. You just fix it and keep on going." The philosophy appears to apply even when the failed experiment is very expensive. In 1986, when Fournie was medium section mill superintendent, he championed the installation of a $1.5 million arc saw for cutting finished beams. Not only did the magnetic fields attract any small unattached pieces of metal for yards around, including pens and watches, but the engineers were never able to refine the equipment to the point of effective operation. Since promoted to vice president of operations, Fournie is somewhat amused to find that visitors "can't believe you can make a mistake like that and not get crucified." He tries to take the same attitude toward those who work for him: "You don't start them out on $1.5 million projects, but you have to give them freedom to make mistakes. The reward for having ideas is getting to carry them out. You give them bigger and bigger chunks and evaluate." Operators on the line have a consistent view of the atmosphere for innovation: "You have an idea—good, bad, indifferent—just spit it out and we'll talk about it. Someone may laugh [at it] but you'll laugh back next week."

A potential hazard of this positive attitude toward risk is that no one wants to admit a mission is impossible. "Once we say we will do something," Fournie explains, "we hang on and try like nobody's ever tried before. Tenacity makes lots of projects work here that don't work other places—but when it just can't be done, it's hard to call the project off." One of his key criteria for determining if people are ready for promotion is whether they know when to ask for help, when to admit that they are in over their heads: "It's a tough call . . . the hardest decision to learn to make." He speaks from experience. He not only set up the arc saw project; he killed it.

- **Managerial System: Hiring Practices and Career Paths.** The most important managerial system in a learning laboratory is selecting and retaining the right employees. Because employees must be innovators, constantly challenging the status quo, they are selected as much for their potential, their attitude toward learning, and their enthusiasm as for a specific background. Although top managers and a few specialized "gurus" at Chaparral have extensive steel experience, when Chaparral was first set up, management decided not to look for workers with industry experience. Beach explains: "We were looking for bright, enthusiastic, articulate people, and we preferred people who had not been exposed to other companies' bad habits." The company therefore hired (and continues to hire) from the immediate geographic area, seeking ranchers and farmers with mechanical ability but no steel experience. Chaparral looks for "a twinkle in the eye, a zest for life," for "basically conscientious people who can put in a strong day's work and enjoy what they're doing." This love of work is critical in an organization that deliberately runs somewhat understaffed to avoid laying people off during market downturns. A learning laboratory is not a good match for someone who works solely for a paycheck or who equates promotion with increasingly easy work.

Chaparral's original applicants went through six weeks of intensive training with daily evaluations and faced stiff competition. Top performers were given their choice of jobs and an immediate 20 percent pay raise. Highly selective hiring procedures continue to reflect concern that new employees fit into the Chaparral culture, that at least one supervisor be personally committed to their training and progress, and that the team have a stake in their success. Although personnel does some preliminary screening, current applicants undergo one or more days of demanding interviews with at least five employees, including two foremen, before they join Chaparral. Only one out of ten applicants selected for interviews can expect to be hired, and the final decision belongs to the foreman with direct responsibility. These very cautious, resource-intensive selection practices may account in part for the extremely dedicated workforce, which boasts an absentee rate about one-fourth that represented by the National Association of Manufacturers.

Since people are the key resource in a knowledge-intensive organization, keeping them is critical.[17] Beach explains: "From the very beginning, we designed this organization with Maslow's hierarchy of needs in mind [i.e., that once people's basic needs for food, shelter, and belonging are satisfied, they will aspire to a fourth level of need, self-esteem, and finally, the fifth, self-actualization.] I know it's not stylish, but we really believe in that hierarchy, so we constantly look at what will help people become self-actualized, at their ego needs. . . . People like a challenge and a well-defined goal out there."[18]

A strong aid in motivating continuous innovation is a clear path for advancement, not just in salary but in position. Asked what he looks for in a job, a millwright responds: "As long as I am moving forward, I can think for myself, and I feel that I am contributing, I will be happy." Chaparral managers believe that skilled, innovative people will leave an organization if they see no possibility for personal growth. During the 1986–1987 reorganization to address possible stagnation, some maintenance and shipping operations were reorganized with autonomous, self-directed teams with rotating leadership. Management regards this change as an organizational experiment that may provide a model for other parts of the company.

The company invests in crosstraining at a number of levels. Management has learned the necessity of training in advance of the need because of problems experienced early in the company's history, when operators who were automatically moved up into lead and then supervisory positions as vacancies opened proved underprepared. In response, Chaparral management established such idiosyncratic practices as "vice-ing." In other companies, when a foreman or supervisor is absent, usually the foreman from a prior shift stays on in that position. When a similar absence occurs at Chaparral, the prior foreman works the extra hours, at his usual pay level, but in a *subordinate* role. The most senior operator or craftsman is then temporarily promoted to "vice foreman" to cover the supervisory position. Thus the company retains operating experience in the form of the prior shift foreman and simultaneously trains the senior operator for future work as a foreman.

A supervisor explains, "You get a little pay increase and a whole lot more responsibility. I got a chance to see whether I really wanted to be a foreman or not, and it gives you a little respect for the boss's job."

Managers are similarly prepared for the future. Recently the production managers for the three mills were given the title of general manager and asked to learn each other's jobs and to cover for each other. This crosstraining is intended to prepare them for general management of an entire operation, when Chaparral starts up another site.[19]

Subsystem Four: Creating a Virtual Research Organization through Networking. A learning laboratory obviously needs access to the latest knowledge, embodied in the best minds and best equipment available. However, not all companies can afford an internal research organization. Moreover, no company can cover all the technological advances, worldwide, that may affect its future. Therefore the principle behind the fourth and final subsystem is this: create a virtual research organization through extensive networking and alliances—for learning and for economic reasons (see Figure 5).

- **Activity: Integrating External Knowledge.** At Chaparral, employees constantly scan the world for technical expertise that others have already invested in. Managers never hesitate to invent when necessary but only after assuring themselves through extensive searches that no available system will suit their needs. While building the horizontal caster, they made repeated trips to the few other world sites that had somewhat similar equipment. Chaparral also constantly benchmarks its capabilities, not just against immediate competitors but also against best-of-class companies, even those from totally different industries. Three on-site laboratories support production through chemical and physical product analysis, but the company has created a virtual research organization through extensive networking and alliances. Information obtained externally is rapidly incorporated through development projects, flowing through the created network almost as readily as it does inside the walls of the learning laboratory, because in both cases people working directly in production transmit the knowledge.

FIGURE 5

Subsystem Four: Creating a Virtual Organization

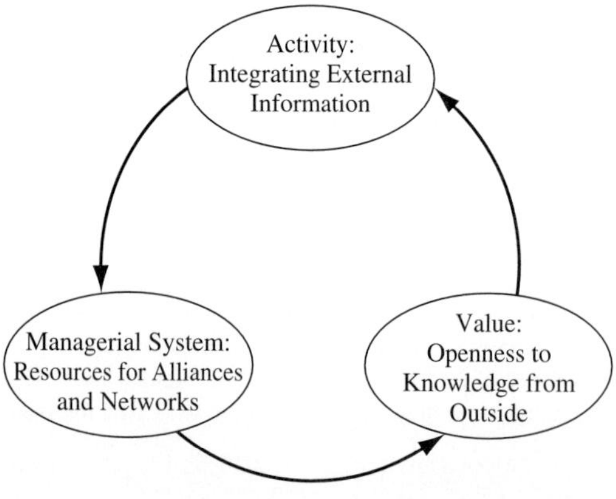

The network was heavily used in the near net-shape project. At its outset, a team of managers and foremen visited Japan and determined that one firm had the most advanced "profile" casting process in the world. Yet it was inflexible, would cost "more than the entire company," and was more labor intensive than desired. Chaparral therefore enlisted the help of German and Italian suppliers. At first extremely skeptical of this small company from the middle of Texas and hence unwilling to try to make the radical mold design, the German suppiler began to believe that seemingly impossible goals could be achieved only when it saw how far Chaparral had progressed with the help of the Italians. Chaparral employees visited the German vendor every few weeks, and ultimately Chaparral's expertise combined with the German's cutting-edge knowledge to produce a patented mold that neither company could have made alone.

Concerned that the relatively thin cast steel emerging from this mold might have undesirable metallurgical qualities, Chaparral reached again into its network for testing. Since it had not yet built its own rolling mill, and none existed anywhere that could receive the novel shape, managers identified a production laboratory in Mexico with very flexible equipment. Chaparral's superintendent of steel-melt technology went to Mexico to direct the simulation of the future mill design, using Mexican equipment and workers. To their delight, Chaparral managers discovered that the samples demonstrated superior characteristics.

The development and testing laboratories in the German and Mexican firms served as virtual extensions of the corporation, for they possessed special equipment and skills that complemented Chaparral's design capabilities. However, knowing such sources of expertise exist would be useless if the factory were not able to tap them, to jointly create more knowledge, and then to absorb that knowledge into the production system.

Chaparral's collegial knowledge network seems more characteristic of a research laboratory than of a factory. Possibly because of Forward's early career as a research metallurgist, the company aggressively pursues the latest knowledge. Chaparral sought the coveted Japanese certification for steel, not because managers thought they would ever sell much steel in Japan, but because they believed the Japanese would go through the company's process carefully—and Chaparral would learn much from the exercise. Forward takes a distinctly nontraditional view of environmental scanning for a factory: "By the time you hear about a technology in a paper at a conference, it is too late." This philosophy explains why Chaparral invests in unorthodox knowledge-gathering mechanisms, for instance, by cosponsoring a research conference with the Colorado School of Mining about a new alloy under investigation. Forward himself treks back regularly to his alma mater, MIT, to consult with university experts.[20]

Very occasionally, this method of R&D has failed the company. The most notable example is the previously mentioned electric arc saw. No one in the plant or in its network of experts, including the vendor, was able to solve the technical difficulties. The mill manager left the saw in place for almost a year to allow time for everyone to attempt solutions and to accept failure, while he sought an alternative technology. It is not clear that an in-house R&D facility would have helped in this case because the required invention lay in the unfamiliar realm of electromagnetic fields. Managers at Chaparral believe that through their virtual R&D organization, they can actually tap a larger variety of knowledge bases than they would be able to support internally, given their size. Still, they will have to continue investing in internal expertise in order to integrate the externally obtained knowledge.

- **Value: Openness to Knowledge from Outside.** Knowledge garnered through such networks can flourish only in an environment that rejects the "not invented here" mentality. At Chaparral, "not *reinvented* here" is the operative slogan. There is no value in recreating something—only in building on the best existing knowledge. People in a learning laboratory value the capability to absorb and use knowledge as much as to create it. They understand that all invention is a process of synthesis. As its practices suggest, a key Chaparral value is global outreach—openness to innovation, whatever its origin. Knowledge is valued not so much for the pedigree of its source but for its usefulness.
- **Managerial System: Resources for Alliances and Networks.** To support information gathering and reinforce global outreach, the company invests heavily in employee travel (and regards the expenses as just that—investments), often sending a team, including foremen and technical staff as well as vice presidents and operators, to investigate a new technology or to benchmark against competitors. Newly acquired knowledge need not filter down through the ranks, because the people who absorbed it are the ones who will apply it. In 1990, seventy-eight people from production, several of them operators, visited a customer site at least once. They also visit other minimills. Asked why he visited a sister plant, an operator states: "To see if I could pick up any new ideas to use here." And did he? "Yes, several." And he points out some small operational changes. Forward elaborates: "We send the people who can best tell us what's going on—whoever they are." Chaparral managers also invest in long-term relationships with suppliers to extend the network. In 1985, while seeking Japanese suppliers, they identified a high-quality company that was not doing much development. As a general manager explains, although it was clear that initially the supplier "would learn more than we" from the alliance, Chaparral spent the money to send employees over to "hand hold" so as to develop a capability that would be useful in the future.

This constant dispersal of mixed employee teams to customers, competitors, and suppliers throughout the world serves a dual purpose. The visits are regarded as learning sabbaticals that keep life "exciting" for employees. They are a source of information for the company as a whole. "We want them to . . . come back with new ideas about how to make improvements or new ways to understand the problem."[21]

Conclusion

The factory, in fact any backroom operation, is not usually regarded as an arena for experimentation and learning. Chaparral Steel challenges this concept of operations. Factories *can* function as learning laboratories (see Table 1). The most important characteristic of such organizations is that they are totally integrated systems. They are difficult to imitate because every employee, from CEO to line operator, is technically capable and interested in learning. Moreover, the whole organization is designed around the creation and control of knowledge. The four subsystems described

Table 1 Comparison of Traditional Factory with Factory as Learning Lab

	Traditional Factory	*Learning Lab*
1. Reasearch and development function	Separate and distant from production.	Merged with production (everyone does development).
2. Experimentation	Rare, feared.	Constant, welcomed.
3. Innovation	Exclusive province of engineers.	Everyone's business (but their own methods).
4. Equipment and process	If it works, don't fix it.	Design your own; constantly improve.
5. New technology	Reject: not invented here.	Never reinvent here.

above are not only internally linked but tremendously dependent upon each other. Continuous education depends upon the careful selection of willing learners. Sending workers throughout the world to garner ideas is cost effective only if they are empowered to apply what they've learned. The organization is unlikely to be open to outside knowledge if it does not place a strong value on sharing knowledge or does not give rewards for bettering the whole company's performance. Thus continuous learning depends upon the sense of ownership derived from the incentive systems, upon the pride of accomplishment derived from special education systems, upon values embedded in policies and managerial practices, as well as upon specific technical skills. The line operator appears to take the same perspective on the conduct of daily activities as the CEO. Chaparral is tremendously consistent.

Paradoxically, the system's interdependence is also a potential weakness, as competencies often are.[22] A learning laboratory may have trouble recreating itself. Any organization is likely to be somewhat limited by its "congenital knowledge" and by the stamp placed upon it by its founders.[23] A significant challenge for Chaparral is how to grow. Forward has noted that "to stand still is to fall behind," and the company's credo to provide growth for its skilled people requires some forward momentum. Therefore, it must either grow larger where it is or clone itself in a new location. How will it transplant the deep worker knowledge, the motivation, the commitment, and the informal systems of knowledge sharing?

For other companies interested in creating factories as learning laboratories, the questions are: Can it be done in a plant within a large corporation, where many of the managerial systems have already been set corporatewide and therefore are not at the plant manager's discretion? Can an existing plant be transformed when plant managers may not have the luxury of selecting people as freely as Chaparral did? Can a company less geographically isolated hope to reap returns on investing in its employees' intellectual advancement, or will they be lured away by other companies?

Similar questions were raised a decade ago when U.S. manufacturers first began to understand how Japanese companies were competing on the basis of quality. Initially overwhelmed by the difference in activities, values, and managerial systems implied by the total quality approach, many U.S. managers were pessimistic about their ability to change their operations to the extent needed and reluctant to invest in employee education. Yet today, many U.S.-based factories achieve quality levels never even aspired to in 1980.[24]

For many of these improvements, the Japanese were our teachers, having learned from Deming and Juran.[25] According to some researchers, Japanese managers may also be ahead of their U.S. counterparts in creating learning laboratories, since "Japanese companies are usually adept at organizational learning."[26] As the references at the end of this article suggest, many of Chaparral's managerial practices and vaues also characterize best practices in Japan, such as investing extensively in formal and informal education; searching worldwide for the best technology and methods and absorbing that knowledge into home operations; and valuing employee empowerment, problem solving, and risk taking.[27] Apparently, such learning systems are not uniquely Japanese.

Of course, creating a learning laboratory in a "greenfield" site is easier than in an existing plant. As not all U.S. factories can start up from scratch, creating a learning laboratory implies big changes. But do we really have any choice? No financial formulas, corporate reshuffling of departments, or exhortations by corporate management to integrate across functions will foster the creativity and productivity needed for international competition.

Experts on change management suggest that three critical elements are required for altering current practices: (1) dissatisfaction with the status quo; (2) a clear model of what the changed organization will look like; and (3) a process for reaching that model, that vision of the future.[28] Examples such as Chaparral can aid all three. One way of stimulating dissatisfaction with current practices and hence motivation to change is to observe other companies where an alternative management style appears to be yielding superior results. By benchmarking against such companies, managers can derive principles to incorporate into their own particular visions. Chaparral offers a model of a factory as a learning laboratory, and if the specifics are not transferable, the principles underlying the Chaparral vision are.

The precise process for implementing these principles will differ markedly from company to company. It is possible to interrupt a factory's current systems by introducing new equipment, new learning skills and activities, new knowledge-creating management systems, or new values. But interrupting a current system is only the first step. As the Chaparral example demonstrates, learning skills, management procedures, and values are interrelated. Values unsupported by management systems are vapid; management systems that run counter to values are likely to be sabotaged; learning activities unsupported by values and management practices will be short-lived. If a learning capability is to be developed, the whole system must eventually be addressed.

References

The author is grateful to the employees of Chaparral Steel for their patience, to Gil Preuss for help in field work, and to colleagues Chris Argyris, David Garvin, and Steven Wheelwright and two anonymous reviewers for comments on earlier drafts. The study was supported by the Division of Research, Harvard Business School.

1. G. Forward interviewed by A.M. Kantrow, "Wide-Open Management at Chaparral Steel," Harvard Business Review, May-June 1986, pp. 96–102. Organization theorists would see Chaparral's culture as an appropriate response to such an environment. Scholars theorize that learning will not happen without a certain amount of stress and that complex, uncertain environments require decentralized, laterally

linked organizations. See: C.M. Fiol and M.A. Lyles, "Organizational Learning," *Academy of Management Review* 10 (1985): 803–813; and R. Duncan and A. Weiss, "Organizational Learning: Implications for Organizational Design," *Research in Organizational Behavior* 1 (1979): 75–123.

2. Senge argues persuasively that successful leaders are systems thinkers, able to see "interrelationships, not things, and processes, not snapshots." See: P. Senge, "The Leader's New Work: Building Learning Organizations," *Sloan Management Review,* Fall 1990, pp. 7–23. Other theorists similarly note how interrelated are strategy, structure, and culture in creating learning environments. See: Fiol and Lyles (1985).
3. I assume that learning occurs if "through its processing of information, the range of [an organization's] potential behaviors is changed." See: G. Huber, "Organizational Learning: The Contributing Processes and the Literatures," *Organizational Science* 2 (1991): 89. That is, beyond contributing to an accumulation of formal knowledge bases, learning creates "capacities . . . for intelligent action." See: G. Morgan and R. Ramirez, "Action Learning: A Holographic Metaphor for Guiding Social Change," *Human Relations* 37 (1983): 21. A growing literature on the topic emphasizes that organizational learning is more than an aggregation of individual learning. See, for instance: B. Hedberg, "How Organizations Learn and Unlearn," in *Handbook of Organizational Design,* eds. P. Nystrom and W. Starbuck (New York: Oxford University Press, 1981), pp. 3–27. While the four critical activities proposed here have not been previously combined into a framework, each has been identified as characteristic of a learning organization. On problem identification and solving, see: E. Hutchins, "Organizing Work by Adaptation," *Organization Science* 2 (1991): 14–39. On integration of internal information, see: Duncan and Weiss (1979). On experimentation, see: R. Bohn, "Learning by Experimentation in Manufacturing," (Cambridge, Massachusetts: Harvard Business School, Working Paper No. 88–001, 1988). On acquisition and use of external information, see: Huber (1991), pp. 88–115.
4. Chaparral managers have verified the accuracy of the descriptions of activities and events offered here, but they are not responsible for the characterization of learning laboratories in general or for the way I have analyzed their organizational culture.
5. Other researchers also imply the utility of stretch goals as stimuli for learning. Such goals may be thought of as "performance gaps" deliberately induced to motivate knowledge generation. See: Duncan and Weiss (1979). Itami suggests that "overextensions" and "dynamic imbalances" created to challenge the organization characterize the most successful Japanese manufacturing companies. See: H. Itami and T. Roehl, *Mobilizing Invisible Assets* (Cambridge, Massachusetts: Harvard University Press, 1987). Moreover, the goal for this particular project fits Senge's prescription of a blend of extrinsic and intrinsic visions, in that both outside competition and improvement over prior performances are invoked: Senge (1990).
6. Morgan and Ramirez (1983) suggest that a "holographic" organization (which epitomizes a learning organization for them) is designed so that "the nature of 'one's job' at any one time is defined by *problems facing the whole*" (emphasis in original, p. 4). Similarly, from his study of knowledge creation in some of Japan's top firms, Nonaka concludes that "every single member of the organization ahould be able to suggest problems . . . and . . . solutions." See: I. Nonaka, "Managing Innovation as an Organizational Knowledge Creation Process" (Rome: Technology Strategies in the Nineties Conference Paper, 21 May 1992), p. 44.
7. Quoted by B. Dumaine, "Chaparral Steel: Unleash Workers and Cut Costs," *Fortune,* 18 May 1992, p. 88.
8. Argryis and Schon point out that "espoused theory" does not always influence behavior; "theory in practice" does. See: C. Argyris and D. Schon, *Organizational Learning* (Reading, Massachusetts: Addison-Wesley, 1978). See also: S. Kerr, "On the Folly of Rewarding A, While Hoping for B." *Academy of Management Journal,* December 1975, pp. 769–783.
9. Von Glinow argues that the most effective organizational reward system to attract and retain highly skilled people is an "integrated culture" that combines a concern for people with very strong performance expectations. Chaparral's system appears to fit her description. See: M.A. Von Glinow, "Reward Strategies for Attracting, Evaluating, and Retaining Professions," *Human Resource Management* 24 (1985): 191–206.
10. In a macro-analysis of sixteen studies using forty-two different data samples that estimated the effect of profit sharing on productivity, the authors conclude that "these studies taken together provide the strongest evidence that profit sharing and productivity are positively related." See: M.L. Weitzman and D.L. Kruse, "Profit Sharing and Productivity," in *Paying for Productivity* (Washington, D.C.: The Brookings Institution, 1990), p. 139.
11. This advantage was confirmed in a study by: K. Clark and T. Fujimoto, "Overlapping Problem Solving in Product Development," in *Managing International Manufacturing,* ed. K. Ferdows (North Holland: Elsevier Science Publishers, 1989). Huber (1991) suggests a reason for the advantage: "When information is widely distributed in an organization, so that more and more varied sources for it exist, retrieval efforts are more likely to succeed and individuals and units are more likely to be able to learn. Therefore information distribution leads *to more broadly based* organizational learning" (pp. 100–101). Fiol and Lyles (1985) similarly cite research showing that learning is enhanced by decentralized structures that diffuse decision influence.
12. Nonaka (1992), after describing very similar policies at Kao Corporation, observes: "Asymmetrical distribution of

information destroys the equality of relationships and leads to unilateral command instead of mutual interaction" (p. 48).

13. For an understanding of the impact on communication patterns of physical proximity and centrally located common facilities, see: T. Allen, *Managing the Flow of Technology* (Cambridge, Massachusetts: The MIT Press, 1977), ch. 8.
14. Interview by A.M. Kantrow (1986). A remarkably similar philosophy was observed in four Japanese companies, where "employees are trained from the first day on the job that 'R&D is everybody's business.'" See: M. Basadur, "Managing Creativity: A Japanese Model," *The Executive* 6 (1992): 29–42. Itami (1987) similarly proposes "'6 excessive' experimentation in production" since "experimentation and learning do not take place only in the lab" (p. 95).
15. Descriptions of Japanese best practices reveal similar strong emphasis on both on-the-job training and formal education. See, for example: J. Sullivan and I. Nonaka, "The Application of Organizational Learning Theory to Japanese and American Management," *Journal of International Business Studies* 17 (1986): 127–147.
16. For an interesting contrast in stimulating learning, see the carefully constructed routines in the "learning bureaucracy": P. Adler, "The 'Learning Bureaucracy': New United Motor Manufacturing, Inc.," in *Research in Organizational Behavior,* eds. B.M. Staw and L.L. Cummings (Greenwich, Connecticut: JAI Press, forthcoming).
17. Clearly a tradeoff exists between stability in the workforce and the diversity needed to stimulate innovation. March proposes that "a modest level of turnover, by introducing less socialized people, increases exploration and thereby improves aggregate knowledge." See: J. March, "Exploration and Exploitation in Organizational Learning," *Organization Science* 2 (1991): 79. However, others point to the "insistence on selection of company members at an early point in life and avoidance of the introduction of new people at higher management levels" as an important influence on an information-sharing culture: I. Nonaka and J. Johansson, "Japanese Management: What about the 'Hard' Skills?" *Academy of Management Review* 10 (1985): 184. Simon observes that turnover can become a "barrier to innovation" because of the increased cost of socialization: H. Simon, "Bounded Rationality and Organizational Learning," *Organization Science* 2 (1991): 125–134. Chaparral's management takes pride in its low turnover rate.
18. Interestingly, Hanover Insurance CEO William O'Brien made a similar reference to Maslow's hierarchy: "Our traditional hierarchical organizations are designed to provide for the first three levels [of the hierarchy] but not the fourth and fifth. . . . Our organizations do not offer people sufficient opportunities for growth." Quoted in Senge (1990): 20.
19. Such experience and knowledge sharing even at managerial levels is noted as a characteristic of Japanese organizations, which are "able to cover for an absent individual quite easily, because other individuals have a relatively greater understanding of the requisite information": Nonaka and Johansson (1985): 185. See also the discussion of designing organizations with redundant skills: Morgan and Ramirez (1983): 5.
20. Again the parallel with Japanese practice, at least as described in literature, is striking. Nonaka and Johansson describe how Japanese firms consult with outside experts, not as troubleshooters, but as educators on the general topic. It is up to the company's own personnel to translate that newly acquired intelligence into application. See: Nonaka and Johansson (1985).
21. Kantrow (1986): 101.
22. I have argued that core capabilities almost inevitably have a flip side, core rigidities, that hamper nontraditional projects and can hobble an organization in moving to new competencies. See: D. Leonard-Barton, "Core Capabilities and Core Rigidities in New Product Development," *Strategic Management Journal* 13 (1992): 111–126.
23. Huber (1991) discusses this limitation. See also: J. Kimberly, "Issues in the Creation of Organizations: Initiation, Innovation, and Institutionalization," *Academy of Management Journal* 22 (1979): 437–457.
24. This includes Japanese transplants such as New United Motor Manufacturing, Inc., whose employees are mostly rehires from the same United Auto Workers workforce that had one of the industry's worst labor records. See: R. Rehder, "The Japanese Transplant: A New Management Model for Detroit," Business Horizons, January-February 1988, pp. 52–61.
25. See the profiles of these two men in: O. Port, "Dueling Pioneers," *Business Week,* 25 October 1991, p. 17.
26. Nonaka and Johansson (1985). In fact, some management practices now being imported into the United States were advocated by younger U.S. contemporaries of Deming such as Chris Argyris, whose early books were translated into Japanese within a year of their publication in the United States. See: C. Argyris, *Personality and Organization* (New York: Harper Brothers, 1957) and *Integrating the Individual* (New York: John Wiley & Sons, 1964).
27. See, for example, R. Rehder and H. Finston, "How Is Detroit Responding to Japanese and Swedish Organization and Management Systems?" *Industrial Management* 33 (1991): 6–8, 17–21; R.T. Pascale, *Managing on the Edge* (New York: Simon & Schuster, 1990), ch. 9; and G. Shibata, D. Tse, I. Vertinsky, and D. Wehrung, "Do Norms of Decision-Making Styles, Organizational Design, and Management Affect Performance of Japanese Firms? An Exploratory Study of Medium and Large Firms," *Managerial and Decision Economics* 12 (1991): 135–146.
28. See, for example, M. Beer, *Organization Change and Development* (Santa Monica, California: Goodyear Publishing Company, 1980), ch. 3.

Reading 1.3
What Really Makes Factories Flexible?

Harvard Business Review

David M. Upton
Harvard Business School

Manufacturing managers in a broad array of industries agree that achieving low cost and high quality is no longer enough to guarantee success. In the face of fierce, low-cost competition and an army of high-quality suppliers, companies are increasingly concentrating on flexibility as a way to achieve new forms of competitive advantage. The flexible factory, they hope, will enable them to respond to customer orders quickly, provide a broad product range, or introduce new products to the range effortlessly. The push to make factories more flexible has been spreading throughout manufacturing and currently is even permeating industries such as chemicals and paper, in which the assumption for decades has been that the plants with the longest production runs are typically the most competitive.

Having acknowledged the importance of flexibility, managers in industry after industry are finding it frustratingly difficult to improve. Some have organized cross-functional teams in the hope that new ways of working will generate greater agility. Many have collectively invested tens of billions of dollars in hardware and software in the hope that computer-integrated manufacturing (CIM) will transform their factories into highly flexible operations. Time and time again, managers have been disappointed and frustrated because they have not understood exactly why enhanced flexibility has eluded them. Was it the degree to which they automated their operations? Was it the software, whose complexity and cost they had underestimated? Or was it the inability of employees to take advantage of the new technologies?

In a quest to help manufacturing managers understand why the improvement of flexibility has been so elusive, I embarked on a study of 61 factories in North America that manufacture fine paper. At first glance, the paper industry may seem like an unusual starting point for the study of flexibility. It is rarely characterized as flexible and has not, until recently, given high priority to improving flexibility. However, there are some less obvious characteristics that make the paper industry an ideal place to begin.

Unlike most industries in which different plants make different products—and, indeed, in which the same plant may make different products—the paper industry's products are comparable across plants and are always manufactured by the same fundamental process. There are only a small number of ways in which one type of paper differs from another type—the most fundamental being the basis weight, or area density, of the paper. These characteristics, or grades (each particular pulp and weight combination is a grade), are straightforward to measure. Those facts enabled me to develop concrete measures of both the range of products that a given plant could produce and the time it took a plant to switch from making one product to making another product.

My findings turn much of the conventional wisdom on its head. In the plants I studied, there was little direct correlation between the degree of computer integration and the degree of operational flexibility. I found that large plants were not inherently less flexible than small plants. Contrary to what many believe, newer, bigger processes were typically better able to perform quick changeovers than older, smaller machines. And although experienced workers provided powerful advantages in some situations, they impeded a plant's ability to be flexible in others.

The primary revelation of my research concerns the role of people—both managers and operators. The flexibility of the plants depended much more on people than on any technical factor. Although high levels of computer integration can provide critically needed advantages in quality and cost competitiveness, all the data in my study point to one conclusion: Operational flexibility is determined primarily by a plant's operator and the extent to which managers cultivate, measure, and communicate with them. Equipment and computer integration are secondary.

At many of the plants in my study, however, managers embraced computer integration as the solution to the growing need to forge new capabilities. In reality, computer systems were often a quick fix that helped managers avoid the tremendously difficult task of defining precisely what kind of flexibility they required from a plant and then setting goals, revamping measurement and compensation systems, building training programs, and overhauling work practices in order to achieve that flexbility.

To state it simply, most managers put too much faith in machines and technology, and too little faith in the day-to-day management of people.

What Is Flexibility?

Ten or 15 years ago, quality was much like flexibility is today: vague and difficult to improve yet critical to competitiveness. Since then, managers and academics have studied and experimented with ways to improve quality; as a result, there is currently an enormous variety of quality-improvement techniques and a plethora of textbooks and gurus from which to choose.

Flexibility is only beginning to be explored. If managers aim to improve a plant's flexibility, where should they start? The first problem is one of definition. Flexibility means different things to different people. At the plant level, flexibility is about the ability to adapt or change. But there are many ways to characterize such an

Source: Reprinted by permission of Harvard Business Review. From "What Really Makes Factories Flexible," by David M. Upton, July-August 1995.

ability. One manager might be talking about the cost of changing from one product to the next. Another might be talking about the ability to ramp production volumes up and down to fit the demand of the market. Yet another might be talking about the ability to increase the range of available products. All these abilities might be called flexibility, but they require very different courses of action to develop.

Ten or 15 years ago, quality was much like flexibility is today: vague and difficult to improve yet critical to competitiveness.

The type of flexibility a given company should emphasize should be determined by its competitive environment. Whether one is referring to products, production volumes, or manufacturing processes, flexibility is about increasing range, increasing mobility, or achieving uniform performance across a specified range.

Product range can mean different things. For example, a plant can have the ability to make a small number of products that are very different from one another, or it can have the ability to produce concurrently a large number of stock-keeping units that are only slightly different from one another.

Mobility means a plant's ability to change nimbly from making one product to making another. It is this kind of flexibility that is associated with quick response times—mobility minimizes the need for long runs and allows production to follow demand without excessive inventory.

Finally, there is uniformity of performance. Plants always have one product that they would much rather make than others because that product maximizes productivity, quality, or some other measure. When a plant moves away from its favored set of parameters, performance falls off. If it falls off steeply, many managers will label the plant inflexible. According to that definition, a flexible plant is one that can perform comparably well when making any product within a specified range. It is this capability that is most improtant to a full-range manufacturer; the uniformity of performance across the range, more than the size of the range, will protect a plant from cherry-picking predators. Uniformity of performance also applies to production volumes: Some plants are able to work productively over a broad range of output volumes, while others are unable to increase or decrease volumes without incurring considerable penalties.

Once managers have defined the kinds of flexibility they want to develop, they face another set of issues. First, because flexibility is not easy to measure, improvement in flexibility is also difficult to measure. For example, is a plant that can produce 200 different colors of paint 100 times more flexible than a plant that can make 2 products (say, a car and a truck) on the same line? Clearly, simply counting products won't suffice in defining flexibility because it doesn't account for the "differentness" of the products concerned.

Second, the products that a plant actually makes do not necessarily reflect its flexibility. Those products that a plant *could* make matter, too. For example, a factory can have the flexibility to machine titanium without a chip of titanium ever being cut. Unlike most other sources of manufacturing advantage, such as low cost or high quality, flexibility is sometimes about a *potential* ability rather than a demonstrated one.

Third, it is often unclear which general features of a plant must be changed in order to make its operations flexible. For instance, many management experts preach that cross-functional empowered teams and computer integration help make plants more flexible. But do they always translate into operational, competitive capabilities that the company can sell? The answer is often no. Similarly, one type of equipment may be capable of making a large range of products, while another may possess quick-changeover capabilities and both may be called flexible. Yet the latter will be of little value if the market demands great variety. The same holds true for training programs, information systems, incentive schemes, and measurement methods. They, too, must be aligned to provide precisely the form of flexibility that is needed from the overall system. Few factories have actually aligned those elements.

Surprising Discoveries

Managers at many of the plants I studied deemed an astounding 40% of flexibility-improvement efforts to be unsuccessful or disappointing. In the vast majority of those cases, the cause could be traced to a failure to identify precisely what kind of manufacturing flexibility was needed, how to measure it, or which factors most affected it.

A paper mill usually consists of a pulp plant, which provides pulp for a group of two to ten converting plants that turn the pulp into paper. The paper-making process is essentially a water-removal procedure. Water is removed from a pulp slurry by use of gravity, squeezing, and heating. The pulp slurry is laid onto a moving fabric belt from which water drains off. The pulp web is squeezed and heated by a series of rollers until it is strong enough to support its own weight. Further heating then takes place until the paper's moisture content is just below that of the ambient atmosphere. The paper is then collected on a reel and sliced into roll sizes that are convenient for customers and sheeting machines.

By collecting production data and interviewing managers, supervisors, and operators, I was able to measure the breadth of paper grades that each plant was capable of producing and the changeover time that each plant required to switch between grades. In addition, I measured the experience of the workforce by length of service, the vintage of the technology, the level of computer integration, the scale of the plant, and the degree of emphasis that management placed at each plant on various types of flexibility. In order to assess the latter, I asked the shop-floor people responsible for running the plant how much emphasis *they* thought that managers were placing on a whole range of factors, from safety to quality to the various types of flexibility. (See the table "Plant Flexibility: Factors and Findings.")

I found large differences in the flexibility of the plants. Changeover times varied from one minute to four hours, even for similar types of change on comparable equipment. I found that the

Plant Flexibility

Factors and Findings

Description	Units	Mean	Minimum	Maximum
Changeover times	minutes	15.40	1	240
Range of plant	pounds	64.6	1	161.0
Change frequency	changes per month	29.02	0	500
Date of last major rebuild	year	1973	1919	1991
Width of paper web	inches	159.8	76.0	348.0
Speed	feet per minute	1,466	250	3,200
Net output	tons per day	236.8	18	1,200
Average crew service	years	17.6	1	25
Degree of computer integration (general)	scale of 0.0 to 8.0	5.16	1.83	8.0
Break frequency	breaks per week	16.0	1	60

The data are drawn from a study of 61 fine-paper plants in North America.

potential range of product variation in the plants differed by a factor of 20. This discrepancy is not surprising given the wide variety in the capacity and vintages of the plants in my study. What is surprising, however, is that very few of my assumptions about the reasons for such disparities—the same assumptions, incidentally, that prevail within the industry at large—turned out to be correct.

Just because a plant is flexible in one way doesn't mean it will be flexible in another.

First, I found that the degree of computer integration in a plant was not in itself associated with either increased range or improved changeover times. (See the chart "Computer Integration Reduces Flexibility.") Indeed, even computer hardware and software designed specifically for the purpose of improving the changeover process, such as automatic grade-change systems, did not decrease the time required to make grade changes in the plants and actually inhibited the ability of plants to produce a broad range of grades. This discovery is especially important because managers often have difficulty justifying computer-integration projects on the basis of cost savings or quality improvements and therefore justify them on the basis of the improved flexibility they will provide.

Second, I discovered that there were no de facto relationships among the various forms of flexbility. In other words, just because a factory is flexible in one way does not mean it will naturally be flexible in others. To explore whether there are generally flexible and generally inflexible plants, I looked at the relationship between two forms of flexibility in each paper plant: its mobility (measured by changeover times) and its range (the breadth of product characteristics it could produce). I found no relationship between the two types of flexibility. (See the chart "There Is Little Relationship Between Two Types of Flexibility.")

Many companies that have installed software-based CIM to improve flexibility have been misled by the *soft* in *software*.

Third, while there was a clear relationship between the scale of an operation and the breadth of product it could produce (larger plants made a smaller range of product characteristics), I found no clear link between the scale of an operation and its ability to change swiftly between products. Although it is true that the cost of downtime resulting from changeovers is higher in bigger plants (because the lost output and investment involved are much larger), it is also true that the advantages of quick changeovers, which provide the ability to undertake just-in-time production, could be accrued at a proportionately larger scale.

Fourth, workforce experience, an important factor in building flexibility, affected different types of flexibility in different ways. In general, the more experience a workforce had (as measured by length of service), the greater the range of products the plant could make. Surprisingly, however, changeover times, or mobility, were worse in plants with more experienced workforces.

Finally, much of the variance in plant flexibility (both range and mobility) could be explained by managerial action—the extent to which managers emphasized the importance of a particular

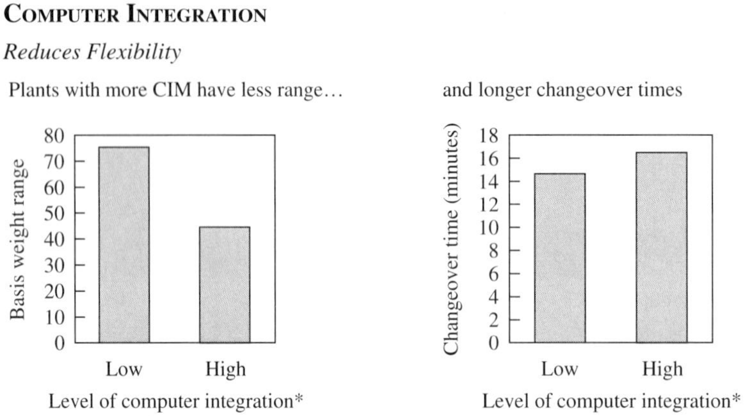

**The data have been corrected for factors other than CIM, such as scale and workforce experience.*

kind of flexibility to operators—rather than by the structural characteristics of the plant. Plants whose managers had not made flexibility a clearly understood goal were much less flexible than those whose managers had.

The Unfulfilled Promise of CIM

Why is it that huge investments in computer-integration systems have failed to increase flexibility? Let's think about how computer integration affects the range of work that a plant can do. Imagine a plant designed to manufacture a range of products when the process is computerized, any competent engineer will ask, Over what range of conditions should this system work? After the possible operating range has been identified, the engineer will come up with a dollar figure for the computer systems needed to coordinate the process. When the company's manager is taken aback by the price, the engineer will point out that the extremes of the range are pushing up the price. The manager will then decide to sacrifice the extremes on the grounds that they are seldom needed.

But look at what has been given up. Although handling the extremes was difficult when the plant was being operated manually, an experienced and skilled crew could coax a plant to make those products. With the processes computer integrated, the plant runs into the "there isn't a button for that" problem. At such plants, the range of products often fell by 20% to 30% after the mill made a major investment in CIM. What's more, the operator skills required to perform such acrobatics typically atrophied from lack of use.

Plants become more flexible when managers stress to workers the importance of flexibility.

Some of these problems were recognized 30 years ago when "hard" automation reigned supreme. The hope was that software-based CIM would solve such problems. Although it is true that computer integration may be changed to accommodate new requirements, many companies have been misled by the *soft* in *software. Soft* implies easily changeable to malleable. Experience shows that manufacturing-integration software is often anything but. As one manager pointed out, "I'd be better off with a flame cutter and a hacksaw than I would with a team of software engineers. At least I'd be able to see what was taking so long to change."

One of my more surprising findings was that computer integration did not decrease the time needed to switch from making one product to making another one. While the worst manual-change system took much longer than the slowest computer-integrated system, the best manual-change teams were much faster than the computer. Many teams of operators had developed routines and tricks that enabled them to change the plant over efficiently, and the best teams took great pride in the fact that they could beat the computer. "The computer is really slow; it gets kind of boring sitting and watching the plant run, so we figured out a better way to change manually," one young operator in the Midwest said, echoing a sentiment expressed by many people in other plants. "We all practiced and worked out who should do what and when, and now we can always beat the computer and do changeovers more quickly than we used to do them with the computer."

The operators' ability to outperform computers can be explained as follows: One of the primary risks in making any change is that of catastrophic failure, such as a paper break, which is akin to a tool breakage in machining. When the paper running in the machine breaks, it must be rethreaded, which is always time-consuming and expensive. To avoid downtime, the methods for changing between products under computer coordination are conservative. People, however, may be a little more daring. They have access to sensors that are unavailable to computerized systems, such as the sound of the process or the feel of the pulp. Some managers might fear that such practices will result in higher breakage

THERE IS LITTLE RELATIONSHIP BETWEEN TWO TYPES OF FLEXIBILITY

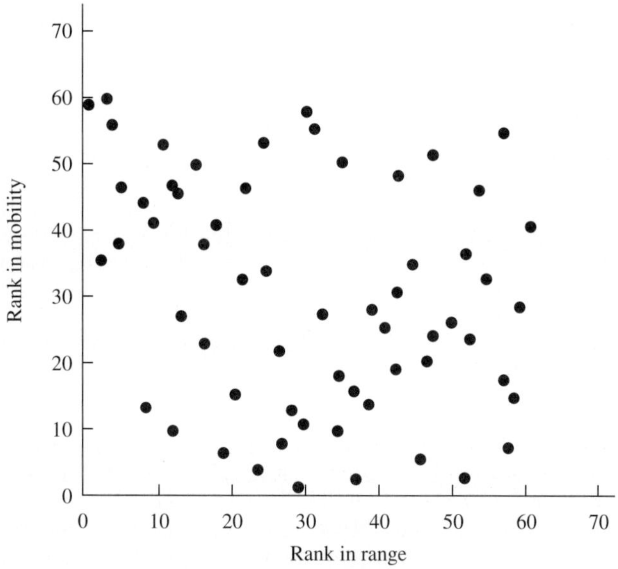

● Relationship between range (breadth of product characteristics) and mobility (changeover times) at each paper plant

rates and more downtime. My data show that, to the contrary, the computer-integrated systems not only were slower than the manual systems but also resulted in a higher breakage rate in the average plant.

Why would managers invest in such equipment if it is not particularly effective in boosting flexibility? The answer is that they often do not know how effective—or ineffective—it is. In most traditional industries, managers do not routinely benchmark flexibility. While flexibility can provide a powerful means for increasing revenues, most managters still focus on comparing cost figures—if they make any comparisons at all. Even those paper-plant managers who did routinely benchmark flexibility were unable to collect data from a sufficiently large sample of plants; as a result, they were unable to see that computer integration was failing to generate a clear advantage in process flexibility.

It is important to point out that many computerized systems have been installed for reasons other than increasing plant flexibility. Those reasons include facilitating the tracking and delivery of information and improving the quality of the process. And computerized systems used to those ends have succeeded. Judicious use of computer integration can provide valuable new sources of competitive advantage when closely aligned with the competitive needs of the business.

Still, when I have shown my study to managers in a variety of industries, particularly in process industries such as glass, chemicals, and steel, they have sighed. They, too, have discovered that computer integration does not necessarily guarantee greater flexibility. In many cases, the additional complexity resulting from computer integration has been a competitive millstone around their companies' necks, demanding new and expensive skills that have not translated into clear advantages in the market. More data is not the same as more information. Systems often sidetrack manufacturing organizations from the tasks they should be doing well. Many continue to hail computer-integrated manufacturing as a panacea. But unless it is treated as a tool rather than as an end in itself, CIM may exclude most people in an organization from the process of shaping and improving the way the operation runs.

The approach of "investing one's way to flexibility" through computer integration can be damaging in two ways. Not only is computer integration not the panacea for flexibility problems but it also comforts managers with the thought that they are doing *something,* when all along they should have been doing something *else.* The implication for managers, as Robert S. Kaplan pointed out in "Must CIM Be Justified by Faith Alone?" (HBR, March-April 1986), is that the blanket acceptance of computer-integration projects on the grounds that they will improve flexibility must be replaced with a healthy skepticism.

Bulding the Right Workforce

It is not a revelation that people can often become less flexible in their work patterns as they become more accustomed to a particular way of doing their jobs. But my research raises a more complex question. Setting aside work practices, how does experience affect those capabilities that can more directly contribute to a company's success? How does experience affect operational flexibility? While my research shows that a workforce's experience is an important factor in determining the flexibility of an operation, it also shows that experience affects different types of flexibility in different ways: Long service had a positive effect on the range of manufacturable products and a negative effect on the ease with which the plant could switch between those products. This finding makes sense. In order to work on unusual jobs at the extremes of a plant's range, people need experience to know how to get the job done. Experienced people know how to run the plant without breaks when making very thin paper or how to get the paper dry without burning it when making very thick paper.

Plants become more flexible when managers stress to workers the importance of flexibility.

A few theories could explain why plants with more experienced workforces were less mobile. First, many of the more experienced crews held the view that each individual plant was made to do one thing well and that changeovers that went beyond those established parameters were making improper demands on the plants. Less experienced crews, however, had learned papermaking at a time when flexibility has started to become critical. Many of these crews had developed their own novel ways of making changeovers and saw them as a welcome switch from the tedium of machine tending. In particular, they often veiwed change as a defining part of their jobs. I don't think that it is a coincidence that one outstandingly mobile plant had the following rule when selecting its operators (although not its supervisors): You must never have worked in this industry before.

Pursuing Flexibility

How, then, should managers approach the challenge of making their manufacturing operation more flexible? The first step is to ask, What form of flexibility does the company need from its plants? For example, a manufacturer that wanted to excel at customizing products would need to develop the ability to carry out a large range of jobs in the plant, while one that wanted to use quick response as its primary competitive weapon would need to focus on building quick changeovers into the manufacturing process.

Once they have identified those capabilities, managers must determine the type of workforce or equipment they need to enhance flexibility. For example, the manufacturer emphasizing customized products would require an experienced workforce that could make even the most difficult products. The company hoping to be flexible through quick response would be better off with a workforce less steeped in the traditional ways of operating, a workforce that is ready to adapt to an environment of continuous change.

Next, managers should find ways to measure the type of flexibility sought and should emphasize the importance of those measures to the workforce, letting them make industry comparisons and providing them with incentives to underscore the point. Most managers at the plants I studied were still clinging to measures that had no connection to flexibility. For example, plant managers and their superiors tended to focus on measures that made sense when their plants were big enough to be able to compete on cost: capacity utilization and tons of output per hour. To expect a plant both to be flexible and to continue to focus solely on the utilization of its equipment and the cost of production makes no sense. The plants that were flexible in terms of range and mobility tended to have clear, nonfinancial measures of the flexibility they were trying to develop—changeover times, lead times, or process range.

Training is an important mechanism for building operational flexibility. On the surface, training simply provides the skills people need to carry out new tasks. Its ancillary roles, however, are much more critical. First, training plays an important part in moving people beyond a "this is the way we've always done it" mentality. In industries in which people are trained primarily by serving as understudies to those with more experience, the new mind-set is particularly critical. Second, training builds confidence: Many people in manufacturing are reluctant to try new approaches or techniques because they are afraid of exposing their ignorance. Third, it builds an esprit de corps and helps emphasize the growing competitive need for the right kind of flexible operation—providing a sense of common purpose from a common experience.

Increasing flexibility may be costly in the short run, but it gets easier over time. Plants become more flexible because their managers emphasize the importance of flexibility and because they *practice* being flexible. A self-reinforcing process then begins. Because such plants are flexible, they are assigned varied and quick-response work, which in turn makes them still more flexible. And just as a plant with mediocre quality can be stretched by challenging it to produce higher-quality goods, previously inflexible plants can be made more flexible by changing the work assigned to them.

Once manufacturing managers have determined the type of flexibility they need, they have to give careful thought to how they will develop that capability. Designing the right mix of machines, computer systems, and people, and figuring out the most effective way to orchestrate them are hard work. But turning only to machines, hardware, and software as the solution will not suffice. People count more than machines.

Combining Computers and People to Build Flexibility

Managers and engineers have long preached that computer systems should complement, rather than replace, the skills of operators. In the end, however, most have embarked on a path to computer integration that caused them to place machine over man and resulted in less flexible rather than more flexible factories.

What accounts for the disparity between idea and execution? The problem rests with how—or whether—managers define the type of flexibility to pursue and then choose the appropriate computer systems, work practices, training programs, incentives, and measures. To create a highly flexible computer-and-people-integrated manufacturing system, managers of each individual factory have to come up with their own unique formula. The managers of Mead Corporation's Escanaba Mill, in Escanaba, Michigan, which makes coated fine paper and employs 1,300 people, did just that.

In the early 1900s, the Escanaba complex, like the paper industry in general, was struggling: The market was in a deep slump, there was a glut of capacity, and price cutting was rampant. To make matters worse, the mill was facing intensifying competition from a growing number of competing mills with bigger and newer machines. Finally, a spurt of imports from Europe raised concerns that foreign producers, which had never posed a threat, might be planning a major assault on the North American market.

Like most of their peers in the paper industry, Escanaba's managers had long believed that the key to competitiveness was to achieve the lowest possible costs through long production runs and few product changeovers. But suddenly many in the industry were forced to rethink those assumptions and look for new ways to distinguish themselves from competitors.

The answer for Escanaba, its managers decided, was to be more responsive to customers than their competitors were. They would accomplish that goal by being highly mobile—able to change production schedules quickly so that they could fill orders much faster than competitors could. At that point, the industry's customers might have to wait as long as two weeks to receive an order, even in slack times. Escanaba's managers set out to slash that time to one or two days.

The new strategy of emphasizing responsiveness demanded a new way of working in the mill: It required faster product changeovers and nimble decision making. When managers analyzed how the mill had been operating, they came to a sober conclusion: The plant was slow in switching from one product to another and in changing schedules to accommodate new customers' requirements. It took too long to execute a production-schedule change, and grade changes often generated product-quality problems or mistakes. They also realized that the plant was not inherently incapable of quick changeovers; changeovers hadn't improved, because improving them had not been a high priority. Managers had been judged primarily on their success in maximizing the mill's capacity utilization and product quality, and they simply focused on long production runs to achieve their performance targets. The long runs meant that operators did not have to learn how to improve changeovers. They also produced a culture that placed very little value on responsiveness to customers. "We make it; you [the sales force] sell it—that's the way the manufacturing people used to see things in this mill," recalled Henry Swanson, the mill's manager of process control and information.

The mill's managers also realized that computerization in itself was no panacea. Some of the machines in the mill were relatively highly computer integrated. "Even though we had a lot of computers on the plant, sometimes we'd make a couple of hours' worth of production before someone realized that something was wrong," Swanson said. "It was just too easy to trust that the computers had got it right." In addition, the opaque computerized system prevented workers from learning how they might improve operations.

So one of the first steps that managers took to transform Escanaba was to rip out the old millwide computer systems. They replaced them with a new system called QUPID (for Quality and Information for Decisions). Unlike the turnkey systems that previously coordinated manufacturing processes, QUPID was custom designed to support operators in each operation; the operators controlled the manufacturing process and would be free to make changes, depending on what they saw happening on the production line. In other words, the system was designed from the outset to help workers make better decisions rather than to cut them out of the decision-making process.

To that end, the mill's managers insisted that operators be intimately involved in the system's design and development. "If the previous system taught us anything, it taught us that we didn't want black-box computer integration that only the vendor really understood," said Glendon Brown, the mill's vice president of production technology. "We needed an architecture that we were part of and that was much more open and easy to change."

Rather than depending on a single supplier, the mill bought the system's building blocks from several different sources. There were two fundamental criteria for choosing those sources: Each had to be a leader in its area of expertise and be willing to customize the system for Escanaba.

Significantly, Escanaba designed the system's interfaces in-house. Each function designed its own interface to ensure that its people got the information they felt they needed to do their jobs and in the format that was easiest for them to understand. "If we were going to succeed in the longer term, we needed the ability to make changes 24 hours a day, seven days a week—times when most managers aren't around," Brown said. "We needed to provide useful information to our operators at the lowest level in the organization, so that *they* could make decisions."

The mill also overhauled its training programs. But it took several attempts before managers hit on the right approach. First, Mead made a common mistake: It used technical people to explain to the operators how the new computer system worked. "It was a disaster," Swanson said. "The technical guys knew too much about the systems and told people more than they needed to know about technical issues and not enough about the business issues and why we needed to work differently."

The mill's managers dropped that program. In its place, they created one designed to help workers understand what they had to do to satisfy customers. The program explained why quick changeovers were critical to the mill's long-term success. And it emphasized that the systems were only a tool to help them perform their jobs better. Operators who had long assumed that keeping the machines running was all that mattered began to look at their work differently.

That program has since evolved. Originally, a professional trainer conducted classes in a seminar-like setting. However, managers came to believe that learning and training should be part of all their employees' jobs. With that goal in mind, managers trained a team of the mill's supervisors and operators to teach others on the floor and in classrooms.

The result of the training program is a plant that now excels at learning. Armed with an intimate knowledge of different jobs and their challenges, the new trainers have played an instrumental role in helping operators become increasingly adept at carrying out quick changeovers and responding to the demands of new customers. Operators are no longer mere machine tenders.

Managers also wisely realized that the mill's measurement and incentive systems had to bolster its new strategy. Accordingly, measures and incentives aimed at maximizing capacity utilization and output and minimizing costs have given way to measures and incentives aimed at maximizing responsiveness and customer satisfaction.

Each year, the mill surveys customers to identify what it is doing well and how it needs to improve. Using the survey, a team drawn from managers and employees throughout the mill then generates specific operating goals. For example, the team might challenge Escanaba to reduce changeover losses by 25% within six months. (This kind of focused goal is much more effective than simply declaring, like managers at all too many factories still do, "We should strive to improve flexibility.") The compensation of managers and superintendents is based on the mill's success in achieving those goals.

The results have been astonishing. The mill's responsiveness and customer satisfaction have increased dramatically. In 1983, it sold the most paper in its history. Escanaba is now the most productive mill in Mead's fine-paper group and has dramatically increased its market share. By emphasizing what workers need to do and providing them with the information they need to do it, Mead has proved that a factory can increase flexibility and, at the same time, boost productivity and lower costs. It's just a question of figuring out precisely what kind of flexibility one wants to achieve and giving people the support they need to achieve it.

Building Flexibility You Can't Sell

When factory managers don't carefully assess their strategies before embarking on a flexibility program, the results can be competitively destructive. One U.S. mill, which I'll call Kildare, had just such an experience. A complex equipped with medium-size paper machines, Kildare made relatively high-volume fine papers and employed about 1,500 people.

In the early 1990s, Kildare's markets were in the grip of a severe slump, and mills with giant, lower-cost machines had come to dominate the business. With Kildare losing market share and struggling to stay in the black, its managers soberly concluded that they could not stand still and wait for the market to rebound. Several competitors had launched reengineering programs aimed at improving their responsiveness, that is, their ability to fill customers' orders quickly. Deeply worried, Kildare's managers decided that they had better follow suit and began to reengineer the order-to-delivery process to make the mill even more responsive than the competition.

To that end, the mill created a training program to teach operators the skills that would enable them to make the swift product changeovers required by the new strategy. In addition, the computerized recipes and other procedures used to produce each grade of paper were carefully documented and routinized. The only aspect that management overlooked was the measurement and reward system, which continued to focus on the plant's capacity-utilization rate. Delivered in the form of praise and criticism as well as pay, the message to supervisors was to keep utilization high. With its emphasis on fewer and longer production runs, that message flew in the face of management's contention that responsiveness was now what mattered most.

Nevertheless, by mid-1993, after an intense, year-long effort, Kildare had cut its lead times in half and had achieved its target of becoming world class in responsiveness. But managers then made a painful discovery. The mill's vastly improved responsiveness had enabled it to hold on to existing customers, who where delighted by Kildare's ability to fill their orders more quickly. But for two reasons the mill had not been able to increase its market share or sales significantly. First, competing mills had made similar strides, and many of them had the advantage of bigger machines. Second, there was simply not enough business in Kildare's particular market.

Kildare's managers realized that they would have been better off had they focused on expanding the range of products that the mill made. A series of opportunities to make a challenging array of relatively high-volume products that were outside the mill's traditional product range sparked the epiphany. The main opportunity was a chance to supply papers to be used to make lottery tickets sold in Central America. The papers had to be difficult for would-be counterfeiters to forge, yet cheap and sturdy.

Kildare's experienced crews were able to manufacture these difficult products. But they found that the systems and procedures installed to improve responsiveness, which assumed a core set of standard, unchanging papers, were a hindrance. For example, the emphasis on perfecting a routine and documenting the procedures and computerized programs for making each paper grade was counterproductive when no two orders were exactly alike and when success depended on excelling at figuring out how to make a particular order during trial runs rather than on delivering that order quickly. In other words, the systems and procedures designed to make the plant more responsive discouraged the development of the skills required to excel at experimentation. Conversely, the trial runs, whose duration was impossible to predict, produced the kind of constant scheduling headaches that one tries to minimize or eliminate if responsiveness is the top priority.

About six months after managers had targeted the new markets, they finally recognized the problem and dismantled many of the changes they had instituted to increase the mill's responsiveness. For example, operators no longer have to enter the programs for making every paper grade into the computer that controls production. Other practices introduced to improve responsiveness—for example, a streamlined order-processing system—did not have to be dismantled. In addition, the mill changed its measurement and reward system to relax pressure on operators to keep the machines running and make fast changeovers. As a result, although responsiveness is no longer the top priority, Kildare has been able to hang on to its traditional customers and win new ones. The mill's sales and profits are soaring. Admittedly, the end of the industry's recession is one reason for its success. But, more important, the mill now excels at a type of flexibility that it can sell.

C H A P T E R

2 PRODUCT DESIGN

2.1 Introduction

The strategic impact of improving a firm's product design process is well established. For example, the majority of a product's manufacturing costs are determined at the completion of the design stage.[1] Moreover, competitive pressure to shorten the design cycle and to increase product variety has intensified significantly in many industries. Shorter design cycles, in particular, reduce the time to bring a product to market and the cost of both redesign and engineering changes. This, in turn, can lead to significantly higher profits over a product's life cycle.[2] Increased variety, on the other hand, can increase the frequency and overall cost of new product development. Finally, it has become clear that lasting improvements in product quality and heightened expectations for product performance, which are mandated by market forces, can only be achieved by "building quality in," i.e., through effective product design.

As a result of these developments, many markets today are characterized by a diverse and rapidly changing mix of products with reduced product life cycles, segmented target markets, enhanced product performance, increased product value, and superior levels of product quality. Our central concern in this chapter is with the management of design and development processes in this volatile environment. We are particularly interested in the role of technology to support product design efforts and the linkage between design and the effective use of automation throughout the firm.

---

[1]See, for example, Fabrycky (1987), who estimates that up to 85 percent of life cycle costs are committed by the end of the concept definition and preliminary design steps. More recently, Ulrich (1993) conducted a field research study to verify this claim. He demonstrated that at least 50 percent of costs (for a class of mature products) are design determined and that up to 70 percent of costs can be affected by manufacturing process decisions.

[2]House and Price (1991), for example, quote a study by McKinsey which showed that the products that were launched within budget, but were six months late, resulted in 33 percent less profit over a five-year life cycle, when compared to products that were within budget and on time. In contrast, products that were 50 percent over budget, but launched on time, resulted in only 4 percent less profit.

The process of new product development can be broken down into a collection of stages which include the following activities:

1. Needs analysis.
2. Ideation and concept development.
3. Concept screening.
4. Requirements specification.
5. Product design.
6. Prototype development.
7. Process planning.
8. Design qualification and performance testing.
9. Preproduction.
10. Volume production.
11. Marketing and promotion.
12. After-sales support.

In order to illustrate the process, consider the case of a food manufacturer developing a new type of frozen food. The company has discovered that there is growing interest in healthy food with low fat and low sodium (stage 1). A design team consisting of marketing, research and development, and strategic planning personnel is formed to go through a (possibly formalized) process to generate a set of product concepts, which incorporate these needs (stage 2). In some cases, the concept will involve modification of an existing product (a line extension), and in other cases it may involve invention of an entirely new category of frozen food. Such concepts are usually formalized in terms of graphical, descriptive, and technical data (e.g., a picture of the new dinner after it is cooked, a description of what it will taste like, its aroma, and the price at which it will be sold in the supermarket).

In the next stage (3), the design team reviews the set of concepts that have been generated. In a consumer products firm, this involves market testing, in which the data describing each concept is presented to panels of consumers and their responses (in terms of perceptions and willingness to purchase) are recorded. The design team then reviews these data, and, based on some type of analysis, screens out those alternatives which lack market promise and/or are not feasible to develop. In other industries, where product performance differences can be measured more objectively (i.e., computers), concept screening may occur within the firm and will be based on internal analyses of the pros and cons of the business case.

Figure 2.1 illustrates this selection process as a development "funnel" in which concept alternatives enter on the left and survivors pass through to the right. A recent survey of new product development in the packaged goods industry (EFO Group 1995), indicated that the survival rate of concepts to the next stage of development is 40 percent.

Stage 4 requires translation of the selected concept(s)—expressed in terms of product attributes understood by the customer—into the technical and engineering specifications which are needed to explicitly design the product. For our example of a frozen dinner, this would include factors such as measurements of the fat and salt content and the principal ingredients. Taken together, stages 1 through 4 provide a basis for dis-

Figure 2.1

The product development funnel:
new product survival rates in the packaged goods indusry

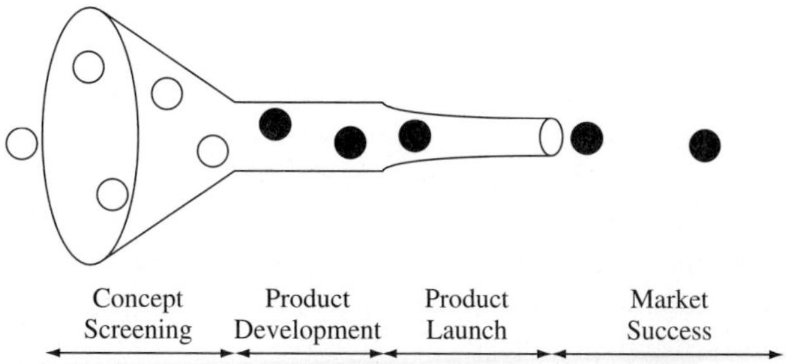

	Concept Screening	Product Development	Product Launch	Market Success
%*	40	48	60	38
CUM %**	40	19	11	4

* "... of those products entering each phase, what percentage survive."
** "... of those products initially proposed at concept stage, what percentage survive."

Source: EFO Group (1995)

cerning the "voice of the customer" and translating the message carried by that voice into specifications for the new product.

Stages 5 through 8 constitute the heart of the development process and lead to the translation of the concept into a tangible artifact, which in our case can be tasted and experienced. This phase also defines the production process which will be required to manufacture the product as well as the mix of ingredients and suppliers for these ingredients. Typically, additional consumer and laboratory testing is required at this juncture (stage 8), and, as a consequence, a considerable amount of iteration among stages 5 through 8 can take place as alternative product specifications are developed, tested and modified. These stages constitute the middle section of the funnel in Figure 2.1. According to the EFO survey results 60 percent of all prototypes developed are actually launched.

Upon completion of this phase of the development process, the firm must turn its attention to how it will produce the product in its plants and how it will market and distribute the product to its customers. Our frozen food manufacturer may carry out pilot production runs and test new equipment and processes (stage 9). Marketing will develop advertising materials and plan the sales campaign (stage 11). Ultimately, the plant begins producing the product in volume and the sales force and distribution functions begin moving the product through the supply chain so that it will reach the supermarket shelves (stage 10). After-sales support (stage 12), in the case of a food product, may involve supporting the relationship with the retailer and assuring quality of the product. In technologically more complex products (e.g., bulldozers, computers, automobiles), after-sales activities include product maintenance and repair. For extended use products that have reached the end of their useable life, the items must be discarded. Increasingly, manufacturers are being held accountable for the manner in which their products and any of its by-products (e.g., packaging) are removed from the environment.

While there is a clear precedence requirement for the execution of each stage noted above, many stages occur in parallel and overlap in time. Moreover, as we have seen, the activities carried on within the development process stages involve simultaneous participation from a wide range of firm functions and disciplines (e.g., industrial design, market research, research and development, product engineering, manufacturing engineering, marketing, production planning, distribution, and customer service). Thus, typically, a high level of coordination is required throughout the development process. Finally, for most products, there is a finite time interval available to complete these activities, since at some point in time the market life cycle ends and demand for the product disappears (due to product obsolescence and/or as a result of competitor actions).

It is interesting to note that most products that are launched fail to be successful in terms of meeting management objectives (e.g., long-term sales, profit, etc.). The EFO survey puts this survival rate in the packaged goods industry case at 38 percent. The cumulative survival rate, from concept to successful product, turns out to be only 4 percent for this sample. Thus, we can see that in spite of its strategic importance and high level of management attention, new product development remains a challenging area for managers.

In general, the product development process described above gives rise to complex information flows within the firm and requires extensive analysis at each stage of the funnel. From a managerial perspective, two critical questions must be answered.

1. How should the firm *organize and manage* its product development activities in order to consistently generate a stream of successful new products?
2. How should product *design alternatives* be evaluated, selected, and prioritized? In particular, what is the impact of each design alternative on the various downstream processes in the firm's value-added chain?

In answering the first question, the firm allocates its scarce design, production, marketing, and other relevant resources to each development project and develops procedures to promote effective use of these resources. The second question requires definition of specific strategic development targets that go beyond market positioning and engineering specifications and that incorporate a long-term, life-cycle perspective. The primary performance metrics for the design process include: unit product cost, time to market, product performance, and overall development project cost. As we shall see, these performance attributes interact over the product life cycle, in the context of a changing competitive environment, to generate the cash flows which ultimately determine the return on the firm's development project investment. In answering the second question, firms also must develop procedures for predicting how development resource investments and design process procedures interact to affect peformance with respect to the metrics noted above.

In this chapter we will consider how to answer these questions and how these answers are linked. Both current practices and the results of current research on these issues will be reviewed. We also will explicitly consider how technology can affect the product development process by focusing on specific procedures and tools which have emerged to support design activities and whose use can profoundly affect both individual project success and the overall competitiveness of the firm.

2.2 Organizing and Managing the Product Development Process

Traditionally, product designs were analyzed from the point of view of meeting market needs at the least cost. This so-called value engineering approach would trade off product function against product cost through mechanisms like design simplification and manufacturing methods improvement. It was also common to carry out a producibility study for a new product to assure the capability of the manufacturing system to fabricate the parts and to assemble the product. Producibility and value engineering were often undertaken well after a design concept had been finalized, and to some extent they represented defensive mechanisms on the part of the various downstream functions of the firm. Figure 2.2 illustrates the flow of events in a typical sequential product design process.

The traditional role of production automation technology, in the process flow depicted in Figure 2.2, is to improve the efficiency of the firm's current manufacturing processes. The effectiveness of such improvements is, to a large degree, based on the stability of the firm's product/process designs. As noted above, product designs are evolving rapidly and in many cases are subject to significant technological change. As a result, the impact of automation technologies on the overall competitiveness of the firm has often been disappointing. In many cases, firms automate to produce yesterday's products.

In addition, use of technology for productivity enhancement (i.e., through manufacturing process improvements) often is not effective in this sequential environment because the associated changes are adopted in response to finalized designs. It may be

FIGURE 2.2

Typical product design practice

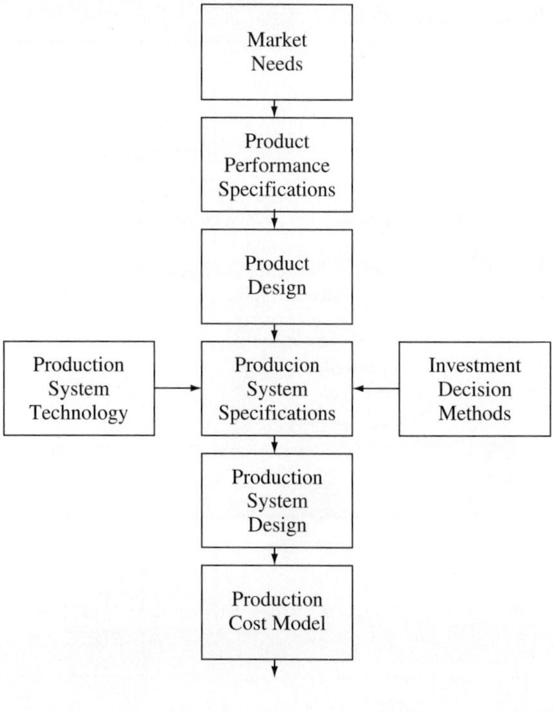

more appropriate to modify product design rather than the production process in order to achieve maximum effectiveness. In other words, *joint* determination of product and process designs can lead to improved competitive performance. Such simultaneous analysis of products and processes is not feasible in the sequential mode.

The *concurrent* approach discussed in this chapter involves a major reorganization of the design process. Figure 2.3 illustrates the logic and information/decision flows associated with a typical concurrent design process. It is characterized by functional cross linkages, multiple feedback loops, and significant activity overlap. The implementation of this concurrent process treats design as only the first step, in an iterative procedure, to bring products to the marketplace. Both design and advanced manufacturing technology must be matched in order to realize the full productivity potential of automation of production processes and the value enhancement potential of product innovations.

We note, in particular, that the traditional, sequential design organizational structure can make the product development process more difficult to manage. This can slow down the overall process and adversely impact product performance and quality. For example, product engineers often focus on generating new inventions and are not concerned with the impact of these inventions on downstream phases (the famous "over the wall" syndrome). Consequently, cost and performance trade-offs associated with manufacturing, marketing, and support of the product throughout its life cycle may not be adequately considered in the analysis of design alternatives. By the same token, in environments where manufacturing has control, the selection of designs on the basis of their producibility may not give sufficient weight to the marketability or serviceability of the product.

Figure 2.4 illustrates the overlapping structure of the product design process in the automobile industry. It is based on the international study conducted by Clark and

FIGURE 2.3

Outline of the strategic product design method

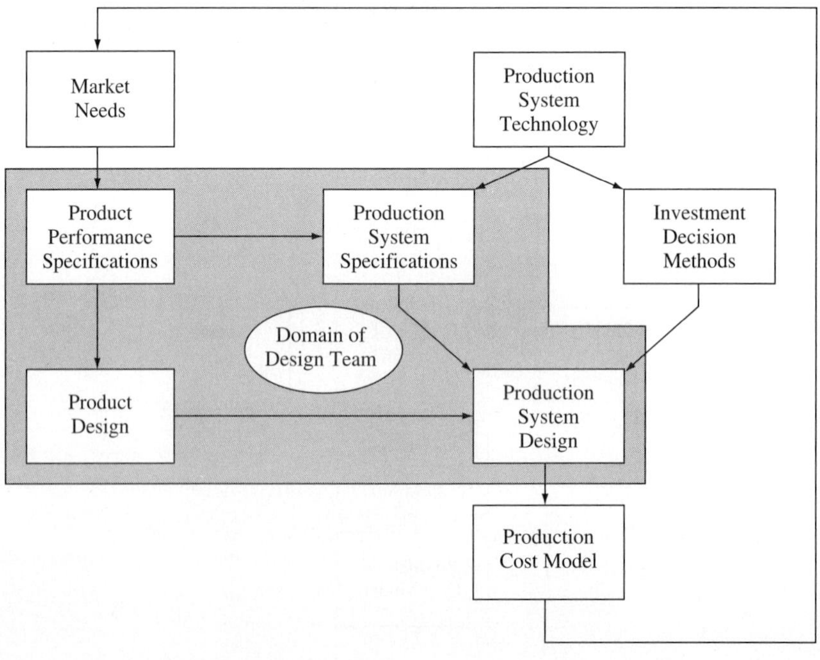

FIGURE 2.4

Average project schedule by stages in the automobile industry

A. U.S. Average

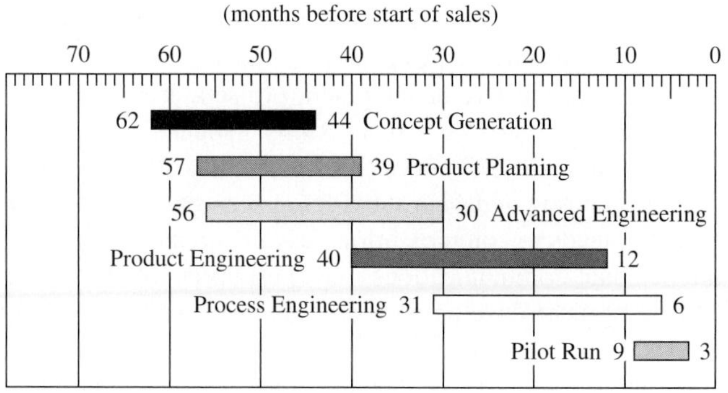

Note: Average lead time of 6 U.S. projects.

B. European Average

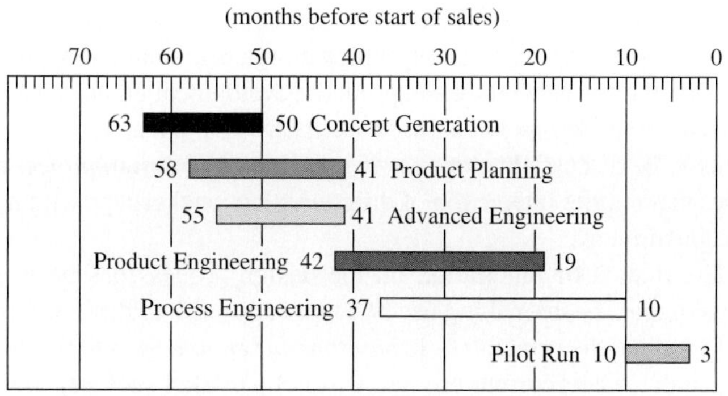

Note: Average lead time of 11 European projects.

C. Japanese Average

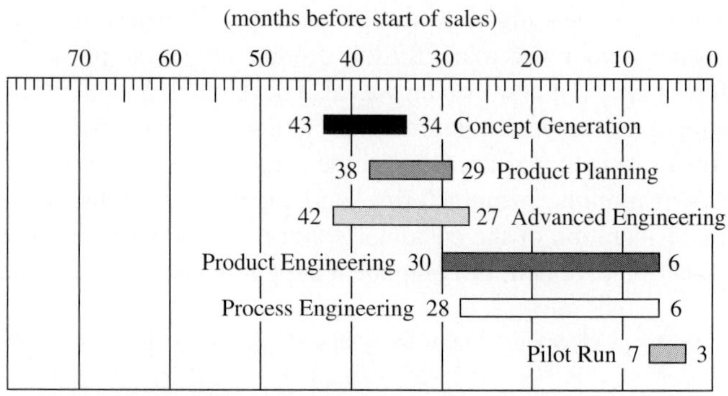

Note: Average lead time of 12 Japanese projects.

Adapted from Clark and Fujimoto (1991)

Fujimoto (1991) in which 29 car development projects carried out by US, European, and Japanese manufacturers were analyzed. The data focused on the resources, organization, and outcomes of each vehicle development project. The data in Figure 2.4 illustrates that the Japanese manufacturers, on average, were able to execute a much faster design process (43 months as compared to 62 or 63 months for the American and European producers respectively). *Concurrence,* the hallmark of the Japanese process, led to faster and more productive development projects. It is important to note that recently American manufacturers have successfully adopted concurrent procedures and, as a consequence, national differences on time to market have decreased significantly.

The restructuring of design activities in accordance with the simultaneous approach of Figure 2.3, leads to complex information and analysis requirements. Data accuracy and interpersonal communications must be improved.[3] This follows, since all participants in the product development process must have access to the same, up-to-date product design information at all times. *Computer integrated manufacturing,* with its common database of product specifications that is shared among product designers, manufacturing engineers, and production staff, is an important enabling technology for concurrent engineering. The use of a common database of specifications has led to significant improvements in cost, quality, and development time (e.g., up to a 35 percent improvement at one major aerospace manufacturer).

It is important to note that significant organizational innovation is needed to support design alternative generation and product development in the concurrent mode. Methodologies such as *design for manufacturability (DFM)* and *quality function deployment (QFD),* to be discussed below, are two examples of design project management methodologies that require interaction of different functional groups within the firm and that promote concurrency.

The impact of integration on the design process has been widely documented.[4] Many companies are making use of a team approach to design. Recently Ingersoll Rand (1991) applied the principles of *simultaneous engineering* to the development of a handheld grinder. Their current product was not a market success, and all previous attempts at introducing a redesigned grinder had been unsuccessful. This project was distinguished by its use of a structured, multidisciplinary team. The team carried out extensive direct interviews with product end users to define ergonomic and performance attributes for a radically new design concept. The new design incorporated plastic composites (which was a first for this product class) and provided improved safety and comfort of use. The new grinder was also developed in record time (one year vs. the average three to four years required in previous grinder redesign efforts).

The organizational structure for the team included early and continued involvement of sales, marketing, manufacturing, and engineering. It also was characterized by explicit consideration of the customer's voice at each stage of the design. Finally, there was direct involvement of component suppliers and external industrial designers. The

---

[3]It has been noted in some firms in the computer industry that there is a greater likelihood for error in the concurrent design structure, since information now travels farther and there is less opportunity for immediate accuracy checking. In a sequential environment, information may only move on to the next phase.

[4]Wheelwright and Sasser (1989) observed that "integrators" with backgrounds in fields like industrial engineering can keep design and manufacturing perspectives in balance. Cross-functional teams require collaboration from the start (Dean and Sussman [1989]).

FIGURE 2.5

Ingersoll Rand power tool division's new product development process

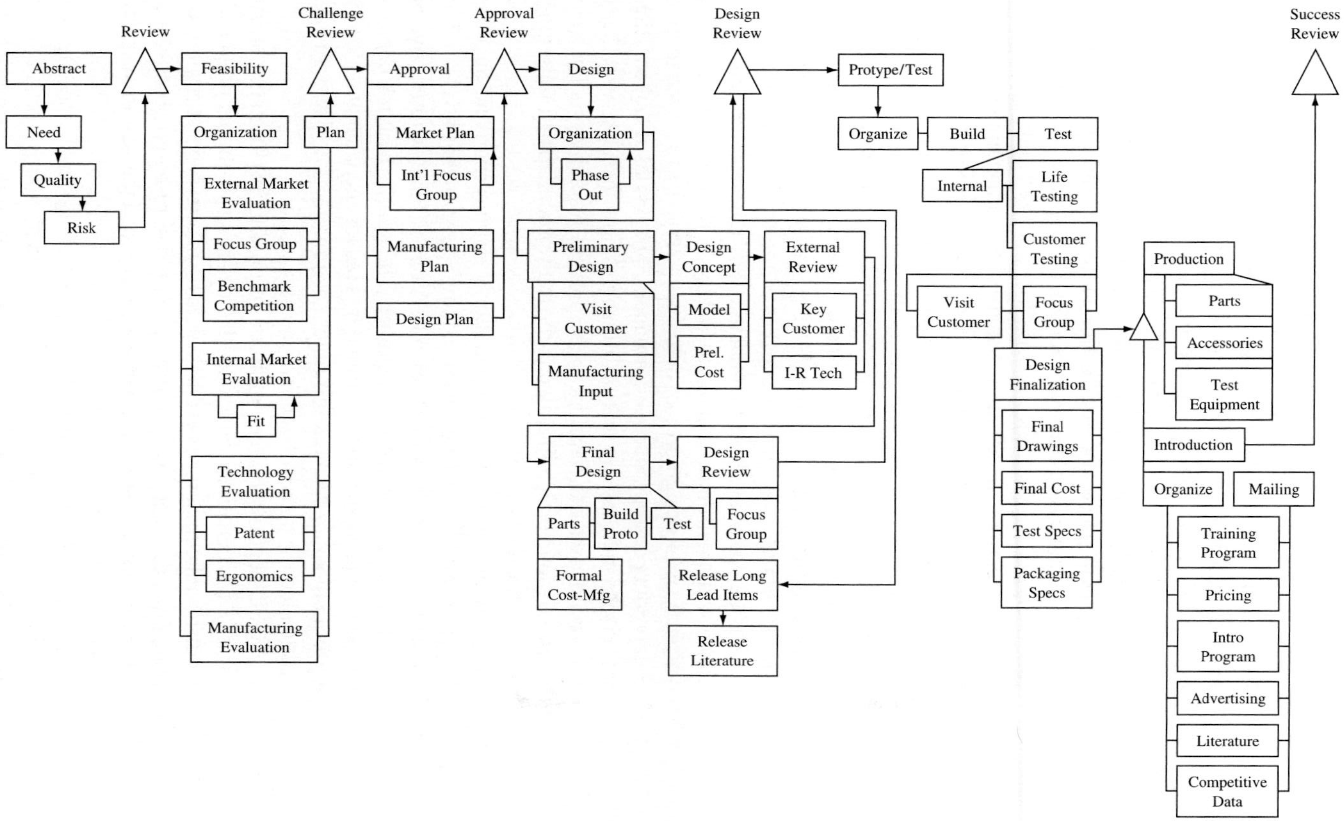

result was a new product design which was perceived to be pathbreaking for this industry. The new design achieved significantly higher market penetration. The organizational multifunctional "road map" (see Figure 2.5) was critical, since the entire design process represented a radical departure from Ingersoll Rand's traditional approach for introducing new products. As a result of their success with the grinder, the company adopted this process road map and the performance target of a one-year product development cycle for subsequent new product projects.

It is interesting to observe that significant difficulties arose at Ingersoll Rand when they attempted to achieve the level of concurrence required to make their new product development processes as effective as that of the grinder on a long-term basis. In a follow-up study, Dougherty and Cohen (1993) (and Reading 2.1 at the end of this chapter), analyzed a set of projects that were organized under the process structure illustrated in Figure 2.5 and discovered that there were a number of ingredients necessary for achieving sucessful innovation. Based on these observations, they developed a framework that identifies three dimensions of concurrence: *functional, institutional* and *portfolio.* The first, functional, refers to the use of cross-functional design teams discussed above. The second, institutional concurrence, involves maintenance of a high degree of consistency in setting design project strategic goals and in setting performance evaluation incentives. Many companies seem to have a disconnect in this area: the goal is to launch an innovative new product in a compressed time frame, while the members of the design group are held accountable to traditional cost and efficiency measures associated with their home functions. Finally, portfolio concurrence involves explicit tradeoff analysis of design project outcome attributes (e.g., time to market, investment cost, unit product cost, product performance). The model of Cohen, Eliashberg, and Ho, to be introduced in following section, illustrates how these performance trade-offs arise. Portfolio concurrence then refers to the process whereby product specific goals and development resources are allocated in the multiproduct portfolio setting which is common in most companies.

The interaction between marketing, engineering, and manufacturing illustrated by the Ingersoll Rand design team requires a significant level of communication between the design engineers and the members of the team who discovered customer needs by listening to the voice of the customer. The *quality function deployment (QFD)* method, introduced in Japan (and described by Hauser and Clausing [1988]), is a recently developed methodology which promotes such interaction. The QFD approach builds a concept called the *house of quality* (see Figure 2.6).

QFD begins with the definition of product performance attributes relevant to consumers. The perceived value of both the firm's existing (or proposed) product and the performance of competitor products, with respect to these attributes, is then estimated. Such perceptual maps are used routinely in market research product positioning studies. The observed gaps between competing products and the customers preferred "ideal point" indicate opportunities for improved product design. The challenge is to communicate these performance requirements back to product designers and engineers who can translate them into product and process specifications. Our research in the food industry suggests that there is often a gap between concept and product in terms of perceived value.

Let us consider the example of automobiles. Consumer performance variables include things like miles per gallon, the acceleration rate, and the number of quality

FIGURE 2.6

House of quality

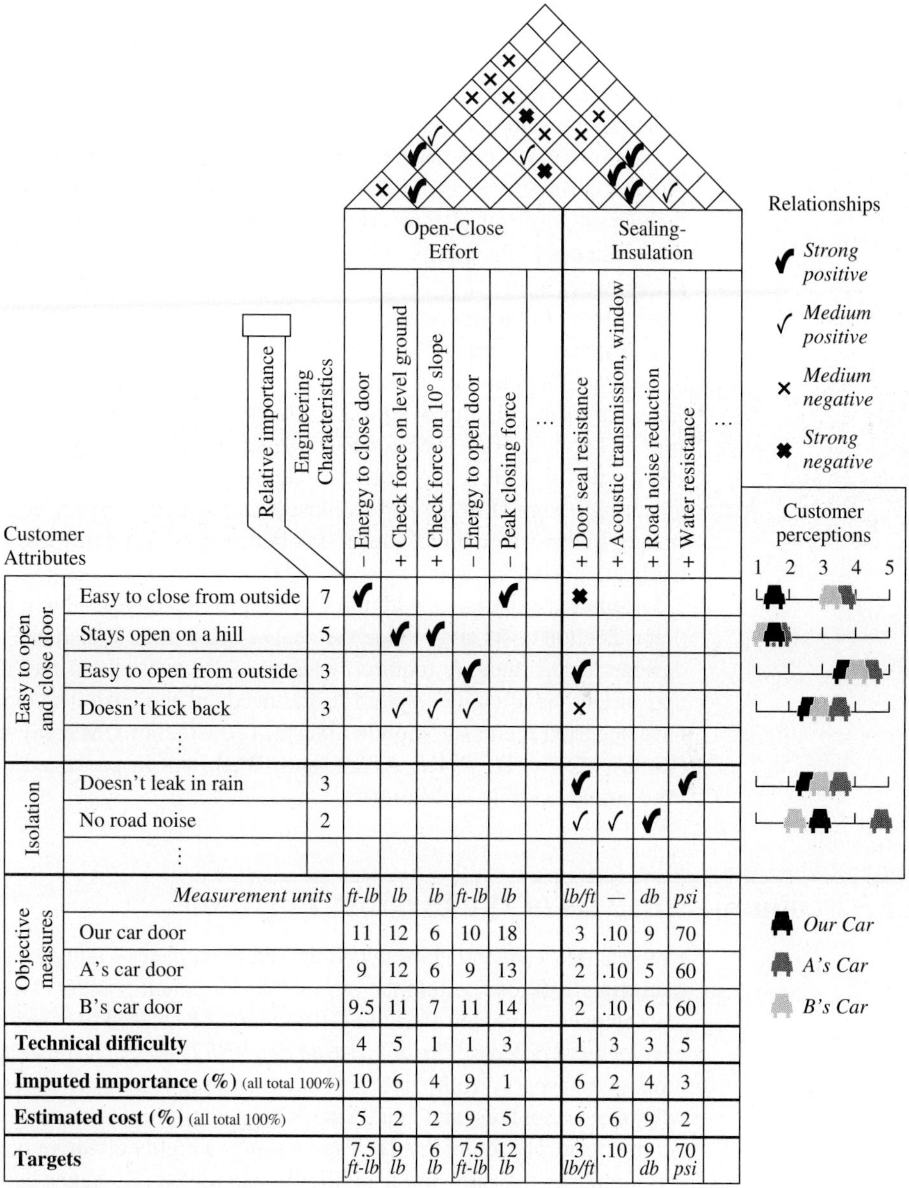

Customer Attributes		Relative importance	− Energy to close door	+ Check force on level ground	+ Check force on 10° slope	− Energy to open door	− Peak closing force	…	+ Door seal resistance	+ Acoustic transmission, window	+ Road noise reduction	+ Water resistance	…
	Engineering Characteristics		Open-Close Effort						Sealing-Insulation				
Easy to open and close door	Easy to close from outside	7	✔				✔		✖				
	Stays open on a hill	5		✔	✔								
	Easy to open from outside	3				✔			✔				
	Doesn't kick back	3		√	√	√			×				
	⋮												
Isolation	Doesn't leak in rain	3							✔			✔	
	No road noise	2							√	√	✔		
	⋮												
Objective measures	*Measurement units*		*ft-lb*	*lb*	*lb*	*ft-lb*	*lb*		*lb/ft*	—	*db*	*psi*	
	Our car door		11	12	6	10	18		3	.10	9	70	
	A's car door		9	12	6	9	13		2	.10	5	60	
	B's car door		9.5	11	7	11	14		2	.10	6	60	
Technical difficulty			4	5	1	1	3		1	3	3	5	
Imputed importance (%) (all total 100%)			10	6	4	9	1		6	2	4	3	
Estimated cost (%) (all total 100%)			5	2	2	9	5		6	6	9	2	
Targets			7.5 *ft-lb*	9 *lb*	6 *lb*	7.5 *ft-lb*	12 *lb*		3 *lb/ft*	.10	9 *db*	70 *psi*	

Source: Hauser and Clausing (1988)

defects. The perceived differences between the performance of competing brands are clearly measurable with respect to such variables and contribute directly to product success. The QFD method can be used to define engineering characteristics of a product design which will improve vehicle performance by reducing or eliminating the consumer's perceived gaps in the company's current product. Chassis weight, engine horsepower, and dimensional conformance are examples of engineering design characteristics which are related to the perceived performance attributes noted above.

Estimation of the contribution of each engineering characteristic to each consumer product performance attribute is the key to QFD. Members of the design team (which include engineering, manufacturing, and marketing representatives) develop quantitative estimates of the marginal impact of each engineering characteristic on each product performance dimension. Figure 2.6 illustrates these linkages, which constitute the "main floor" of the house of quality.

Since engineering design attributes are themselves interrelated, it is also necessary to estimate the correlation between these characteristics (i.e., the "roof" of the house). Thus, for example, improved dimensional conformance could be related to chassis design and weight. Target performance levels of engineering characteristics can then be set in order to achieve the target product performance attribute goals. The resulting engineering design attribute targets must be achieved through specification of the product design. In this way a direct causal link between market value and product specifications has been created.

Reported experience with the QFD approach suggests that it can lower start-up and preproduction costs and reduce the design cycle time. It also can reduce the number of design changes that are required when it is discovered that the launched product does not, in fact, adequately respond to the needs of the customer. Successful applications have occurred in the automobile industry (Toyota, Ford, Mazda), high technology firms (Kodak, Xerox, DEC, HP, AT&T, and IBM), and in packaged goods companies like Procter and Gamble and Colgate.[5]

2.3 Economic Analysis of the Product Portfolio

House and Price (1991) introduced the *return map* (see Figure 2.7) as a methodology for managing trade-offs within the product development process. They report on its significant impact at Hewlett Packard (HP) where it was used to improve productivity of the development process. The *break-even time (BET)* concept, illustrated in Figure 2.7, captures some aspects of the trade-off between time-to-market, investment of product development resources, and market sales in an environment characterized by a fixed product life cycle. HP found that BET reduction is a highly effective performance goal, since it is simple to measure and it implicitly captures the linkages between product performance, quality, and timeliness.

[5]See Griffin and Hauser (1991) for a recent study of the implementation of QFD. They discovered that 83 percent of the projects using QFD reported a strategic benefit. They also documented the role of QFD in enhancing communication within the development process.

FIGURE 2.7

HP's return map: an example of a post-launch decision tool

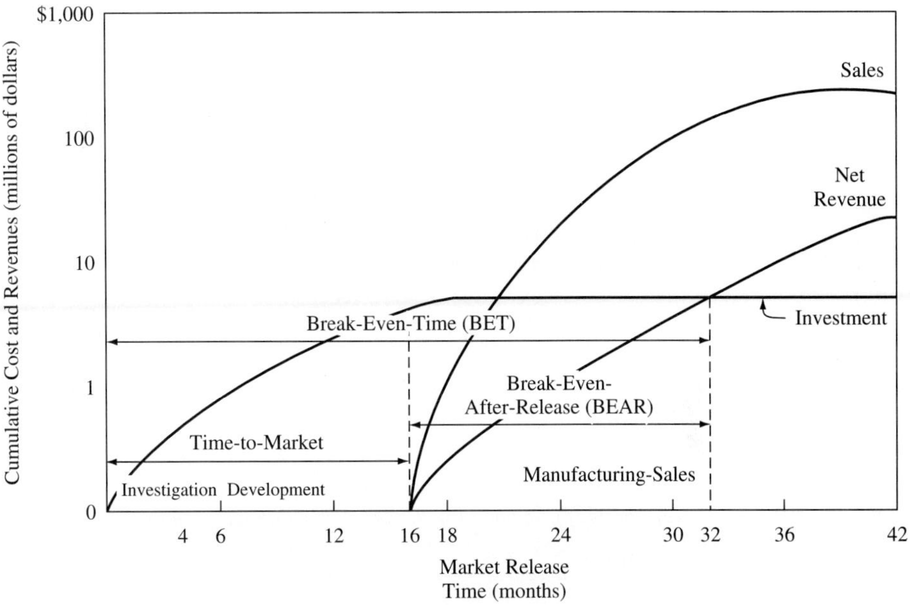

More recently, Cohen, Eliashberg, and Ho (1993) analyzed performance trade-offs directly by developing a multistage, economic model of the product development process. Each development stage impacts the perceived product quality. The output quality of each stage is determined in a cumulative manner (i.e., the quality "produced" at the performance development stage is an input to the process planning stage). Moreover, product quality can be enhanced by activities within each stage by (1) selecting the level of inputs utilized (representing capital, labor, and information) and (2) fixing the duration of the stage. As more time and resources are invested within each, better quality will result. Improved products are generated, however, at the expense of delaying the introduction of the product into the market and/or by increasing the project development costs. Figure 2.8 illustrates how scaling the "performance ladder" can evolve throughout the various stages of a development project.

The managerial question of interest is what should the duration of each stage be and what level of resources should be committed within each stage. Consequently, it is necessary to understand how input resources affect product quality (i.e., by estimating a *design production function*) and how sales are affected by the achieved level of quality (i.e., by estimating a *sales forecast function*). Analysis of the model indicates that determination of time-to-market and the quality level of either existing or anticipated competitor products are key factors. In some cases (e.g., when there is no competition in the market), a firm would be better off to introduce a lower quality product more rapidly. In other instances (e.g., when there is a high quality incumbent in the market), the firm should delay product introduction and invest in product design enhancement prior to

FIGURE 2.8

Product performance improvement process

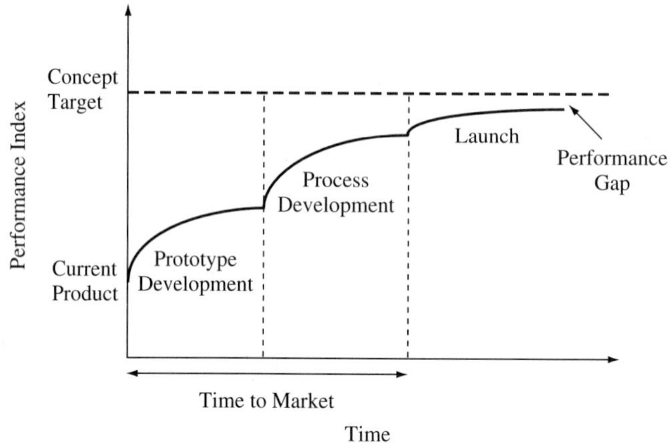

FIGURE 2.9

Variations of profit and break-even time with time-to-market

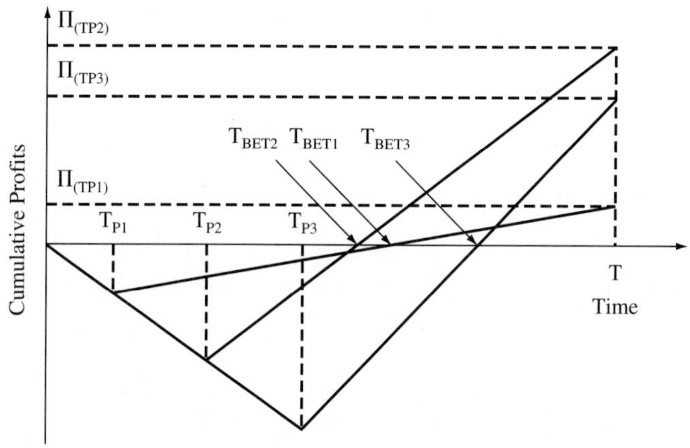

launch. The selection of appropriate targets for these performance metrics is critical for effective product development.

Figure 2.9 illustrates the cash flow consequences of the trade-offs. As the firm develops a project, its cumulative profits are negative. At launch time (TP_1, say), revenues begin accruing based on market share, price, and market size considerations. Delay of launch (to TP_2 or TP_3) leads to higher expenditures (i.e., the firm falls "deeper into the well"). Launch at the later date presumably is accompanied by a better product with a greater share. Hence, revenues rise at a faster rate. Based on the objective of maximum, cumulative profit (as opposed to BET), the best choice in this example is TP_2.

In a recent project carried out in the packaged goods industry, Cohen, Eliashberg, and Ho developed a decision support tool which was based on the model and trade-offs discussed above. The purpose of the support system is to facilitate concept screening and prototype development by providing a framework which makes use of market test data and other relevant project information. The performance metrics generated by the support system at each development stage include updated measures of consumer acceptance, projected sales and cash flows, market share, unit product cost, and development costs. Based on this information, the firm develops a product with specific attributes, which give rise to market acceptance, revenues and, ultimately, profits. The support tool quantifies the trade-off among resource/information inputs, time, material flows and cash flows.

Figure 2.10 contains the results of analysing company data, based on a sample of launched products. The product performance improvement function indicates how two key consumer acceptance attributes, scoreboard (SC) and purchase intent or "definitely will buy" at the prototype stage (DWB_p) are determined by the level of labor inputs

FIGURE 2.10

The firm's new product performance improvement function

$$\text{Log(SC)} = \underset{(0.0245)^{***}}{4.3495} + \underset{(0.0175)^{**}}{0.0348}\ \text{Log}(R_L) + \underset{(0.3302)^{***}}{0.0348}\ \text{Log}(R_E) + \underset{(0.0037)^{***}}{0.0123}\ \text{Log}(R_N)$$

(Adj R-sq = 0.34)

$$\text{Log(DWBp)} = \underset{(0.0245)^{***}}{2.8474} + \underset{(0.1860)^{*}}{0.3263}\ \text{Log}(R_L) + \underset{(0.0923)^{***}}{0.3611}\ \text{Log}(R_E) + \underset{(0.0392)^{**}}{0.1188}\ \text{Log}(R_N)$$

(Adj R-sq = 0.33)

* statistically significant at 10% level.
** statistically significant at 5% level.
*** statistically significant at 1% level.

FIGURE 2.11

The sales forecasting functions

Concept Stage

$$\text{Sales} = \underset{(292.81)^{**}}{-932.54} + \underset{(4.34)^{**}}{14.65}{*}(DWB_C + \text{Taste}) + \underset{(5.27)^{**}}{21.48}{*}\text{Uniq}$$

(Adj R-sq = 0.44)

Prototype Stage

$$\text{Sales} = \underset{(337.13)^{**}}{-931.30} + \underset{(5.16)^{**}}{11.41}{*}(DWB_C + \text{Taste}) + \underset{(5.71)^{**}}{18.51}{*}\text{Uniq} + \underset{(4.63)^{**}}{10.86}{*}DWB_P$$

(Adj R-sq = 0.51)

Marketing Stage

$$\text{Sales} = \underset{(320.93)^{**}}{-872.38} + \underset{(4.90)^{**}}{12.85}{*}(DWB_C + \text{Taste}) + \underset{(5.82)^{**}}{13.37}{*}\text{Uniq} + \underset{(1.48)^{*}}{2.85}{*}DWB_P + \underset{(30.88)^{**}}{79.39}{*}\text{Media}$$

(Adj R-sq = 0.66)

* statistically significant at 5% level.
** statistically significant at 1% level.

(R_L), the experience level of the research and development team leader (R_E), and the sample size of the market tests (R_N). The forecasting functions indicated in Figure 2.11 were generated on the basis of data available at the completion of each development stage. We see that consumer test responses on purchase intent at both the concept and prototype stage (DWB_C, DWB_P), taste (Taste) and uniqueness (Uniq), as well as media spending (Media) contribute to first year sales forecasts.

2.4 Design Support Tools and Methods

There is a strong relationship between product design and the performance of downstream value-adding processes. As a consequence, a family of methods have been developed to exploit the linkages between the nature of the product and the processes involved in its production, delivery, after-sales support, and disposal. These methodologies have come to be known as *design for X (DFX),* where X denotes a wide range of performance attributes such as cost, quality, performance, service, environmental impact, manufacturability, and reliability. DFX methodologies should be invoked early in the design process, and their use neccessitates adoption of a long-term, cross-functional perspective.

In order to illustrate DFX, consider two examples; a hand grenade and a vacuum cleaner. The hand grenade is an item that is used once and must be 100 percent reliable. The vacuum cleaner is more complex, has a range of customer options, and is expected to last for years. As a result of these differences, the parts, production processes, and service systems appropriate for each product will differ. The hand grenade is made of high quality parts that are glued or welded together and usually is not repaired after its manufacture. The vacuum cleaner requires high quality parts that can be screwed together, and the manufacturer must provide replacement parts and a field repair service capability.

Design for manufacturing (DFM), which is the most widely known example of the DFX philosophy, is a general process which explicitly takes into account the manufacturability of proposed designs. The objective of DFM is to develop product designs which are easy to produce, and which have low cost, high reliability, and superior quality. Choice of fabrication methods, materials, processing equipment, and production personnel are considered in assessing each design alternative. *Design for assembly (DFA)* is a closely related process concerned, in particular, with the requirements for the assembly of components into finished products. It is based on time-motion studies of assembly operations for different classes of assembly methods (e.g., manual, dedicated automation, and flexible automation). The objective of DFA is to predict a product's total manufacturing cost, (including component handling and insertion). These forecasts are generated for each design/process scenario and are based on an empirically validated model of the assembly process.[6]

[6]See Dewhurst and Boothroyd (1989) for a description of a software support system for DFA which has become widely accepted.

Ulrich and Eppinger (1993) describe the DFM methodology as consisting of the following steps:

1. Estimate manufacturing costs.
2. Minimize the cost of components.
3. Minimize the cost of assembly.
4. Minimize the cost of supporting production.
5. Consider the impact of DFM decisions on other factors.

They review methodologies for each of these steps and indicate how they can be linked together in an iterative fashion under the DFM umbrella. Of particular importance are the drivers of product cost which include the cost of components, the cost of assembly, and relevant overhead costs. We will return to the issue of costing in Chapter 8 where we consider accounting practices in automated production environments.

Scientific research on DFM and learning from practice has led to the enunciation of certain principles of design (see Stoll [1986]). These principles have been used to derive rules or protocols for formulating efficient designs. In a recent survey, Sackett and Holbrook (1988) evaluated currently available DFA software tools. These systems are based on the application of design rules which are directed towards either minimizing product complexity or towards maximizing the ease of fabricaton and assembly. The rules can be classified as follows:

Minimize System Complexity:

1. Minimize total number of parts.
2. Develop modular designs and group components into subassemblies.
3. Use standard components.
4. Design parts for multifunction use.
5. Maximize parts commonality by standardizing function and style.

Maximize Ease of Fabrication and Assembly:

6. Avoid use of separate fasteners.
7. Minimize assembly directions of insertion (e.g., use of chamfers and location features).
8. Maximize compliance by designing parts for stability.
9. Minimize handling.
10. Try to minimize product weight.

The intent of these design principles is to provide guidance to product engineers and designers so that manufacturability will be taken into account early in the design phase. Their impact can be dramatic. For example, IBM used design for assembly techniques to develop the Proprinter.[7] The resulting design had 65 percent fewer parts and an assembly time which was reduced by 90 percent.

[7]See IBM Proprinter case (March 1989).

The organizational requirements and management mechanisms for successful implementation of DFA have also been studied. Ettlie and Reifeis (1987) analyzed the experience of three companies and discovered that use of a cross-functional design team, compatible CAD software systems, common reporting and database systems, formal DFM programs, and engineering generalists all were correlated with success. DFM is now widely adopted, and successful applications have been reported by many companies (e.g., General Motors and the Saturn, Allen-Bradley's automated connector plant, the General Electric dishwasher and Polaroid cameras).

The diffusion of the DFX method has recently been felt in downstream operations. For example, Lee, Billington, and Carter (1993) report on the concept of *design for localization,* in which customization of HP's ink-jet printer for the European market was facilitated by a design change. This change allowed the company to produce a generic product at its central facility in Vancouver, Washington, and ship them to the distribution center in Italy. At the distribution center, warehouse personnel added power supplies, software, and documentation appropriate for each country market in response to filling local shipment orders. The significant savings realized was based on inventory reductions enabled by the pooling of demand risk over multiple markets.

Another example of DFX is *design for the environment (DFE).* Lenox and Ehrenfeld (1995) define DFE as "the systematic process by which firms design products and processes in an environmentally conscious way based on industrial ecology principles across the entire product life-cycle." The feedback of material flows generated in the production, use, and disposal of products must now be considered explicitly. The perspective of design from "cradle to grave" has now been expanded to that of "lust to dust."

The industry ecology principles mentioned in the definition of DFE include:

1. Loop closing to promote the circulation of materials within the production/distribution/disposal system.
2. Dematerialization to reduce the use of materials by designing products with equivalent functions.
3. Protection of the natural metabolism by restricting the flow of harmful materials.
4. Systematic use of energy by adoption of conservation, extraction of energy from the process, and through the minimization of waste heat flow.

Today there is a wide range of environmental practices that can be adopted to support these ecology principles. We note, in particular, the use of pollution control and prevention devices and reduction of toxic products and waste. Additional practices include use of recycling, remanufacturing and reuse, and promotion of symbiosis among firms, which allows the byproduct outputs of one firm to become useable inputs to another firm.

DFE can be supported by the adoption of a number of management tools. A primary method is *life cycle analysis,* which requires that an inventory of data be collected to document the impact of the product throughout its life cycle. This data is used to identify means to reduce harmful impacts and is often required by government regulation. An approach referred to as *environmentally conscious manufacturing* has emerged in which specific tools and processes are used to minimize the environmental impact of production. Finally, product designs are now directly affected (e.g., the elimination of freon in refrigeration products).

2.5 Technology to Support Product Design

Automation technologies such as CAD, CAE, and CAM are used to support the design process in many firms. *CAD software systems* assist design engineers in interactively defining the geometry and specification of parts and assemblies. These systems are capable of displaying on computer screens two-and three-dimensional images of the product. These images can then be easily manipulated (e.g., rotation, zoom, contour representation) and modified.[8]

A key advantage of CAD systems is that they provide the designer with improved visualization of the design. As a result, product designers can evaluate the impact of design changes in real time. Moreover, with the aid of a CAD system, the designer can retrieve and edit existing designs to rapidly create new designs. This leads to more effective use of the firm's knowledge base (embodied in its library of parts designs). As a result, it is possible, with the aid of CAD, to increase part commonality and enforce design standards. This leads to greater productivity of the firm's design, production, and after-sales service support processes.

CAE systems allow designers to carry out engineering analysis of product designs. For example, a designer can evaluate the strength of alternative material choices using the computer and thereby avoid the need to develop a prototype and to carry out laboratory tests. CAE can also be used to evaluate circuit design performance. In both cases, the CAE system develops a mathematical representation of the product structure and uses this representation to predict the impact of changes in key design parameters on product performance (e.g., *finite element* or *circuit* analysis). It is important to note that CAE systems are idiosyncratic to the product characteristics. Recently, artificial intelligence technology has been used to develop *expert systems* for CAE analysis. It is expected that such systems will proliferate in the future.[9]

Once a design has been developed and analyzed, the next step is to develop the specifications necessary to produce the product. For the case of metal parts, such specification involves definition of machining instructions which are required to generate the product's final dimensions. If the machining process is automated (see Chapter 4, which describes CNC in more detail), these machine instructions take the form of a computer program. *CAM systems* can automatically develop CNC programs with the machine readable CAD part representation as an input. Tool path analysis and simulation of tool movements can be carried out prior to actual material processing.

A rapidly evolving tool to support analysis of design alternatives is *rapid prototyping*. With the CAD file as an input, these systems generate physical models of the product or component. Various computer controlled processes are available to develop such three-dimensional renderings in a matter of hours and minutes.

---

[8]Mitchell (1991) presents a recent review of the CAD, CAM, and related technologies.

[9]The current status of CAD and CAE technology is reviewed regularly by software database services, e.g., the Datapro (1996) series on CAD/CAM/CAE systems.

The collection of CAD representations of products, CAE analyses, prototypes, and CAM machine instructions are a part of the firm's manufacturing database. Full integration of CAD and CAM is often difficult due to differences in interface standards among the various systems and the lack of common data base procedures. We will return to a discussion of information systems challenges in the automated manufacturing environment in Chapter 5.

It is also necessary for the CAD/CAM user to possess a high level of skill in order to realize the full potential of these automation tools. For example, the user must select processing equipment, tooling, and job step sequencing. These choices can have a major impact on manufacturing productivity of the firm and the final cost of the product. Thus, it is very important for management to develop and nurture its human capital through continuous training and skill building programs.

It is also important to note that the benefits of CAD, CAE, or CAM systems are not separable. For example, a firm will not realize the full benefit of its CAD system unless it has adopted and successfully integrated the downstream CAE and CAM systems. Milgrom and Roberts (1990) developed an economic model of this "externality" effect and concluded that as a result firms should engage in "quantum leap" as opposed to incremental investment strategies for CIM.

Recent evidence (see Liker, Fleischer, and Arnsdorf [1992]) suggests that many firms are still finding it difficult to realize the full potential of these design support technologies. We will return to this issue in our discussion of justification and implementation in Chapter 8.

Figure 2.12 illustrates the overlapping stages of the design process and indicates where in this process various support technologies are applied. Chapter 3, concerned with process design, will review *CAPP (computer-aided process planning)* methods, and Chapter 6 will discuss *MRP II (manufacturing resource requirements).* Statistical quality control (SQC) is not discussed explicitly in this text. It is important to note, however, that SQC and experimental design methods are relevant throughout the early stages of process design, pilot production, and volume production.

FIGURE 2.12

Design support technologies

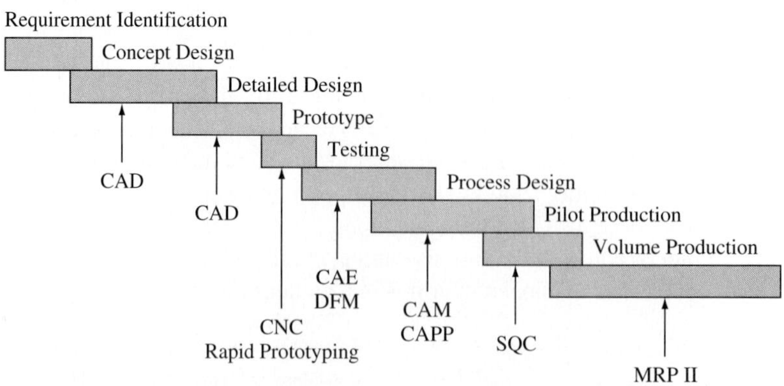

Study Questions

1. Define *concurrency* in product development. How can if be achieved? What are its principal benefits?
2. What are the roles of cross-functional teams in product development? What are the principal benefits and requirements for success in implementation?
3. Define *quality function deployment (QFD).* How does it impact the product development process?
4. What are the principal performance metrics for product development? Explain how each of these metrics affects the economic payoff of a product development project.
5. Give three instances of "X" in the term *design for X (DFX).* Explain how one of these downstrean factors is linked to the product development process.
6. What are the principal advantages of *computer-aided design (CAD)* systems?

References

Boothroyd, G. and P. Dewhurst, (1989), *"Product Design for Assembly,"* Boothroyd Dewhurst Inc., Wakefield RI, 1989.

Clark, K. B. and T. Fujimoto, Product Development Performance, Harvard Business School Press, Boston, 1991.

Cohen, M. A., J. Eliashberg, and T. H. Ho, (1994), "New Product Design Strategy Analysis: A Modelling Framework," *Management of Design: Engineering and Management Perspectives, Design Management,* Kluwer Academic Press. ed. S. Dasu and C. Eastman, pp.45–60.

Cohen, M. A. and T. H. Ho, (1993), "A Teaching Note on Tradeoff Analysis and Portfolio Management in New Product Development," forthcoming, *Technology and Operations Review,* The Wharton School.

Datapro (1996) "CAD/CAM/CAE Systems," Datapro Research Corp., McGraw-Hill Delran, NJ.

Dean, J. W. and G. I. Sussman, (1989), "Organizing for Manufacturable Design," *Harvard Business Review,* January–February, pp. 28–36.

Dougherty, D. and M. A. Cohen, (1995), "Product Innovation in Mature Firms," *Redesigning the Firm,* Oxford University Press, pp. 87–115.

Ettlie, J. E. and S. A. Reifis, (1987), "Integrating Design and Manufacturing to Deploy Advanced Manufacturing Technology," *Interfaces,* November–December, pp. 63–74.

EFO Group (1995), New Product Development Survey.

Fabrycky, W. J. (1987), "Designing for the Life Cycle," *Mechanical Engineering,* January.

Griffin, A. and J. R. Hauser (1992), " The Marketing and R&D Interface," *Handbook of MS/OR in Marketing,* eds. G. L. Lilien and J. Eliashberg, Elsevier Science, Amsterdam.

Hauser, J. R. and D. Clausing, (1988), "The House of Quality," *Harvard Business Review,* May–June, pp. 63–73.

House, C. H. and R. L. Price, (1991), "The Return Map: Tracking Product Teams," *Harvard Business Review,* January–February, pp. 92–100.

Lee, H. L., Billington, C. and B. Carter (1993), "Hewlet Packard Gains Control of Inventory and Service Through Design for Localization," *Interfaces,* 23:4, July-August, pp. 1-11.

Lenox, M. and J. R. Ehrenfeld (1995), "Design for Environment: A New Framework for Strategic Decisions," *Total Quality Environmental Management,* Summer 1995, pp. 37–51.

Liker, J. K., M. Fleischer, and D. Arnsdorf, (1992), "Fufilling the Promise of CAD," *Sloan Management Review,* Spring.

March, A. (1988), "Designing the IBM Proprinter," Corporate Design Foundation, Boston, Case 88-0927.

Milgrom, P. and Roberts, J. (1990), "The Economics of Modern Manufacturing: Technology, Strategy and Organization," American Economic Review, June, pp. 511–528.

Mitchell, F. H., Jr. (1991), "Product Design and Manufacturing," Chapter 9 in *CIM Systems,* Prentice-Hall.

Nevins, J. and D. Whitney, (1987), *Concurrent Design of Products and Processes,* McGraw-Hill.

Polkop L. and P. Lawrence (1991), "The Cyclone Grinder," Corporate Design Foundation, Boston.

Sackett, P. J. and A. E. K. Holbrook, (1988), "DFA as a Primary Process Decreases Design Deficiencies," *Assembly Automation,* Vol. 8, no. 3, August, pp. 137–140.

Stoll, H. W. (1986), "Design for Manufacture: An Overview," *Applied Mechanical Review,* Vol. 39, no. 9, September.

Ulrich, K. T. and S. D. Eppinger, (1995), "Design for Manufacturing," Chapter 9 in *Product Design and Development,* McGraw-Hill, NY.

Ulrich, K. T. and S. A. Pearson, (1993), "Does Product Cost Determine 80% of Manufacturing Cost?" MIT Sloan School Working Paper.

Wheelwright, S. C. and W. E. Sasser, Jr. (1989), "The New Product Development Map," *Harvard Business Review,* May–June, pp.112–115.

Reading 2.1
Product Innovation in Mature Firms

Oxford University Press

Deborah Dougherty and Morris A. Cohen

Launching an innovative product is one thing, sustaining innovation is another. All too often, firms successfully produce one or two innovative products, then fail to replicate their success (for example, IBM's PC or GM's Saturn Project). This is especially true for firms that have been operating in stable markets for some time. Old habits and routines can thwart attempts to become innovative, even when management sets up interdisciplinary teams, encourages coordination and information-sharing across functional lines, and embraces other tenets of concurrent engineering.

How can older organizations transform themselves so they can sustain innovation? Bringing together all the functions involved in developing a new product is important, but insufficient. Firms must also align new products with their current product offerings and their overall corporate strategy. Sustained innovation requires concurrent management, not just concurrent engineering.

The framework introduced here defines at least three critical requirements for becoming innovative. The framework is based on the concept of concurrence, or simultaneity, whereby different functions, projects, and organizational units coordinate their innovation activities. This framework goes beyond concurrent engineering, which focuses on coordination across business functions, to include two other dimensions—concurrence within the portfolio of new and existing products and concurrence with respect to different institutional levels of management.

We apply this framework to the case of MACHCO Machinery, a century-old firm which attempted and failed to become more innovative. By looking at what people at this firm did and did not change, and what happened as a result, we can understand the challenges of becoming more innovative and begin to ferret out what managers can do to transform the firm.

Background of MACHCO

MACHCO is a multibillion dollar machinery and equipment concern that was started more than a century ago with the invention of a steam-driven drill for mining. By 1988, the company had grown to four major business segments with a total of ten groups, with fifty-five plants around the world, and sold everything from huge engines to tiny bolts. Our research concerned the efforts of one of the ten business groups to improve its product development capabilities. This is the Production Equipment Group, which sells a variety of small electrically powered machines (engines, pumps, drills, gauges) used in industrial manufacturing, assembly, and repair processes.[1] For the most part, these were mature, almost ancient markets.

Like other oldline firms, MACHCO's Production Equipment Group had over the years developed a reputation for being "the Cadillac" of assembly equipment in several of its segments. They made profits for years, according to one manager, and, not surprisingly, had settled into a "fat and happy" stasis. The Group's main plant was located in Springton, a small rural town in the Midwest, while the division headquarters (comprised mainly of marketing and sales) was about four hours away by car. MACHCO's factory dominated the small town, since most of its residents worked at the plant. Jobs ran in families, and many shop floor people had fathers and grandfathers who also worked for MACHCO.

Befitting a firm older than Taylorism, Springton historically operated on the principle of segmenting work into the smallest steps that could be measured and controlling work flows and processes tightly. The plant, as well as the firm as a whole, had been organized functionally for years, and this segmentation was carried through to product development, where each group had its own separate task to perform. An engineer explained that marketing always handled all the "external" work on product specifications, while his group did all the internal work. A manufacturing person said: "It used to be, marketing would come in here and tell us what to do. They would throw the specs over the wall. They never talked to the engineers, so both groups hated each other. And manufacturing hated new products, period. They distrusted the whole process and it hurt the numbers."

Relationships between engineering and manufacturing were worse than those between marketing and engineering. Traditionally, according to one manager, manufacturing would wait until a product was completely designed with all drawings final. Then the drawings would go to industrial engineering, who would work out the process, and then on to the shop. They handled the interdependencies by taking plenty of time to work issues out. Not surprisingly, each department also blamed the other for taking too much time. For example, a technical person explained: "Normally, manufacturing would take six to nine months to process parts, so they didn't need all the pieces to fit together all at the same time. In our usual process we would just throw a design over to them and let them take as much time as they wanted." Contrast this explanation with one from a manufacturing person: "It's hard to get engineering to move on things. We work over here on schedule and we are evaluated on keeping to the schedule, but the engineers have a blank check on time."

Source: From *Redesigning the Firm,* Edited by Edward Bowman and Bruce Kogut, Oxford University Press, 1995.

Not only did the people in different departments harbor a degree of contempt for each other, they saw the same things very differently, and did not really appreciate one another's concerns.[2]

The strong separation between the departments did not hamper their work much, since the products themselves were not integrated with each other or with resources. Managing product lines was the functional province of marketing and sales personnel, most of whom were located at group headquarters. Product strategies were not used at the plant to guide decisions. This astrategic approach worked well for a long time, according to one, because changes were rarely made: "Production machinery made god-awful gobs of money, and there was no pressure on innovation as we know it now. We had an 80 percent market share." They developed few new products, although the product line had evolved incrementally over the years. As of 1988, the plant produced hundreds of different kinds or models of small production machinery, an array that had evolved from and was continually upgraded in response to specific customers, through the addition of "specials" or "buy-outs" to fill in lines.

Resources were allocated in a standardized, generic fashion as well, reinforcing the standardized products. Not a greal deal of investment was made in product engineering, and MACHCO had closed its technology center in the late 1970s. Their core product "architecture" was a class of extremely rugged electric motors, used to power many of their machines. MACHCO people invented this motor years ago, and since then had been tweaking it to make different sizes and speeds. As one put it: "We built a company that knows everything about industrial grade electric motors."

The manufacturing technology was also mature, based on the forging and machining of metal parts. Everyone referred to what they did as "making chips" and "cutting bar stock." One engineer who had recently joined said: "This was an excellent machine shop. They would machine something or work on it so they can machine it, even if other technologies are better. They'd keep backing up and beating themselves trying to do what they knew best. It was inbred in people." Manufacturing resources were managed departmentally rather than by different product lines, and controlled by a system of standard costs which emphasized gross volume over flexibility. Manufacturing managers understood that their jobs were to "keep up productivity levels and not show a negative variance." Everyone in manufacturing, from the machine operators to the managers, was judged on volumes shipped, use of equipment, and adherence to standards. An engineer explained: "Once stuff hit the shop floor it was all automatic. They were not set up to make any adjustments." Needless to say, their emphasis was also short term, as this manufacturing manager explained: "We did not look forward. We worried about costs, and we looked at things from month to month. If we had a bad month, we would hope to make it up the next. The corporation looked at things quarter to quarter."

The only strategic guides used in the day-to-day decision-making at the plant were abstracted rules of thumb concerning costs and volume production. This segmented, incremental approach to resource allocations worked well enough for years, but it had an inherently conservative bias. As far as Springton was concerned, keeping costs low was the objective, not spending money was the strategy, and meeting weeky schedules and cost targets were the key performance criteria.

In addition to the compartmentalized approach to work and the departmental approach to resource allocation and development, Springton had a particular, often unspoken, social system. Not surprisingly, the plant and the firm more generally comprised a world with a unique culture and institutionalized practices that people followed without thinking.[3] Three aspects of this culture are important to the story of becoming more innovative.

One important aspect was how people decided what work to do. To sort among the many possibile choices—work on this enhancement, fix that performance problem, respond to this customer's requests—they relied on an informal network of incremental, day-to-day deal-making and negotiating, with long time employees who knew the system playing key roles. One explained that getting work done ". . . depends on the informal system. It comes down to this person having or developing a working relationship with others who have the technologies he needs to access, and his ability to keep a door open to his supervisor so he doesn't have to go through the chain of command every time he needs something."

A product line manager explained how he would follow the network system rather than official procedure as well: "The guys are so busy here. You have to constantly push for your stuff, face to face, even to get something small done. . . . I go over to engineering all the time to work out quality problems or other problems. I should go to manufacturing, but I go ask the project engineer. We have lots of people who have been here for twenty-five years who know a lot of stuff, so it's easiest to go to them." He also explained that since engineering chose the problems they would work on via political nudging and firefighting, he routinely asked engineering for much more than he really needed to manage his line well, just to make sure he was always on people's agendas. This process was incremental and reinforced the scattered sense of product strategy. It was also very much an insider's game, and people who had been at the plant for only a few years felt like outsiders.

A second important aspect of the social system was its hierarchical nature, based on top-down control. Senior management, ensconced at Group headquarters, "called the shots" regarding production and product lines, and Springton executed. There was no real communication or interaction between the strategic and the operating worlds, so that informal networking emphasized day-to-day, piecemeal execution of tasks and trouble shooting. The connection between senior managers and the plant had become a thin thread of control around cost and "invoices." Nearly everyone at Springton, regardless of function, pointed out that they had been driven by costs. One said: "We look for cost reductions on products without thinking of the final effect on the product. "The emphasis on cost control had been so strong for so long that people would not risk $5,000 worth of scrap to experiment with a new design or process. Senior managers were making money while Springton was making production machines, and their very limited interaction was comfortable to both sides as long as these two goals did not conflict.

Third, because of the constricted, top-down, authoritarian relationship, people at the plant felt that they could not discuss matters with management, and routinely hid any problems from them: "If there is a problem, they will more than likely cover it up—you know, in quality, delivery. . . . People are scared to let divisional know about problems because of the backlash. Also, it [hiding problems] comes from the bottom up—people say: 'Hey, I live next door to you . . .'"

Another attributed this hiding to a fear of punishment: "Everyone in manufacturing is thinking they will be brought out to the woodshed if they screw up." He said that failure also was not acceptable: "We don't like to admit that people make mistakes . . ." And an old timer explained that they hide mistakes because they cannot really trust senior managers to react fairly: "We don't hide problems, but we will gloss over them. Upper management says: "We are all one big happy family, so if you have a problem tell us." But one time that we did, the whole top of our head comes off and blue smoke fills the room. . . . You can't turn it on and off, because the next time you will not tell them about problems. Besides, you got your manager sitting there and you don't want to get him in trouble."

As of 1988, MACHCO's production equipment business was based on segmentation: segmentation of departments from each other, of resources from products, and of operations from strategy. They routinely made few tradeoffs, except around "cost" in the abstract. The plant was good at machining and volume production of clearly designed parts. The engineers were good at adjusting and tinkering with established products. The sales people were good at calling on established accounts. The rift between the operating and strategic levels had become large. Changes had indeed occurred during the 1980s. The group had invested in a number of programs to modernize manufacturing, such as improved worker-management relations, which emphasized development of the labor force, quality circles, shifts toward just-in-time inventory processing, and a capital improvement plan. They had also developed cellular manufacturing and implemented a state of the art CAD system. On the engineering side, they actively recruited more college-educated engineers to broaden their skill base. People who had joined the firm during that period said that MACHCO's Springton plant was more advanced than many in other established industries such as automobiles and engines.

But these changes did not improve their product innovation abilities, since it took them an average of four years to get a new product out. Despite the time, most of the new products were me-too designs that did little to increase their market share. MACHCO had not introduced any truly innovative products in some time, and even their "bread and butter" products were being superseded by low-cost foreign competition and innovative domestic competition.

The Fix/Phase I

In late 1987, division management decided that their deficiencies in product innovation were no longer tolerable, since they were facing increased competition in all their markets. An "innovation champion," Jim Fox, a business manager at group headquarters, was charged with the task of devising and implementing a new product development process. Fox researched product innovation, and came up with a process outlined in Figure 1. The project characteristics listed in Table 1 reflect some of the key characteristics that are embedded in the flowchart: (1) interdisciplinary teams with team ownership of the product's development; (2) full customer and supplier interaction throughout; (3) careful product conceptualization and project-objective setting with respect to product performance and development cycle time; and (4) adequate phase reviews so that connections between the project and the rest of the organization were organized around joint problem solving and effective resource allocation. These characteristics are part of proficient product innovation that studies of innovation have uncovered,[4] so Fox had done his homework well.

MACHCO's process was also deliberately tailored to resolve what group managers saw as their major problems. For example, by centering on inter-disciplinary teams, they attacked the historical "over the wall" (sequential) development process. The flow chart also carefully laid out what to do and how, which gave people a path to follow as they went about this very new kind of activity. Management included a very ambitious one-year time frame as well, in part to break through what headquarters saw as Springton's tendency to avoid risk and stick to their usual methods and to respond to emerging competitive pressures. However, the new process did not lay out the roles for senior management or others at Springton, except for participation in review meetings.

A new finishing machine, to be called Venus, was selected as the first new product, because MACHCO's existing machine suffered from a low-market share, a poor performance image, and the effects of a series of unsuccessful attempts to win market share through product redesign. They recognized that if the company did not do something with this product, they would soon be out of that business all together. Management felt that if they could go back to the drawing board and come out with a product that was demonstrably better than the competition's, they could realize growth at the expense of the competition. In addition, this machine was relatively simple, and as one team member put it: "We figured if we couldn't innovate here we couldn't do it anywhere."

In late January 1988, a memo was sent to the function heads in Springton outlining general objectives for the project, which included developing a whole line of machines, doubling market share within three years, and using modular design and design for assembly procedures. A tight schedule of market research, commercial and technical feasibility studies, and extensive customer visits was laid out. By March, a person from each department had been appointed to the team. The marketing person commuted to

TABLE 1 MACHCO's Four New Products Compared

Project Characteristics	*Venus Grinding Machine*
Functional concurrence	
All departments throughout	Yes, but manufacturing not fully involved
Team-owned product concept	Team created
Team in charge of details	Yes
Joint customer visits	Yes
Portfolio concurrence	
Project scope fit 1 year frame	No. 3 models, change to 1 in development
Product strategy clear	Yes, emphasis on clearly superior product
Technical feasibility assessment	Yes, but shell problem occurred
Manufacturing feasibility assessment	No, not thorough
Institutional concurrence	
Good support from departments	Yes, except manufacturing role process not clarified
Access to resources	Yes, funds as necessary
Effective phase reviews	No
Joint problem solving with top management	Strategic input, hands-off management
Project outcomes	
Team satisfaction	High
Time to market target	Met for first model, others delayed
Product and project costs	Overage
Customer acceptance	Very high
Overall project success	High

Springton, but engineering and manufacturing people were co-located and assigned full time.

Working in a multidisciplinary team required a new approach to taking responsibility and understanding one's role, as one recalled: "It was very different from our usual approach. The whole team was responsible to go find out what the market wanted and what the customer needed, and then move the product into design and manufacturing. We never did that before. We had no infrastructure on how it should work."

Despite the uncertainties, the team went right to work evaluating the market and customer needs by visiting customers (including users, buyers, and facility managers) and distributors together. The direct exposure to customers gave them many insights and ideas, and helped to define the product strategy further. By May the team had developed a business plan, which included meeting specific measurable features (for example, durability, repairability, safety). An important insight from all the customer visits was that the machine should be ergonomically sound: safe, easy to use, and comfortable. As shown in Table 1, the product scope was large, consisting of three different models and many accessories. The team also concluded that a significant performance improvement was needed if they were to meet the general business objectives of the project. An incremental improvement would lead to a quick response by their major competitor, who, as market leader, could lower price and/or quickly match incremental product enhancements.

In the next three months, the team brought the product concept alive, working with a design house to develop various mock-ups. By July, the team came to realize that the new finishing machine would have to be radically different from current designs in order to deliver all the customer attributes they developed in the planning phase. The machine's shape needed to fit into the opera-

Mars Engine Starter	*Jupiter Drill*	*Saturn Drill*
No, IE late, manufacturing not involved	No, began as engineering team, others only partly involved	No, began as engineering team, others only partly involved
Team created	No, different ideas in different departments	No, concept changed by management not team
Yes	No, intervention from others	No, intervention from others
No	No	No
Yes, came up with 1 model	No. 5 models, too large	No, multiple models
Yes, enter as newcomer	No, too vague, too big	No, strategy changed, objectives conflict
Yes, decided to use existing product	Partly, but need new tech platform	Yes
No, all parts outsourced	No, major problems late in development.	No, some problems late in development
No, team on its own, marginal staus	No, roles not clear, manufacturing not ready	No, marketing and manufacturing not connected
No, low priority, limited access	Yes, funds as necessary	Partly, funds limited
No	No	No
Team avoided management	Management intervened in details	Management intervened in details
Medium	Low	Low
Late	Very Late	Very Late
Very Close	Close	Overage
High	Very High	Very High
High	Low	Low

tor's hand, which meant that the body had to be made from plastic, not metal. To fit the motor and other parts inside this radical housing, they would have to redesign the machine and add an internal shell made of bored composite to hold all the parts. The group made the decision to take the radical approach, even though working with plastics and bored composite was completely new to the plant.

Over the next six months, the team worked on all the new parts, including the new shell. Throughout the process, senior management maintained a hands-off stance. They did not get involved with day-to-day problem solving, but did make the necessary resources available by holding the MACHCO's normal rules of avoiding costs in abeyance. They also made it clear that the one-year cycle time target could not be compromised.

By January of 1989, however, they ran into trouble manufacturing the internal shell. The part called for production methods that they had no experience with, and one person observed that they did not build the connections with manufacturing very well:

> The flaw in our organization turned up when we moved into manufacturing. The magnitude and scope of the change was too much for the people involved. They had to wear hats they never wore before. A whole lot of decisions had to be worked out at once because as the product moved into the plant, there are a lot of loose ends in production control, planning, shipping. . . . The coordination effort was the hard part. Normally . . . we would just throw it over to them and let them take as much time as they wanted. Here, everything was to be done at the same time. All of our GANTT charts were all laid out on top of one another. We weren't prepared to handle that.

The team decided to source the troublesome part externally for the time being, hoping to solve the problem when they could, but that doubled the part's already high cost. Their inability to produce the

FIGURE 1
MACHCO Division's New Product Development Process

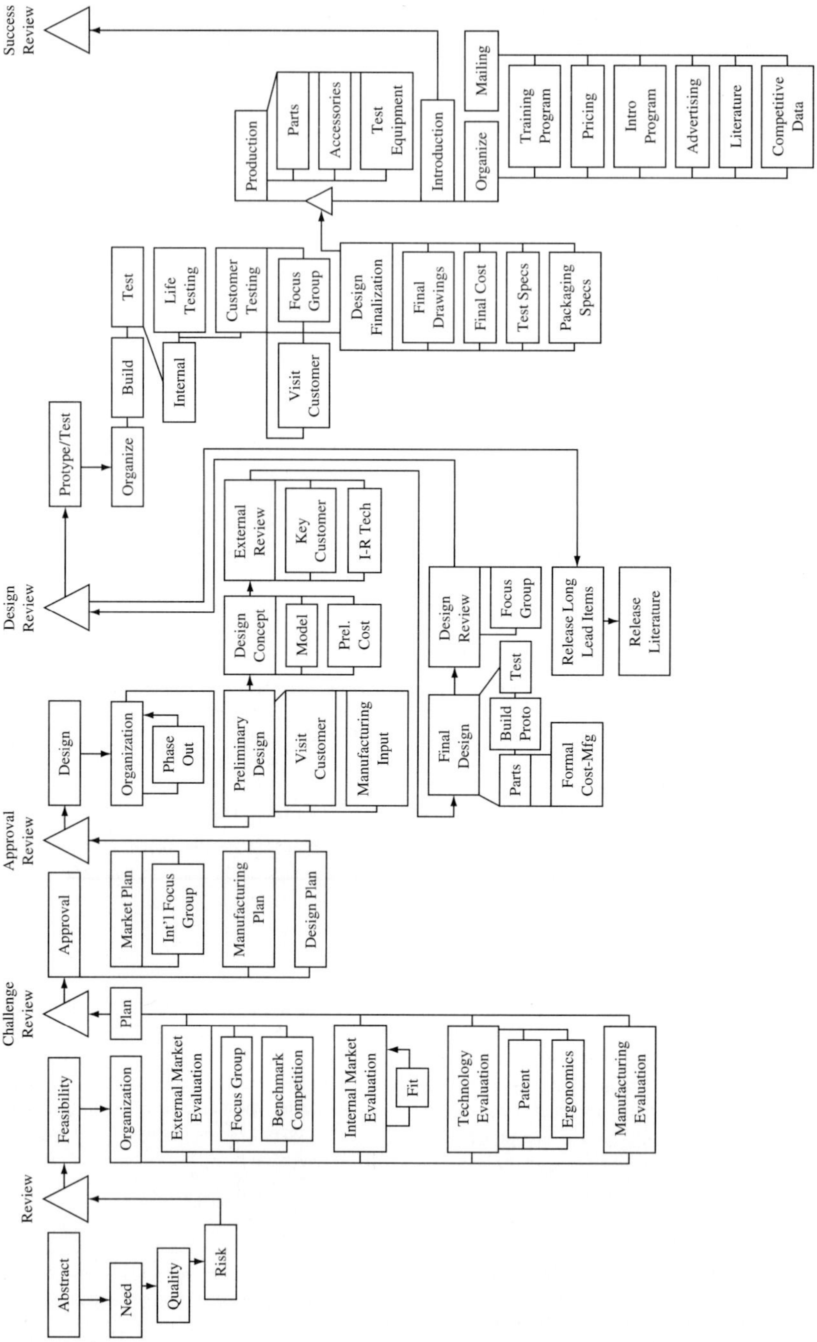

internal shell also had forced the team to reduce the original product scope significantly from three different models to one.

By February, they put prototypes with customers and learned that their newly designed machine was a hit. By June, the machine was formally introduced to the market, only two months later than the date they set the year before. Initial sales ran 20 percent above previous levels even though there was only one model, and they lacked the benefit of advertising or structured promotion. The compressed development time and word of mouth alone generated significant interest in the marketplace. Several big problems remained, particularly that the shell was still being produced externally at twice the planned cost, which significantly increased the cost of the machine (to the point where its profit margin was eliminated). The continued inability to work out this manufacturing problem made the whole team feel burned out. Management declared a victory, however, because the new innovation process had indeed produced a demonstrably superior and commercially viable new product in one-third the usual time. In addition, manufacturing ultimately (after a year) mastered the composite boring process and were able to bring its fabrication in house.

The Fix/Phase II

During the next year production equipment group managers decided to launch three more new efforts with the same development process and the same one-year time frame. Unfortunately, all three were over time and budget, as the status summary in Table 1 indicates.

The Mars Engine Starter (a special motor to turn over industrial diesel engines) started in September 1990. As shown in Table 1, this effort also had many more of the innovation characteristics in Fox's flow chart than the other two projects. The Mars project relied on a multidisciplinary team, but the day-to-day work was carried out in engineering, and there was no participation from industrial engineering or manufacturing initially. The team had little customer input because they felt that they already knew about this market. The product conceptualization was done well. One member explained that the team spent a lot of time up front on the concept development: "We really beat it out." Relationships between the project and the rest of the plant and senior management were handled primarily by avoidance. The Mars team also did not let development problems "go upstairs" to group management. The project was also disconnected from resources in manufacturing, in that all the new parts necessary to produce their elegant design had to be made outside.

The Jupiter Drill project was begun in late 1989 with a one-year development plan, but it was almost three years before two of the five models were "signed off," or approved for sale. Jupiter also has cost problems that have not yet been resolved. The Jupiter drill is a \$1,500 to \$15,000 piece of equipment, depending upon design, used for boring and screwdriving in mass production processes, especially for automotive assembly.

As Table 1 indicates, the Jupiter project did not embody many of the characteristics of an effective innovation. There was no interdisciplinary team in the beginning, and only occasional true teamwork later on. Jupiter's product conceptualization was weak, its scope large. Some said that the project goals and objectives were so broad that each day they would "make one step forward and three back." The product required significant advances in technology for Springton, such as new motor designs and clutches and use of computers, and the basic machine had over 130 parts, vs. 30 for the finishing machine. People at group headquarters disagree, however, and argue that the product was not a major technological advance.

The project ran into significant delays in manufacturing, partly because of a lack of coordination and partly because the product required a new know-how. They were using standard micrometers to test the new parts, but these did not measure the various dimensions adequately. The parts fit the drawings, but, since they were made from composites, they had different properties than metal. Management authorized new testing equipment, and by July the plant had produced parts that both performed and fit together. The project's relationships with senior managers were also strained.

The Saturn Drill project began in early 1990; it was less technically complicated, but still a challenge. This drill would sell for several hundred rather than thousand dollars and was designed for professional repair shops. The development team was managed from engineering again. Neither marketing nor IE was assigned full time or co-located, and the team as a whole made no visits with customers. The product's conceptualization (as summarized in Table 1) only partially embodied the necessary attributes, and the goals were changed by management significantly after about eight months. In December of 1990, management decided to be the only producer of this kind of drill in the United States, and to erect such significant entry barriers—patents and high capital costs—that foreign competitors could not copy the product. The team was given one year to accomphish the new objective. Again, the vertical relationships became quite strained. Management felt that the people at Springton were not working creatively or thoughtfully, so they increasingly intervened to solve problems in the project itself (that is, by requiring the team to submit a weekly schedule).

By July of 1992 it looked like all three new products would finally be going into production. MACHCO people had learned about managing innovation, team work, how to solve problems creatively, and, from the school of hard knocks, where they had remaining weaknesses. By January of 1993, MACHCO (finally) had several very good new products. But everyone was extremely frustrated. Senior managers felt that the delays were at least partly avoidable, while operating people felt that they were being asked to do the impossible. All four innovation projects ran into problems with manufacturing. None of the products were "simultaneously designed" for function and manufacturing. All four competed with each other for resources (the rest of the Venus models were being developed at the same time as the Mars, Jupiter, and Saturn). So, while the official product development process contained a number of the characteristics of effective innovation, often these characteristics were not put into practice.

Let us step back from these particulars for a moment to analyze what happened by drawing on the many studies of innovation and its relationship to organizations. One important insight is that MACHCO's problems are within the range of "normal," because many organizations have difficulty breaking down functional barriers, keeping teams going, getting resources, and so forth.[5]

Another insight is that the implementation of new technologies is more successful when accommodations are made in administrative policy and organizational structure at the same time. Ignoring organizational change happens often, however, one writer estimates that 50 percent to 75 percent of U.S. firms have experienced failure in implementing manufacturing technologies, because they have overlooked necessary organizational changes.[6]

A third important insight is that organizations constitute systems of know-how and culture that are "interlocking, complex, and tenaciously held."[7] The configurational nature of organizations suggests that change will be difficult, and not simply because employees themselves resist change. Innovations introduce dynamics that can get truly gummed up in the interlocked set of organizational beliefs and practices, because unforeseen mismatches between an innovation and the organization's existing procedures inevitably crop up. For example, new equipment may need debugging, but is often put into full operation right away, or a new process may negatively affect measures used to judge the success of certain people or operations, or pressures for continued productivity create maelstroms of conflict when the activities of innovation hinder volume production.[8]

Therefore, to become an innovative organization is to also manage complex, challenging organizational changes. Piecemeal tweaks or incremental shifts, such as putting in a few teams or decentralizing some decisions, is not enough. Managers need to grab the configuration and shift it all at once. This did not happen at MACHCO. In fact, for the three new products that followed the initial effort, we can see that MACHCO's established ways of working, summarized in the beginning, still dominated.

One model of change management suggests some dynamics that we can develop for MACHCO.[9] According to this model, people cannot even begin to change until they see that the current system no longer works and that another system is available. This step is called "unfreezing," which occurs when the worth of the current configuration is disconfirmed; at the same time, people are motivated to change and provided with a sense of "psychological safety" to do so. Change can be a confusing and frightening process, and people are reluctant to change without this kind of help. Change occurs by "cognitive restructuring," or learning, which provides people with both a new perspective and the specific skills to see and feel differently. Finally, the new configuration is "refrozen" by integrating it into a new set of relationships.

A study of a poorly performing automotive plant illustrates this change process. The plant was attempting to change from an autocratic to a quality-oriented, decentralized management style, but for two years not much happened because changes were made in a piecemeal fashion that merely conflicted with the existing culture. They made no effort to provide people with insights into what the new system would look like, nor the psychological safety to begin to change. The people experienced the pressures to change as a "squeeze through the eye of the needle, where the safety of an effective production culture was being left behind but the new culture remained only a poorly understood theory."[10] Change finally took place when people developed metaphors through whch they could envision their new roles and the new system. The metaphors enabled them to see through the eye of the needle, to transform their understanding and understand how to get from "here" to "there."

Toward a Model of Organizational Transformation

The people at MACHCO were definitely feeling as though they were being squeezed through the eye of a needle. The old way of working that did not enable innovation had been disconfirmed, and all but a few people recognized that they needed to change. But try as they might, they could not carry out the innovation activities listed in Table 1 very well. We think that they did not because they were mired in their old segmented system, and the new innovation process did not help them break out of that old system entirely.

We propose that MACHCO needs to change three dimensions of organizational practice all *at once* in order to become innovative. We use the term *concurrent management* to describe the overall idea: "concurrent" because the three changes occur together, and each change requires a high degree of coordination within its specific domain: and "management," to emphasize that firmwide change is needed. The three dimensions that managers need to change at the same time, in order to unfreeze and change the three kinds of segmentation that trapped MACHCO are: the horizontal dimension to integrate the various sets of functional expertise (*functional concurrence*); the new to old dimension to connect the new product with the existing patterns of products and resources (*portfolio concurrence*); and the top to bottom dimension to align strategic with operational orientations (*institutional concurrence*). These three concurrences "make sense" individually at MACHCO, and so can be useful metaphors for change. The project characteristics associated with each dimension are listed in Table 1.

Functional Concurrence. The need to integrate the functions around new product development, is widely recognized, and team development has been widely discussed,[11] so we will not belabor this point here. However, MACHCO did not fully achieve functional concurrence in any effort. They did achieve the close, trusting team relationships on the Venus and Mars projects, which shows that, despite their highly functionally segmented work before the innovation, they could indeed create teams. But in no project did they achieve simultaneous engineering, wherein the design and manufacturing are worked out together, up-front. And in two cases marketing was disassociated enough from the projects that the initial designs were rejected. It is obviously hard to create complete functional concurrence, in which people jointly conceptualize the product and jointly bring it alive. To do so, people need roles that encompass the entire project rather than only a step in it, a rich understanding of one another, considerable trust, and the ability to make decisions for their particular function based on the whole.

Comments from Jupiter participants highlight the fact that MACHCO's historical lack of appreciation and trust between manufacturing and engineering was carried forward into this project. An engineer attributed most of the problems in manufacturing to the industrial engineering department: "We had involved industrial engineering, and they were familiar with the parts, but they did not pay attention to how critical the new parts were in the design so they didn't focus up front. They assumed standard processes would be acceptable." Yet an industrial engineering (IE) person explained that engineers did not give them good specifications:

> There were so many stumbling blocks, it wears on you. In December we wanted it so bad, but it seemed like we took three steps backward every time we solved a problem. We spent Christmas and New Year's in here building tools and testing them. We gave it all we got, and we found out uh oh, it's not here. The engineers were somewhat tight on design, and they didn't know if they needed those parts that tight. It was a downer. We would ask the engineers: "What do you have to have," and they would say: "We don't know." They wanted the parts in this condition to validate them, but they were not sure of the design. We didn't know what to give them.

In the Saturn project, a manufacturing engineer explained a major rift between marketing and design: "Marketing along with their consultants wanted a mechanism in front of the tool because it would be easy to operate and innovative. The engineers said that would increase costs because it would make the handle longer. Marketing wanted someone to come up with a mechanism that would work on the front, and engineering already had a solution that they preferred. They felt it was a waste of their time to work on the valve when they had other problems to solve." His words reflect a lack of appreciation and trust and hint that each group was digging in, not opening up.

Many more examples could be listed. The point is that, even though effective teamwork could and did happen, it often did not. We suggest that the lack of concurrence in the other two dimensions, in fact, made functional concurrence almost impossible, especially the more rigorous concurrence required by simultaneous manufacturing; to do that requires that people fundamentally alter the way they normally work. Putting people on teams and even building team relationships will not help those people decide between innovation vs. routine products, learn how to participate in designing effectively, allocate resources among competing businesses, or align innovation with strategic issues. The failure to unfreeze these other two dimensions led to a situation where the functional work, we think, kept refreezing and so day-to-day work returned to its normal, functionally disconnected mode.

Portfolio Concurrence A new product is not developed in a vacuum. Many new products in established firms fit into an existing line as an enhancement, addition, or replacement. Even entirely new categories of products need to connect to the existing portfolio because they may be sold to existing users, use an existing distribution channel, or build on an existing technology. Sometimes demand for products may be correlated, so the introduction of a new product can affect the sales of existing products. In addition, product development is resource intensive, but key labor and capital resources are limited and must be allocated over multiple projects. Moreoever, learning can be accumulated over time from the various design efforts. Early versions of a product may "fail" by market or financial standards, but the insights that the developers gained into technology and market issues can contribute to the success of subsequent generations. These interdependencies should be handled systematically, or the inherent ambiguities of product innovation will lead to arbitrary or politically motivated choices—that is, chaos. Chapter seven in Bowman and Kogut (1995) summarizes the challenges in developing more systematic decision criteria, and also illustrates what happens when firms fail to do so.[12]

At MACHCO, while each of the products had a strategy, there was no evident connection among the four new products, nor among them overall and the set of current products. The formal process (Figure 1) contained the decisions to be made at a project level, but did not provide the decision criteria to be used to make choices and tradeoffs across product lines. Moreover, tradeoffs within a project—time to market and product performance—were not considered explicitly in setting the project's strategic objectives (see Cohen, Eliashberg and Ho [1993] for a normative model framework which analyzes such tradeoffs explicitly). MACHCO's existing control "system" was based entirely on rigid and very general cost control, and this system was not changed. The new products competed with each other and with the mature products for scarce tooling, industrial engineering time, and production capacity. For example, even though they had outsourced all their complex parts, the Saturn team planned to have several of its simpler parts made at Springton. But manufacturing decided that the pilot runs of these parts would also be done outside because of resource constraints that the team, at least, felt could seriously hurt the product. As this person notes, they could not work out any alternatives: "The team felt it was important to make these pilot lot parts here, so the plant could get an understanding of the process. But because of the schedule the manufacturing manager said we should put all the parts out. The tool room is so backed up they couldn't handle the load. . . . We suggested that we farm out the tooling for the parts but make them in here, but they didn't want to do that. Having someone else's tools in here is too radical for them."

A manufacturing manager vividly described the more general chaotic conflicts that a lack of portfolio concurrence caused, as the new products conflicted with all the other work people had to do:

> I don't see a universal strategy for new products. Everything is on a micro basis. Is anyone managing the new products overall as a sucker of resources from Springton and the division? No. They micromanage each project with no foresight on what it will do to the others. Plus we are implementing an MRPII program, becoming ISO9000

> certified, and we've got 400,000 square feet of shop floor in the back here producing mature products, and that doesn't happen by itself. . . . There is a need to stand back and look at what we have on our plate and determine what we really want to get done. . . . rather than try to do a hundred things on an inferior level. Things get too segmented on a day to day basis. You have too many hats to wear. Every five minutes you put a new hat on and take the old one off. They only time we can work on something for more than five minutes is when management is coming to town to look at our costs, and we need to come up with another program to reduce people . . .

People at MACHCO had no framework to focus and orient their choices over resource allocation and prioritization. Such a framework is essential in innovation management, because problems and choices come up all the time and cannot be planned for at the project level only. Without any decision frame, MACHCO innovators could not renegotiate targets when problems cropped up, work out conflicts, or solve common problems in a systematic fashion. The existing resources, including time, were fixed, which hurt the innovations in three important ways.

One negative effect was that the fixed time schedule made people feel that the projects were doomed to fail—not a healthy premise. An engineer said: "You need a good handle on the scope of the project and on the technical risk before you set the time period. Otherwise, experienced people who know better will not buy in; they would have no enthusiasm because they know it will fail."

The sense that success was impossible made people "back off" from aggressive pursuit of goals, as this manufacturing person said: "On the development end, if I am given an unrealistic goal, well, I don't stop working, but I take liberties in the back of my mind because I know we will not meet the target date."

Second, the lack of portfolio concurrence—understanding and making tradeoffs among or links between the many issues involved—led to "micromanagement." Upper managers attributed the myriad problems the innovators had to failure on the innovators' part to stick to a schedule or make good decisions, not to the lack of a framework to make choices. Management, therefore, reached down into particular problem areas where they did not have expertise and created even more problems, as this person explained: "They make decisions that they should not be involved in. They are not in here day to day seeing all the problems. So they come in here and say: 'What is the hold up?' We explain the situation and all the ideas we have had, and they say: 'Well, why don't you do this and this.' They don't know that the engineers have already said they don't like that idea or that manufacturing has these kinds of problems."

Third, the inability to make their own choices, as well as the micromanagement, robbed the teams of a sense of ownership, which affected their ability to be teams: "Now everyone has their heads in, and the team is not able to make its own decisions. We trade off an enormous amount to get the product out on time, but then every technical decision is reviewed and criticized by people who aren't on the team, and still get no relief on the time."

We do not suggest that the technical choices he referred to were necessarily correct, or that management's concerns were not well motivated. However, all choices would always be suspect without a system to compare the choices against, and innovators would often make unstrategic choices if they did not know what the strategy was. It should not be surprising that in this very confusing and complicated situation people stayed with what they knew how to do and were rewarded to do. It should also not be surprising that without a more general framework to connect the particular projects to the rest and to set explicit project performance targets people worked with their heads down rather than up and, in so doing, fell into every innovation trap in the book.

It is interesting to note that many firms have adopted reduced time to market or project breakeven times as their standard for product development. The one-year goal for the first phase of the Venus project is a case in point. Motivated by their success at meeting this target (albeit with a product whose main component was being outsourced at a severe cost penalty), MACHCO attempted to use this standard on all subsequent projects. As we have noted, only some of these targets were met. The cause of this failure can be traced in part to a lack of portfolio concurrence. The time/performance tradeoff was not evaluated for each project. Resource allocations across projects also did not necessarily reflect the strategic priorities of each project relative to the company's portfolio of new and existing products.

Institutional Concurrence One reason that no general objectives or strategic rules of thumb were developed to guide the innovation process is that MACHCO had no way to arrive at them. Even if they have a rational decision-making framework to allocate resources and make project tradeoffs, innovators also face many ambiguities and so must make qualitative judgments and choices, renegotiate deadlines, or reconceive product packages and strategies. Decisions require a framework against which the choices are thought about and played out. Just as a continual dialogue between lateral functions is necessary to design and produce a product, a continual vertical dialogue between managerial levels is necessary to create and recreate a project's connection with the firm's strategy.

Institutional concurrence refers to the vertical alignment of the firm as an institution with innovation and assures that what people down at the plant are doing and thinking about innovation accords with what people up at headquarters are doing and thinking. Institutional concurrence provides the social context that enables the creation and ongoing maintenance of vertical interaction and communication. Such communication is essential if both operating- and strategic-level people are to work together under fundamentally ambiguous conditions. For example, both sides need to appreciate and respond to the shifts in market or technology needs that often occur during a project (for example, the need to go "radical" with Venus, to alter the goals with Saturn), even though the decision points are not precise, and be able to judge how the project is doing, even though the milestones may be blurry. Chapter two in Bowman and Kogut (1995) describes how different organizational paradigms regarding "governance" fail to acknowledge the realities of how decisions are made and people are motivated, and thus may not produce effective governance. In the same fashion, divergent paradigms or beliefs about "strategy" and "operations" can disrupt the necessary communications, and thus disrupt innovation.[12]

In all three subsequent innovation projects, MACHCO's established vertical connection of one-way, top-down control and authority did not unfreeze. Indeed, people at Springton hid any exceptions and surprises they encountered from senior managers, in their innovation efforts, even though such inevitable hitches are the primary means to learn. The following comments indicate that people at Springton recognized the need to change the institutional relationships before other change could happen, but, in their usual fashion, they were waiting for senior managers to take the lead:

> There haven't been any real changes. The teams are good, but from a decision-making standpoint we operate as before. We let upper management get too involved in the details, and they do not let the teams do their work. It's a cultural problem at MACHCO, not just at Springton.
>
> We need a new philosophy. It is not enough to just set up teams and say OK, now this is how we will do business. This requires a whole change in point of view, all the way down to the shop floor supervision. This is just like our attempts to use quality circles. They were never a part of the business strategy, and never a real part of our profitability or cost improvement, and so we never really implemented them.

The major indicator of a lack of institutional concurrence was the failure on everyone's part even to recognize that MACHCO and Springton were institutionalized, configured *systems,* with many tightly interconnected practices and procedures. A simple, seemingly obvious, example came up time and again in people's stories. The need to create new parts and produce prototype quantities violated the usual performance measures based on standard costs (which, in turn, were premised on volume and optimization of labor). One reason for the extended delay of the Jupiter project was that Jupiter's new parts were treated like usual parts and standardized in the usual fashion. One manager explained the problems that usual procedures created on the shop floor: "I was asked to get parts for the Jupiter through. They had already made decisions as to the time it takes to run them and the cost. For example, we would tell the operator to make ten of these an hour. But it turned out that the rates were wrong. We have a pay incentive system. The operator knows for sure that he can run the machine with this other part at a certain rate, but we tell him to run the machine for the new part at ten an hour and he can't. So he comes back and says: 'Hey the rate is wrong, can you fix it?' And we say: 'No, that's it.' What's wrong with this picture?"

Parts for the Saturn drill were also delayed in manufacturing because they got caught up in the system, not simply because the team failed. This IE manager explained:

> We were set to run one of the seven new parts on April first. The plan was to run a pilot lot of 100 to 150 pieces through the cells and make sure everything was right. You need to check the gauging, adjust the tooling, and so forth. Well, those parts are still not in stock (three months later). First the tooling wasn't ready, and then the department foreman ran a production lot through right behind the pilot lot. He was supposed to run the pilot lot all the way through first, but he did a 4,000 lot right behind it to minimize setup time. He is graded on how quickly he uses the tools. So now I am managing a 4,000 part production, not just a pilot lot, but I have no authority over that. Fred (Division Manager) gave explicit orders that he wants all the pilot parts in stock as soon as possible, but the foreman is the scheduler of that department, and he has people above him who told him to run the production lot.

Fred, the manager, reacted to this problem by requiring the IE person to submit a weekly schedule, rather than acknowledging that the existing social system, including manufacturing's performance criteria, had to change. Fred obviously thought that the problem was a lack of proper attention to detail. It is not that the people are stupid and overlook what we see as an obvious conflict. Rather, senior managers did not recognize the configured (and not necessarily rational) nature of Springton plant's existing social system.

The major reason for the lack of appreciation, indeed of awareness, was senior manager's detachment from operations, and vice versa. A person at Springton explained the separation of thinking from doing: "The thinking is the easy part. You can envision how the changes in manufacturing can happen, just like you can envision teaching your kid to play baseball. But actually doing it, implementing change, is always harder than you thought. Managerially, people envision this happening at a faster and faster pace than it actually can."

Because of the very abstracted and limited view of each other, upper management saw no conflict in asking Springton people to change radically how they work but also keep volumes up and costs down. Springton saw nothing but conflict and felt that they must have more resources to carry out what they saw as a significantly increased work load. Neither side could believe the other, which meant that there was no basis for trust and, thus, none for unfreezing and changing the system. Consider the lack of understanding between levels this comment suggests: "We are trying to take a quantum leap beyond the competition, and you have a tendency to fall when you do that, which becomes very stressful for everyone, right down to the shop floor people who are working very hard for you. From the corporate structure, people start raising hands and saying why does this have to happen? But they are not in here every day. . . . It is difficult to quantify to a corporate VP, who certainly has other things on his mind, so that he can understand why we can't make this machine do what we want."

Each side of the strategic divide continued to look at the other through the knothole of standard cost and other abstracted performance indicators, which did nothing to provide a common understanding of what they were about, as this manager explained:

> At our annual meeting, corporate management talks about cost management, profit, and market share. We don't relate to that. We talk units and how many units. I remember when a corporate VP came here. He was very honest about the fact that when he looked at tools he has no idea what they are. But he knows our contribution to the

> corporation's bottom line down to the last five digits. . . . We try to share more information within the factory with people about market share, but it's difficult, and I don't understand it really. There is also a comfort level. If I meet with you weekly, I can understand your questions and learn to respond. But if it's once a year or once a lifetime, I don't understand you . . .

This engineer was more scathing in his view of senior management, but the lack of appreciation between the operating and strategic levels is quite clear: "Management is driven by dollars, and some are reverting to Theory X management. They think if they beat us up hard enough, we'll see the errors of our ways. . . . Upper management is bored and confused by lines on a drawing. They want a return on investment."

Without the resource allocation system of portfolio concurrence or the strategic guidance system of institutional concurrence, particular decisions end up creating more disruption. For example, we asked one manager if they were getting better at anticipating the kinds of problems that plagued the Jupiter and Saturn projects, and he said: "Yes, now there is a certain amount of anticipating. But we still try to reduce the leadtime. That is the area we go after. Say we say that the time frame for a part will be two weeks. Management says can you get that to seven days or even five days, and we say yeah, we can. But ten days means we will be doing all the modifications to make the a part work, but when you move it down to five days, all the working time for modification is gone."

We can see how such constant pressure on the details eliminates the room to maneuver, experiment, and learn. We can also imagine that it precludes attention to the whole project or to the portfolio of projects. Nobody can really feel in charge or can take charge. No one can quite see through the eye of the needle.

What Now?

MACHCO managers have brought themselves to the brink of transformation to an innovative organization through their efforts in the last three years. We suggest that more deliberate management of all three concurrencies at once would help them go reach that goal. They began innovation by emphasizing functional concurrence, which is necessary, but we think not sufficient. The lack of teamwork occurred not only because people did not know how to work together, but because the total system of management reinforced the segmentation. We suggested two additional dimensions of organizational change that would correct the total system. To perform effectively, the teams need to take joint responsibility for the product's ultimate development and make discontinuous leaps in method and approach. To work in this new way, the current system of resource allocation and decision making must be freed up. To give the teams the ability to make their own decisions, management must back off, move out of the day-to-day work, and, instead, provide a strategic system that frames and guides daily work. To assure that the teams will make the right decisions, the current system of strategy making must be changed.

MACHCO has made good progress with functional concurrence, but they need to work out the design to the manufacturing link more clearly. In addition, the breakdowns between design and marketing show that there was still a tendency to lobby separately for one's own favorite technology or product appearance. To fix this problem, the teams need to be more clearly multi-functional from the very beginning. If marketing people continue to be located at headquarters, then they must participate actively in the conceptualization process from the outset. The whole team should own the product conceptualization and the product's overall integrity, not any specific function. The team needs to make joint visits with customers throughout the process, no matter how familiar they think the market is. Indeed, if people knew so much about the market, then why did the old product start losing share?

MACHCO needs to start early on portfolio concurrence. A portfolio system is premised on a business strategy that provides a common sense of the firm's domain of operation and character. This strategy focuses attention on markets the organization wishes to compete in, the key factors of competition, and the technologies that the firm can leverage. Within this general strategic frame, managers must first select effective development performance targets for each project (for example, time to market, product performance, unit cost, price point). They can, however, build on changes in manufacturing and engineering that have already been made. In addition, some firms find it useful to set up separate pilot production facilities to avoid conflicts with high-volume products. Experimenting with other kinds of manufacturing control systems (for example, Just-in-Time) could help to unfreeze the long-held standard cost system. MACHCO seems reluctant to "waste money" on such facilities, but they have surely spent in cost overruns or scrap parts at least as much as a true pilot process would have cost. Perhaps it is time to acknowledge this particular cost of innovation and include it directly on the books, rather than in variance. Other options, such as continuing to develop a better base of external suppliers or adding new capacities in molding and composite processing, can also be evaluated.

Several managers recommended other ideas that can be tried to develop a portfolio system, such as more flexible capabilities in tooling or manufacturing that can be deployed to support both innovation and mature products. Cummins Engine, for example, has developed an innovative product/process classification scheme, which accounts for the inherent conflicts associated with manufacturing new and old (as well as high-and-low volume) products within the same facility. Their solution led to the development of a series of alternative process layouts (including transfer lines, cellular and Flexible Manufacturing Systems) to accommodae this variety. In chapter six of Bowman and Kogut (1995), the impact of product variety on manufacturing is explored for the case of the auto industry.

A number of the engineers at MACHCO recommended a more deliberate technology strategy. Instead of re-inventing motor enhancements for each product, for example, they could set out several target technology development projects that could be used across the product lines. We suggest that a more systematic look at other aspects of the value chain should also be carried out (for example, distribution purchasing, parts and after-sales service).

A portfolio system could combine these various functional developments to form a framework of decision criteria for each innovation effort. This framework cannot become rigid, but neither can individual product efforts proceed in a vacuum. Particular criteria to judge priorities for time to market, degree of innovativeness, relationships within and across product lines, and feasibility on technology, distribution, and/or manufacturing capabilities are needed to assess a given product's likelihood of success. The nature of the market and the product's strategy (aggressive, preemptive, or defensive) would dictate the particular time frame for each product. Of course, many new products might encounter new uncertainties during development if expected technologies do not materialize or unexpected competitors do materialize. Additional choices along the way might need to be made to change the plans or perhaps even shelve the product idea for the time being. Considering a particular effort within the context of the entire portfolio of products would help make these choices.

None of these changes would stay in place without better instituional concurrence. Leadership from the top should initiate the strategic dialogue and orient operating-level people in a given direction, and then let the operating levels determine how to move in that direction. MACHCO needs to break down its fundamental communication barriers between the strategic and operating levels. The people at the plant truly do not understand group management, while group management truly does not seem to appreciate the realities of operations. They need to develop a strategic conduit or medium of exchange that is capable of a rich, complex exchange of knowledge, not just bits of information. One place to start would be to build on Springton's networking ability and craftsmanlike orientation to work, which would emphasize the good qualities of the existing system. A second obvious fix is to align the performance criteria, reward systems, and the processes through which activities are evaluated and monitored to support innovation. A third fix would be to develop "metrics" that convey the same message from top to bottom, despite the very different sensemaking worlds at the different levels. Worrying about costs in the abstract ignores the context of investment and opportunity. These concerns need to be translated into operation-level activities that crystalize the firm's improvement of its value and ability to generate cash flow. Finally, a process of participatory strategy making is necessary. Here, strategic and operating people discuss possibilities for business domains or moves within those domains, and try them out.

As indicated in Table 1, all four projects failed to satisfy the three dimensions of concurrency simultaneously. As a result, each project encountered difficulties. The outcomes of these projects reflect these difficulties.

We suggest that unfreezing and changing all three dimensions of organizational action at once are necessary to transform a stable old firm into an innovative one. Together, the three concurrencies capture the configuration of issues that are important to innovation management, from team management to strategic guidance. Making complete changes in each dimension is not possible. The innovative organization is not a stable state, so managing innovation never stops. Moving along these paths together is a continuous process of making connections, trying new ideas, evaluting the outcomes and choosing next steps.

Notes

The authors would like to acknowledge the comments of Daryl Brewster of Campbell Soup Company.

1. The company and products are disguised.
2. For an analysis of how people in different departments think very differently, see Dougherty, Deborah. 1992. Interpretive barriers to successful product innovation in large firms. *Organization Science* 3:179–202.
3. Years ago, people wrote books about factories as institutions, with real live cultures and shared norms. See, for example, Gouldner, Alvin, 1954. *Patterns of industrial bureaucracy.* New York: Free Press; Crozier, Michel, 1964. *The bureaucratic phenomenon.* Chicago: University of Chicago Press. Recent works by Shoshana Zuboff. Larry Hirschhorn, and Robert Thomas also capture the institutional aspects of factories and work.
4. Along with literature mentioned in note 1, see Clark, Kim, and Takahiro Fugimoto. 1989. Lead time in automobile product development: Explaining the Japanese advantage. *Journal of Engineering and Technology Management* t:25–58; Cooper, R. 1983. A process model for industrial new product development. *IEEE Transactions on Engineering Management* 30:2–11.
5. For summaries of these problems see, for example, Cooper, Robert G., and Elko J. Kleinschmidt. 1986. An investigation into the new product process: Steps, deficiencies, and impact. *Journal of New Product Development* 3:71–85; Dougherty, Deborah and Trudy Heller. The illegitimacy of successful product innovation in established firms. *Organization Science* (forthcoming).
6. See discussion in Adler, Paul S. 1990. Shared learning. *Management Science* 36:938–57; and Majchrzak, Ann. *The human side of factory automation.* San Francisco: Jossey-Bass.
7. Westley, Frances R. 1990. The eye of the needle: Cultural transformations in a traditional organization. *Human Relations* 43:273–93.
8. For a summary of these kinds of problems, see Leonard-Barton, Dorothy. 1988. Implementation as mutual adaptation of technology and organization. *Research Policy* 17:152–267.

9. See Schein, Edgar H. 1961. Management development as a process of influence. *Industrial Management Review* 59–77; Schein, Edgar H. 1987. *Process consultation.* vol. II. Reading, Mass.: Addison-Wesley; Lewin, Kurt. 1947. Frontiers in group dynamics. *Human Relations* 1:5–41.
10. See note 9.
11. See Dougherty, Clark and Fujimoto, Leonard-Barton.
12. Chris Ittner and Bruce Kogut, in "How control systems can support organizational flexibility," describe different approaches to developing systematic control systems that also encourage flexibility, and point out that standard control process, in fact, discourages flexibility.

References

Adler, Paul. 1990. Shared learning. *Management Science* 36:938–57.

Clark, Kim B., and Takahiro Fugimoto. 1989. Lead time in automobile product development: Explaining the Japanese advantage. *Journal of Engineering and Technology Management* 6:25–58.

Cohen, Morris, Jehoshua Eliashberg, and T. Hua Ho. 1993. New product design strategy analysis: A Modeling Framework. *Design Management.*

Cooper, Robert G. 1983. A process model for industrial new product development. *IEEE Transactions on Engineering Management* 30:2–11.

Cooper, Robert G. and Elko J. Kleinschmidt. 1986. An investigation into the new product process: Steps, defiiciencies, and impact. *Journal of New Product Development* 3:71–85.

Crozier, Michel. 1964. *The bureaucratic phenomenon.* Chicago: Chicago University Press.

Dougherty, Deborah. 1992. Interpretive barriers to successful product innovation in large firms. *Organization Science* 3:179–202.

Dougherty, Deborah. 1992. A practice-centered model of organizational renewal through product innovation. *Strategic Management Journal* 13:77–92.

Dougherty, Deborah, and Trudy Heller. 1994. The illegitimacy of successful new products in large firms. *Organization Science.*

Gouldner, Alvin W. 1954. *Patterns of industrial bureaucracy.* New York: Free Press.

Ittner, Christopher, and Bruce Kogut. 1994. "How control systems can support organizational flexibility." *Redesigning the firm.* New York: Oxford University Press.

Leonard-Barton, Dorothy. 1988. Implementation as Mutual Adaptation of Technology and Organization. *Research Policy* 17:251–67.

Lewin, Kurt. 1947. Frontiers in group dynamics. *Human Relations* 1:5–41.

Majchrzak, Ann. 1988. *The human side of factory automation.* San Francisco: Jossey-Bass.

Schein, Edgar H. 1961. Management development as a process of influence. *Industrial Management Review* 59–77.

Schein, Edgar H. 1987. *Process consultation,* vol. 1. Reading, Mass.: Addison-Wesley.

Westley, Frances. 1990. The eye of the needle: Cultural transformation in a traditional organization. *Human Relations* 43:273–93.

C H A P T E R

3 Process Design

3.1 Introduction

As firms successfully apply the principles of new product development to respond to competitive pressures, they tend to increase the frequency of product introductions and the variety of products that they deliver to the market. The adoption of automation technologies promises to provide such firms with an important means for dealing with the inherent difficulties of managing the complexity of producing and supporting more products with relatively shorter life cycles and smaller market volumes. As we shall see, these technology developments have given rise to new ways of designing the production process, i.e., of organizing and using the manufacturing resources of the firm. In this book we consider process design by organizing it into three broad classes of activities: (1) *the design of the manufacturing process itself* in terms of selecting the type of process,[1] choosing the production equipment, and designing the plant floor layout, (2) *the development of process plans* for transforming input raw materials into processed, finished products through actions of the various stages of production, and (3) *the management of the execution of production process plans* for the set of products being produced in the factory.

Figure 3.1 illustrates a decision hierarchy which connects the management of these three classes of process design activities. These activities can critically influence the overall performance of the production system in terms of its cost, quality, flexibility, cycle time, and customer service level (usually measured in terms of product availability).

[1]Traditionally, discrete manufacturing processes are classified into broad process categories which are based on the dominant production technology. The principal classes include (1) job shops, (2) batch shops, (3) transfer lines, and (4) continuous flow systems. The job shop is characterized by individual jobs with unique processing requirements visiting an arbitrary sequence of work centers. The batch shop produces in lot sizes greater than one and typically has some degree of standardization with regard to the routing of jobs. The transfer line imposes a common routing on all products which pass through it. Continuous flow processes can be found in oil refineries and chemical plants.

FIGURE 3.1

Process design decision hierarchy

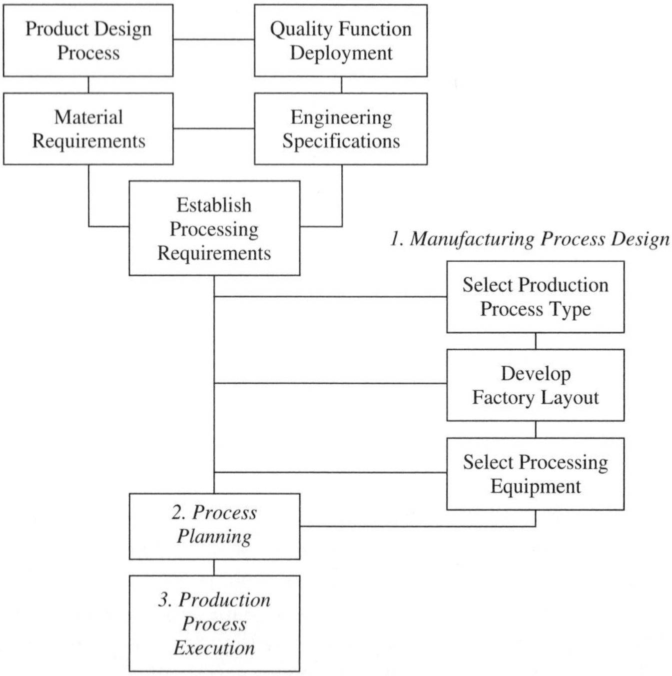

The design of the production facility determines its potential for meeting target levels for each performance attribute. The linked hierarchy of design choices together determine the physical and organizational capabilities of the factory.

In this chapter, we will focus on two aspects of the hierarchy: factory layout and process planning. We will assume that the factories in question are producing discrete items in batches using either a job shop, batch flow, or transfer line process. The issue of plant floor layout will be considered specifically in the context of a restructuring approach, referred to as *group technology (GT),* which has been associated with the implementation of automation. GT leads to the definition of product classifications (groups) and the assignment of these groups to dedicated production resources (manufacturing cells). The principal managerial issue associated with the selection of processing equipment is justification of the investment required for its purchase, installation, and use. Methods for evaluating the capital expenses associated with the redesign of production facilities will be discussed in Chapter 8.

Process planning consists of those activities germane to the preparation of detailed instructions for the fabrication of a part and/or for the assembly of parts into a finished product. In general, process planning includes the following activities: (1) selection and sequencing of processing operations, (2) assignment of the product to specific pieces of equipment with the capability of executing these operations, (3) choice of a materials handling system, (4) establishment of setup (or changeover) procedures for the initialization of production batches, (5) design of jigs and fixtures, (6) selection of machining and cutting tools, (7) calculation of various process (control) parameters, and (8) the definition of tool paths, including creation of numerical control (NC) part programs. The nature of a process plan and the degree of detail required for specifying it vary considerably, depending on the dominant process type in place, the production methods being used, and the skill level of the employees. For example, in a job shop environment with nonautomated equipment and experienced machinists, a detailed plan may be unnecessary. A simple instruction, such as "produce as per blueprint," may suffice. On the other hand, production of a product in an automated transfer line environment, using computer controlled machine tools, may require a process plan that specifies each of the activities mentioned above.

Process planning has traditionally been performed manually by process planners who have extensive experience with factory floor operations. Technology developments of the past few decades, such as *numerically controlled (NC) machines, NC part programs, CAD systems, GT,* and the overall increased use of *computers* in manufacturing, have led to the automation of many process planning activities. *Computer-assisted process planning (CAPP)* systems are now widely used in manufacturing as a way to make firms more productive and less vulnerable to the turnover and retirement of process planners. CAPP is also critical for the integration of CAD and CAM functions and the eventual achievement of a true CIM environment (Chang and Joshi [1990]).

The management of those activities associated with the execution of process plans within the production facility (stage three in the process design hierarchy) is highly dependent on the configuration of processing equipment, the infrastructre defining the material and information flow linkages, and the process plans selected for each part and product. These production management decisions will be discussed in detail in Chapter 6. In that chapter we consider issues concerning scheduling and planning from the long-term (e.g., annual) to the short-term, (e.g., real-time control of material, equipment, tooling and workers) horizon.

We begin, in this chapter, with a discussion of the layout decisions associated with the placement and connection of individual pieces of processing equipment on the factory floor. These layout decisions determine the work flow, productivity, and quality characteristics of the factory. This discussion provides a backdrop to the subsequent sections on GT and cellular manufacturing. These concepts that have emerged as important principles for making process design decisions in automated environments. The relationship of these concepts to computer-aided design, and to other elements in the CIM framework, will thus be analyzed. In this context, we will review the innovative approach followed by Cummins Engines in implementing cellular manufacturing (Venkatesan [1990]). The chapter ends with a brief discussion of recent progress in the application of CAPP technologies and its pivotal role in linking product designs to the manufacturing production process.

3.2 Facility Layout

Facility layout decisions concern the placement of machines, stock-holding points, and design of the material handling system within the production facility. The general objective is to arrange these elements in a manner that enables a smooth work flow and thereby improves performance—especially in terms of reduced *work-in-process (WIP)* inventory, shortened throughput time, and better utilization of the production equipment.

Traditionally, facilities have been organized according to two basic types of layouts: the *process layout* and the *product layout* (See Figure 3.2). In a process layout (also called a functional or job-shop layout) similar pieces of equipment are grouped together, for example, lathes with lathes, drilling machines with drilling machines, and so forth. As a part is processed, it moves from one group of equipment to the next according to its process specification (or routing). The requirement to visit each group of equipment gives rise to the principal disadvantage of the process layout approach, that is, excessive movement (and material handling) of parts as they zigzag their way through the factory floor. As a consequence, production leadtimes increase, and unduly large WIP inventory accumulates. There are also some benefits, however, to process layout, for example, economies of scale through ease of supervision, better utilization of equipment, and availability of skilled employees within each machine group.

To overcome the drawbacks of a process layout, manufacturers have turned to an alternative solution, i.e., *product layouts.* In this approach, machines are arranged so that each collection (or cell) can accommodate a specific range of products. To the extent that equipment is arranged sequentially in the order of the various production process steps associated with a dominant product, this layout is also called a *flow shop* (e.g., each part follows the same path through the cell). Under a product layout, material moves smoothly from one machine to the next without having to wait long or to travel on a convoluted path. Consequently, a dramatic improvement is generally observed in both throughput time and the level of WIP inventory. A small number of multiskilled workers typically tend each cell. Defining a suitable grouping of machines for the cell and the assignment of parts to that grouping are among the most important issues associated with product layout. GT is one approach for developing such groupings and cellular manufacturing is the principal mechanism for using these groupings to reconfigure the factory layout.

3.3 Group Technology

The concepts of *group technology (GT)* were developed in the former USSR in the early 1950s and were adopted in both Japan and Europe by the 1970s (Burbridge [1975], Hyer [1984], and Suresh and Meredith [1985]). Manufacturers in the United States were slow to implement GT principles initially. With the increasing acceptance of CAD and other manufacturing automation technologies, the importance of GT principles has been recognized. Indeed, some observers feel that GT is essential for the success of CIM. As a consequence, GT became well established in the United States by the mid-1980s.

The approach is based on the principle of clustering "similar" parts into part families. The objective in creating a part family is to realize the benefits associated with economies of scale that are present in almost all stages of the product life cycle without losing the capacity to deal with significant product variety. Use of group technology

Figure 3.2

Facility layouts

A. Process Layout

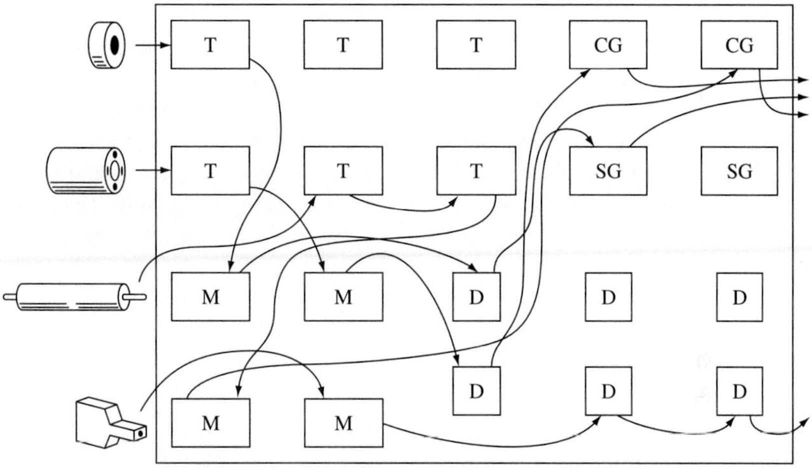

T Turning CG Centerless Grinding
M Milling SG Surface Grinding
D Drilling

B. Product Layout

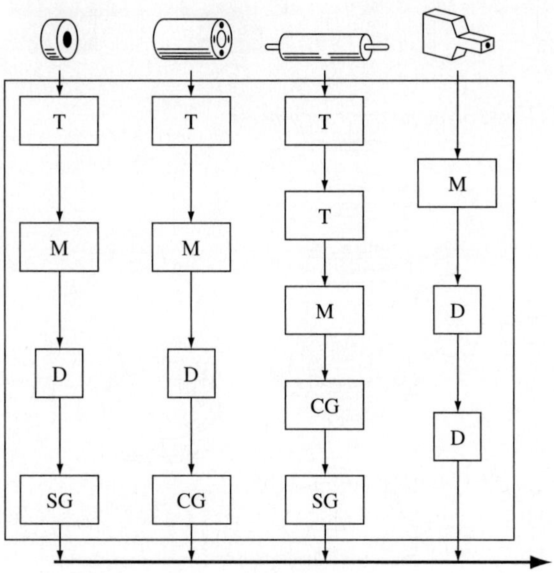

Source: Adapted from Hyer and Wemmerlov (1982)

principles, therefore, can lead to improved product and process designs, thereby enhancing manufacturing operations performance.

The power of GT, and some of the difficulties associated with its implementation, emanate from the fact that the notion of "similarity" is not constant. Indeed, similarity can vary with the stage of the manufacturing life cycle in which GT is applied. Thus, a parts family based on design similarity would include parts that are similar in geometry (shape and size), whereas a parts family based on manufacturing similarity would consist of parts that undergo similar sequences of manufacturing operations or use similar processing technologies. The simultaneous consideration of all of these notions of "similarity" to create suitable groupings is one of the most challenging aspects of GT. In a later section, we discuss the approach taken by Cummins Engines Company to deal with this challenge. Figure 3.3 shows two parts families. Figure 3.3A shows a family of parts

FIGURE 3.3

Using group technology to create part families

A. Similar parts based on shape

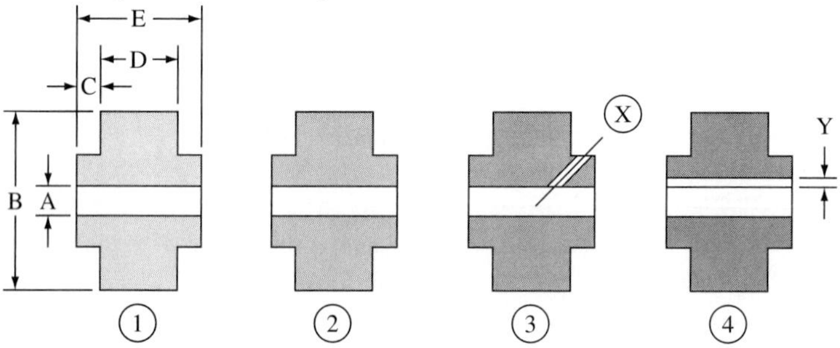

B. Similar parts based on manufacturing process

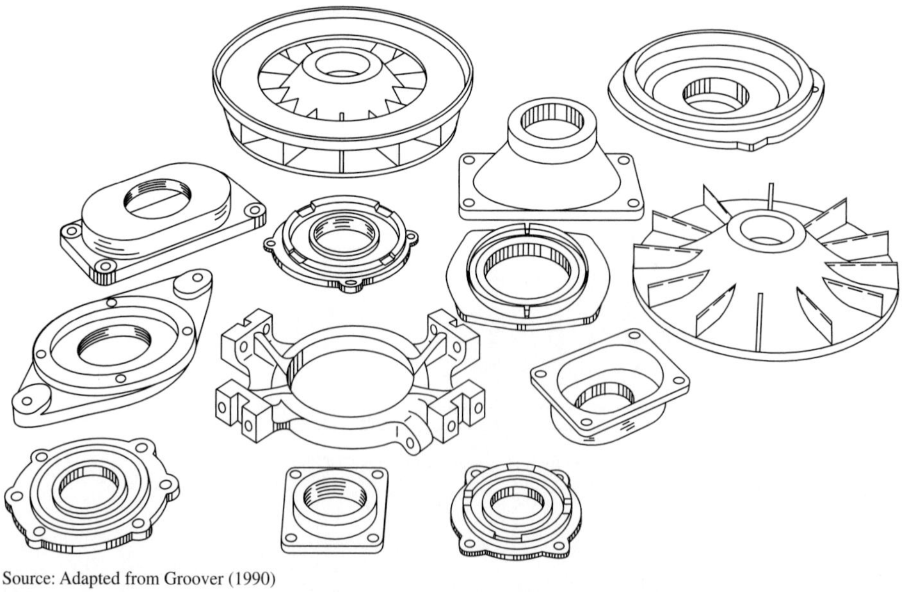

Source: Adapted from Groover (1990)

that are similar in shape, but that may entail very different manufacturing processes. In contrast, Figure 3.3B shows 13 parts that have similar manufacturing process requirements, but have quite different shapes and design attributes.

Numerous classification and coding schemes are available today for the implementation of GT, for example, BRISCH (UK), OPTIZ (Germany), MICLASS (Netherlands), and CODE (USA). Most modern classification schemes are based on both design and manufacturing attributes. Indeed, a classification and coding scheme can be compared to a sieve which sorts parts by their particular attributes (see Figure 3.4). The design attributes commonly used for such sorting include: external and internal shape, length-to-diameter ratio, physical dimensions, tolerances, surface finish, and material used. The manufacturing attributes grouped together include: production processes used, operations sequence, machine tools and fixtures used, production time, batch size, and annual production volume. Several classification schemes have been developed that allow for the use of a database for storing and retrieval of parts designs based on classification code similarity.

MICLASS (metal institute classification system) is one of the most popular systems used in the United States. In MICLASS, a coding number 12 to 30 digits long is assigned to each part. The first 12 digits represent design attributes, whereas the remaining digits represent manufacturing (process) attributes. In classifying a part using MICLASS, the user responds to the questions posed by the software package. Based on

FIGURE 3.4

Classification and coding scheme as a sieve

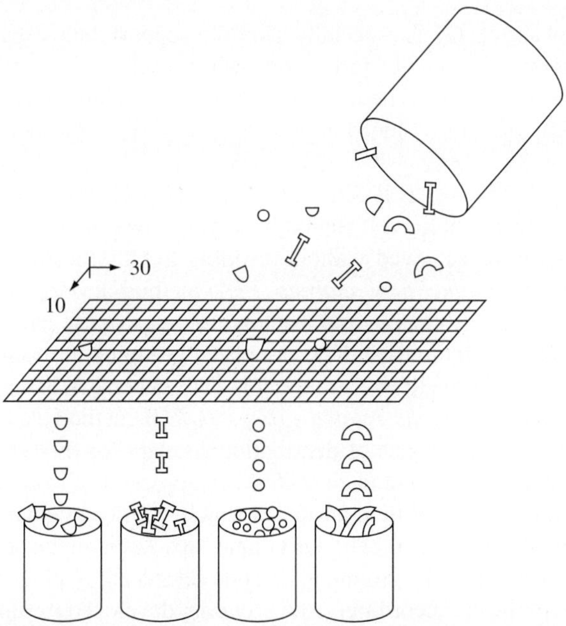

the user's answers, MICLASS then assigns the appropriate classification code to the part.

Once a classification scheme is in place, the next step involves defining collections of parts families. Many statistical procedures are available for grouping objects which are each characterized by a set of attributes. Green and Tull (1978) describe a variety of procedures (including factor analysis, cluster analysis, discriminant analysis) which are used in marketing contexts.

The creation of suitable parts families is clearly a data intensive task. As a consequence, the start-up, or preparatory, work involved in implementing GT (i.e., to use the chosen classification scheme to develop a computer coding for all of the firm's existing parts) is time consuming, tedious, and expensive. However, once this start-up work is completed and, as a consequence, the up-front investment is made, experience suggests that manufacturers are assured of productivity improvement at all value adding stages (i.e., from design through manufacturing to distribution and after-sales service).

For example, when a GT-based code links to a CAD system, it may not be necessary for a designer to create a new part from scratch every time a product is being developed. Instead, by querying the automated, design storage and retrieval system, the designer can locate the parts family which is "closest" (in attribute space) to the proposed new part. With this match, the designer can then create the new part design by suitably modifying and editing the design of an existing part drawn from the matched family (see Rock et al. [1993]). Designer productivity will be substantially improved in this manner. GT can be used to support process design in a similar fashion (i.e., by retrieving and editing a matched process plan). We discuss this approach further in the section on CAPP. Table 3.1 summarizes the benefits observed in the application of GT, whereas Table 3.2 indicates some of the major costs associated with the implementation of GT.

As noted above, GT is especially useful to support the reorganization of manufacturing operations into manufacturing cells, where each cell specializes in the processing of a particular part family. The main benefits of such cellular manufacturing include reduced handling of parts, reduced in-process inventory, and a shortened production lead time.

The proper choice of a classification and coding scheme is clearly important for the success of group technology. It should be kept in mind, however, that identification of part families can be achieved without resorting to classification and coding schemes. For example, the *product flow analysis (PFA)* method developed by Burbridge (1971) analyzes operations sequences and machine routings of the parts produced in a given shop to identify part families and the associated grouping of machines.

In fact, the GT principle of grouping is very general in its applicability. This is illustrated by the *operations related groups (ORG)* method developed by Ernst and Cohen (1990) for an analysis of distribution strategy for the service parts division of General Motors. ORG uses statistical clustering procedures to create groups of parts by considering a full range of item attributes which are of particular significance to after-sales service (see Table 3.3). The ORG approach is distinguished by its linkage of a managerial objective to the grouping, i.e., one selects a grouping to achieve a particular management purpose. Accordingly, the grouping developed to support repair and diagnosis of equipment may be inappropriate for use as a grouping to support inventory management, facility layout, or market segmentation.

Table 3.1 Benefits of Group Technology

Benefits	*Number of Firms Noting Response as "Most Important"*	*Number of Firms Reporting Response as a "Significant Benefit"*
Lead Times		
Reduced total lead times	8	11
Reduced throughput times	8	13
Reduced queuing times	4	6
Reduced setting times	3	7
Product Design		
Design retrieval improved	4	9
Parts standardization introduced or improved	2	8
Reduced number of new parts/year	2	7
Reduced design effort	2	7
Inventories		
Reduced work-in-process inventories	3	7
Reduced finished goods inventories	0	4
Labor		
Reduced labor costs	3	9
Increased operator job satisfaction	0	7
Miscellaneous		
Process planning simplified	5	10
Quality control improved	5	10
Tooling and fixture expense reduced	3	9
Material handling cost reduced	2	10
Space needs reduced	1	7
Cost estimating more accurate	0	7

Source: Hyer (1984b)

Table 3.2 Costs of Group Technology

Expenditure	*Number of Firms Reporting This Cost as "Significant"*	*Number of Firms Reporting This Cost as the "Most Significant Cost"*
1. Planning the GT program	12	4
2. Additional computer software	11	4
3. Training coders	11	2
4. Additional computer hardware	10	4
5. Coder salaries	10	2
6. Machine tool rearrangement	9	2
7. Coding consultants	9	0
8. Management education	7	1
9. Training cell operators	5	0
10. Additional machine tools	2	1
11. Consultants (general)	2	0

Source: Hyer (1984b)

TABLE 3.3 Variables Used in the ORG Grouping Analysis

Variable	Definition
1. FY	= The first year that the part was used (year)
2. LY	= The last year that a vehicle model used the part (year)
3. DLRNET	= The unit cost of the part to the dealer (S)
4. ECD	= Excess cube indicator (1 = large, 0 = small)
5. DYTDRET	= Number of units returned by the dealers in the current year (units)
6. DYRIRET	= Number of units returned by the dealers in the previous year (units)
7. WYTDRET	= Number of units returned by the wholesale distributors in the current year (units)
8. WYRIRET	= Number of units returned by the wholesale distributors (WD) in the previous year (units)
9. BUYLT	= Lead time specified the part's supplier contract (weeks)
10. AGEMEAN	= Mean of the units sold to the Allied Government Export (AGE) market in the last 24 months (units)
11. CANADAMN	= Same as (10) for the CANADA market (units)
12. DEALERMN	= Same as (10) for the DEALER market (units)
13. TARGETMN	= Same as (10) for the TARGET (consisting of high volume parts subject to special promotions) market (units)
14. WDMEAN	= Same as (10) for as WD market (units)
15. AGESTD	= Standard deviation of the units sold to the AGE market in the last 24 months (units)
16. CANADSTD	= Same as (15) for the CANADA market (units)
17. DEALESTD	= Same as (15) for the DEALER market (units)
18. TARGESTD	= Same as (15) for the TARGET market (units)
19. WDSTD	= Same as (15) for the WD market (units)
20. AGESF	= Seasonality factor for the monthly sales in the AGE market over the last 24 months. It is equal to the difference between the maximum and the minimum of the two year average sales for each month.
21. CANADASF	= Same as (2) for the CANADA market
22. DEALERSF	= Same as (2) for the DEALER market
23. TARGETSF	= Same as (20) for the TARGET market
24. WDSF	= Same as (20) for the WD market
25. BOCPI	= Fraction of sales of the part used in the body style BOC (Buick, Oldsmobile, Cadillac) market
26. CPCI	= Same as (25) for the body style CPC (Chevrolet, Pontiac, Canada) market
27. TRKP1	= Same as (25) for the body styles TRUCK and BUS market
28. CHEP1	= Fraction of a particular part used in the Chevrolet Division market
29. PONP2	= Same as (28) for the Pontiac Division market
30. BUIP1	= Same as (28) for the Buick Division market
31. OLDP1	= Same as (28) for the Oldsmobile Division market
32. CADP1	= Same as (28) for the Cadillac Division market
33. LT1	= Actual lead time (weeks)
34. LT2	= Standard deviation of (33) (weeks)
35. LT3	= Weeks late lead time; difference between actual and promised delivery dates (weeks)
36. LT4	= Standard deviation of (35) (weeks)
37. LT5	= Requested lead time from supplier (weeks)
38. LT6	= Standard deviation of (37) (weeks)
39. CRIT	= Criticality index of the part (0-1)
40. NUMCAR	= Number of vehicles in which a part is used

3.4 Cellular Manufacturing

Cellular manufacturing is based on the principles of product layout. It specifies a set of machines (in a cell) which has the capability to produce a "part family" as defined by the group technology approach. Cellular manufacturing applications can be found today in a wide range of industries, including metal fabrication and computer chip production. Machine cells are sometimes classified in one of the following categories, depending on the number of machines included and the degree to which the material flow is automated: (1) a single machine cell, (2) a group machine cell with manual handling, (3) a group machine cell with semi-integrated handling, and (4) a flexible manufacturing system (FMS) (Groover [1990]).

The first category needs no discussion, while the last category, FMS, is discussed in detail in Chapter 4. The second category of machine cells, involving manual handling, is often arranged in a U-shaped layout to facilitate the movement of parts by multifunctional workers (who may operate/supervise more than one machine). This layout can also accommodate variations in work flow of parts made in the cell. The third category involves mechanized movements of parts with occasional support by manual handling. The machines in this category are arranged in either an inline layout (when parts have nearly identical routing) or a loop layout (which allows for different routings for different parts). Figure 3.5 illustrates the different layouts used in designing a manufacturing cell.

FIGURE 3.5

Layouts for cellular manufacturing

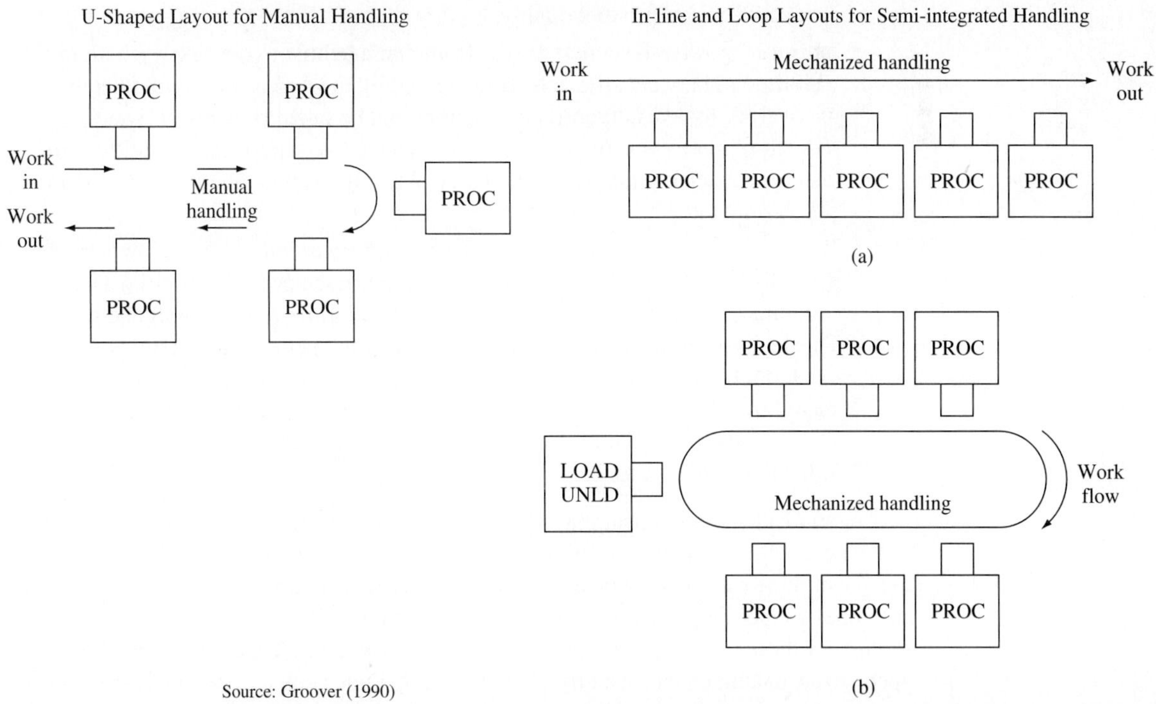

Source: Groover (1990)

The advantages and disadvantages of cellular manufacturing, as compared to those of traditional manufacturing, include:

- *Reduced material handling and production lead time:* As noted above, move distances, as well as move times, can be very small in cellular manufacturing due to the physical proximity of machines with one another in manufacturing cells (e.g., an order of magnitude reduction is not uncommon).
- *Reduced set-up time:* Set-up times typically are dominated by the changes required to reset part holding devices (e.g., jigs and fixtures) and tooling. As a cell is designed to handle a family of parts which are similar in size, shape, and manufacturing and tooling requirements, cell operators can complete set-up changes in a fraction of the time required under a functional layout (through learning and adaptation of the devices). The waiting time for parts will also be reduced significantly.
- *Reduced WIP inventory:* Reduced material handling and throughput time lead to a proportional reduction in the level of WIP inventory in the cell.
- *Reduced makespan (i.e., throughput time):* A reduced need for material handling and set-up changes, and the associated savings of time, greatly reduce the average throughput time of parts in the system.
- *Improved operator expertise and morale:* Since cell operators handle only a limited number of parts, the learning curve effect helps to improve their expertise. Manufacturing cells lend themselves naturally to team-based organization structure, which generates higher levels of worker morale, productivity, and product quality.
- *Improved quality:* As a part moves from one machine to the next with almost no wait time in between, the defects generated by the process are immediately uncovered. As a result, corrective actions can be taken right away. The reduction in lot sizes, made possible by reduced set-up times, also allows for faster feedback on quality problems. As a result, product quality improves in a cellular layout.
- *Lower machine utilization and higher capital requirements:* These two are the main disadvantages of cellular manufacturing. Each cell, by definition, must contain all the machines required to produce a family of parts assigned to it. This leads to the purchase of more machines (one per cell that requires the machine) than that required under a process layout where similar machines are shared by all parts (e.g., there will be more excess capacity in the cell system). The investment requirement, therefore, goes up and the average machine utilization rate is reduced.

In an ideal setting, a firm may simply be able to rearrange its production machines to create a cellular manufacturing layout. In many cases, however, one finds that either (1) one or more parts cannot be easily associated with any part family or (2) one or more machines cannot be placed in a single manufacturing cell due to the specialized nature, or cost, of these machines (e.g., heat treatment equipment). These unassigned parts, or specialized machines, are usually placed in "remainder cells." Obviously, the advantages of cellular manufacturing discussed above will be diluted when remainder cells are required.[2]

The success of cellular manufacturing ultimately depends on how the group technology principles are used to create an "optimal" grouping of parts into part families. Hence, the fundamental issue, as alluded to earlier in the discussion of GT, is the selection of an appropriate criteria for GT that enhances the advantages of cellular manufacturing while reducing its disadvantages.[3] In implementing cellular manufacturing an important task, therefore, is to create a plan for smooth transition from a process layout to manufacturing cells layout. Rearranging machines into cells based on part families is also a major undertaking requiring both considerable time and expense.

Cummins Engine Company, a manufacturer of diesel engines, has made an innovative use of cellular manufacturing principles to significantly reduce material handling, WIP inventory, and production leadtime. They have done so in a manner which also reduced labor requirements and improved delivery performance (Venkatesan [1990]). In the mid-1970s, Cummins produced a stable set of products with large sales volumes concentrated in relatively few product lines. The factory layout was fashioned after a traditional process layout, and WIP inventory was large. Since product quality was high and the firm enjoyed excellent on-time delivery performance and a strong financial position, senior management was not concerned with the WIP inventory level or the production leadtime.

This picture changed dramatically over the next decade. Faced with increased federal regulation on emission standards, fragmentation and volatility of its markets, and competitive pressure from global players like Komatsu, Cummins had to increase its product variety and reduce its leadtimes, while simultaneously reducing prices by 20 percent to 40 percent in order to maintain its market share. However, increased product variety and smaller production volumes played havoc with the efficiency and leadtime performance of its production facilities due to the need for frequent set-up changes as production was switched from one product to another. Cellular manufacturing was adopted as the solution in 1985. The firm's major production facility was reorganized in accordance with the principles of cellular manufacturing. Each cell was designed to produce a small group of closely related parts. As a result of this change, the situation improved initially. As product proliferation continued, however, the manufacturing cells began to choke with the changeover time requirement generated by the increased level of product variety.

Cummins went back to the drawing board and realized that the creation of manufacturing cells based on product design alone was insufficient to address its problem. A refocus, using both product and volume considerations, was essential. The product line was divided into four categories of engine components, each of which was further subdivided in five generic classes based on sales volume, predictability of demand, and sta-

---

[2]See Appendix A for an illustration of the grouping / layout problem and the use of a common, "interchange" algorithm.

[3]See Ernst and Cohen (1990) for a methodology in which the definition of groups can vary with the purpose of the grouping exercise, the nature of the products, the market environment, and the production technology.

bility of product design. Manufacturing capacity was then realigned with this new classification in mind. Under the new structure, cells, traditional layouts, and flexible systems were all used. For example, there were 160 types of flywheel housing parts, and 11 comprised 85 percent of volume. These 11 parts were assigned to a refocused high-volume line. The remaining 149 were assigned to a flexible machining cell. As products evolved through their life cycles, membership in groups and assignment to a production subfacility could change. The resulting hybrid/adaptive system showed dramatic improvements over the purely cellular solution.

The Cummins approach defied conventional GT wisdom. For example, it is generally recommended that flexible machining cells (and flexible machines) should be used only to handle many different kinds of parts simultaneously and not for high volume parts (Jaikumar [1986]). Cummins violated this rule in some of its cells, e.g., flexible cells were assigned some high-volume parts, when their designs were particularly unstable. Cummins' success with its hybrid approach illustrates an important issue: Frequent design change can be just as disruptive to a production facility as high product variety; both require the use of flexible machines to accommodate frequent changeovers. Using flexible machines (as compared to less flexible ones) for products with frequent design changes also shortens the time-to-market, which is of significant strategic value (see Chapter 2). Cummins' experience teaches one additional lesson: Increasing market volatility requires that a dynamic product/process grouping strategy be followed—parts classification and manufacturing cell alignment should be continually reviewed and revised as needed.

3.5 Computer-Aided Process Planning

Although one of the earliest discussions of using computers in process planning can be found in Niebel (1965), the industrial use of computers in supporting process planners in their clerical and report preparation activities began only in the 1970s. A typical example of such use is Lockheed's CAP system (Tulkoff [1981]). In addition to providing clerical and documentation support, *computer-aided process planning* provides several advantages including:

1. Reduction in the time and cost required for process planning leading to improved productivity of process planners,
2. Improved accuracy, performance, and consistency of process plans, and
3. Most importantly, creating a bridge between product design and manufacturing activities, and thereby moving towards the ultimate goal of CIM.

This last point has been discussed in Rock et al. (1993), where they note that CAPP can facilitate the sharing of electronic planning knowledge across multiple manufacturing sites and processes.

Computer-aided process planning is being used today in a wide range of metal fabrication, aerospace and electronic industries. Examples of the types of parts being produced using CAPP include turned and prismatic parts, sheet metal, and electronic assembly. A survey of the industrial use of CAPP can be found in Haas and Chang (1988). A review of concepts and literature can be found in Chang and Wysk (1985).

There are two basic approaches to computer-aided process planning: *variant* and *generative* process planning. Under the variant approach, a process planner uses a computer to retrieve an existing process plan for a similar part. This plan is then suitably modified to create a variant plan to produce the required part. In contrast, the generative approach creates a process plan by using the available manufacturing planning logic, without referring to any existing plan.

The *variant process planning* approach is based on group technology principles, i.e., similar parts are likely to require similar manufacturing processes. In implementing a variant system, the classification and coding scheme used in CAD is also used as the foundation of the CAPP system. The CAPP specifications of the manufacturing process can then be used subsequently to create programmed instructions for the automated manufacturing equipment on the factory floor. Thus, CAPP closes a critical gap between product design and product manufacturing activities.

A two-stage approach involving preparatory and production stages is generally used in developing a variant process planning system. In the preparatory stage, existing parts are coded, classified, and grouped into families based on process similarities. A standard process plan is then assigned to each family and is stored using predefined operation codes, where each operation code signifies a series of processes to be carried out on certain machines. The preparatory stage can be very labor intensive. However, the ultimate benefits of a variant system depend on the efforts spent in the preparatory stage in appropriately determining families and assigning operation codes. During the production stage, a new component to be produced is first coded. This code is then used as the key in searching for the family to which the new part belongs. The "closest" associated standard process plan is then retrieved and edited to create a suitable process plan for the new component.[4]

Limitations of a variant system include its inability to handle components that are dissimilar to the ones coded earlier in the preparatory stage and its dependence on experienced process planners to make the necessary modifications. Indeed in most companies, process planning is a labor intensive job employing many planners who generate a wide variety of plans. It typically takes several days to convert a new product design into a detailed process plan. In spite of these limitations, a variant system is generally easy to develop, install, learn, and use. In many batch manufacturing situations, where similar parts are produced repeatedly, variant systems can prove to be very cost effective. Obviously, the stability of the product mix and the capacity limitations of the manufacturing process must also be taken into account. Accordingly, many firms have developed multiple, independent systems for site-specific requirements.

The *generative process planning* approach involves using computers to automatically generate a process plan for producing each and every part. The advantages of gen-

---

[4] Many CAPP systems provide the interface needed to support access to and retrieval of process related data. They do not, however, provide assistance in resolving downstream scheduling conflicts or for resolving standards violations generated by the interaction of the product design, process plan, and process capabilities.

erative process planning include the ability to produce new as well as existing parts with equal ease, and the potential to create consistent, accurate and optimized process plans. In fact, generative process planning is considered by some to be a key missing element in achieving the vision of true computer integrated manufacturing. Generative systems conceptually include three elements:

- *Part Description Scheme:* The way parts are described is critical to the gathering of relevant information necessary in creating manufacturing process plans. Earlier generative systems were based on group technology codes. The main limitation of GT codes is that in the coding step the detailed information on size and shape of parts is generally lost. Hence, human intervention is still needed in converting part design codes into a process plan. Li et al. (1987) describe a semigenerative system CORE-CAPP that uses GT codes.

  The limitations of using GT codes have led to the development of special descriptive languages that can specify geometric shapes or features such as cylinders, cones, holes, grooves, and so forth. An implicit assumption in the use of these languages is that a process planner is on hand to convert the original design into these descriptions. With advances in CAD, increasingly sophisticated interfaces are becoming available to directly input CAD specifications into the process planning system.
- *Manufacturing Databases:* In addition to the information on the parts to be produced, it is essential that information concerning the capabilities of manufacturing systems also be known. This information may be made available in the form of various databases containing detailed specifications of machines, tools, fixtures, materials, machinability, standard times and costs, and so forth.
- *Process Planning Logic:* The most difficult elements in a CAPP system are (1) representation of the process planning logic and (2) knowledge that a planner uses in making choices of machines and machine tools, cutting tool speeds, process sequences, and so forth. Use of decision trees or decision tables is possible for simple situations. Recent advances in artificial intelligence (AI) and expert systems have made it possible to represent highly complex process planning logic, but this approach is still emerging. A discussion on the possibilities of using expert systems and other knowledge representation schemes can be found in Chang (1989) and Kumara et al. (1986). Rock et al. (1993) describe applications of AI to CAPP and note that it can bring the following benefits to CAPP systems:

—Leveraging of domain independent tools within a firm.

—Improved system performance in terms of execution time and the ability to respond to new events with reactive planning.

—Regrouping of actions to reflect flexible, cellular manufacturing with operators managing overlapping operations.

The generative approach is much more difficult to implement than the variant approach. In fact, relatively few real generative process planning systems are in use today.

The systems that exist may be best described as semigenerative, where most of the process planning functions are carried out automatically, with the remaining functions being performed by human process planners.

Significant recent trends in the progress of CAPP system include: (1) a steady increase in the level of "intelligence" embedded in the systems and (2) an increasing level of automation in the operation of these systems. During the 1990s and beyond, major developments should focus on the use of a CAPP system as an element that seamlessly integrates design and manufacturing activities. The technological challenges of making such integration a reality are considerable, but (we believe) surmountable. The main stumbling block to this effort appears to be organizational resistance. Cooperation of different departments, such as design, engineering, manufacturing, and marketing will be critical for the success of CAPP systems.

Study Questions

1. What are the principal product and process attributes in forming groups in the *group technology (GT)* approach?
2. Explain the rationale of the GT concept. What are the major advantages and disadvantages of GT?
3. What are the main differences between a *product* and a *process* layout?
4. How do market and competitive factors affect product grouping and factory layout decisions?
5. What are the principal benefits of cellular manufacturing? What are the requirements for its successful implementation?
6. What are the differences between generative and varient process planning methods?
7. How can *artificial intelligence (AI)* enhance *computer-aided process planning (CAPP)*?

References

———(1986), "Note on Group Technology and Cellular Manufacturing," Publication No. 9-686-098, Harvard Business School Press, Boston, MA.

Burbridge, J. L. (1971), "Production Flow Analysis," *The Production Engineer,* April.

Burbridge, J. L. (1975), *The Introduction of Group Technology,* Heinemann, London.

Chang, T. C. (1989), *Expert Process Planning Systems,* Addison-Wesley, Reading, MA.

Chang, T. C. and S. Joshi, (1990), "Computer-Aided Process Planning," in D. Cleland and B. Bidanda, ed., *The Automated Factory Handbook,* TAB Professional and Reference Books, Blue Ridge, PA, pp. 367–401.

Chang, T. C. and R. A. Wysk, (1985), *An Introduction to Automated Process Planning Systems,* Prentice-Hall, Englewood, NJ.

Ernst, R. and M. A. Cohen, (1990), "Operations Related Groups (ORGs): A Clustering Procedure for Production/Inventory Systems," *Journal of Operations Management,* Vol. 9(4), October, pp. 574–598.

Groover, M. P. (1990), "Group Technology," Chapter 16, *Automation, Production Systems and Computer Integrated Manufacturing,* Prentice-Hall, Englewood Cliffs, NJ.

Haas, M. and Chang, T. C. (1988), "Survey on Process Planning Systems Usage in Industry," School of Industrial Engineering, Purdue University, West Lafayette, IN.

Hyer, N. L. (1984a), "Management's Guide to Group Technology," *Operations Management Review,* Winter.

Hyer, N. L. (1984b), "The Potential of Group Technology for U.S. Manufacturing," *Journal of Operations Management,* Vol. 4(3), pp. 183–202.

Hyer, N. L. and U. Wemmerlov, (1982), "MRP/GT: A Framework for Production Planning and Control of Cellular Manufacturing," *Decision Sciences,* Vol. 13, pp. 681–701.

Kumara, S. R. T., et al. (1986), "Expert Systems in Industrial Engineering," *International Journal of Production Research,* Vol. 24(5), pp. 1107–1125.

Li, J., C. Han, and I. Ham, (1987), "CORE-CAPP: A Company-Oriented Semi-Generative Computer Automated Process Planning System," in *Proceedings of the 19th CIRP International Seminar on Manufacturing Systems,* Pennsylvania State University, University Park, June, pp. 219–225.

Niebel, B. W. (1965), "Mechanized Process Selection for Planning New Designs," ASTME Paper 737.

Rock, D., S. Washburn, D. Purdon, and A. Houtzel, (1993), "Sharing Knowledge Bases in Industry," *AI Expert,* Vol. 8, no. 6, p. 24.

Suresh, N. C. and J. R. Meredith, (1985), "Achieving Factory Automation Through Group Automation," *Journal of Operations Management,* Vol. 5(2).

Tulkoff, J. (1981), "Lockheed's GENPLAN," in *Proceedings of the 18th Numerical Control Society Annual Meeting and Technical Conference,* Dallas, TX, pp. 417–421.

APPENDIX A

Process plans can be represented (in part) by a table that indicates required machine part combinations. A "1" in the table indicates that the machine designated by the row label is used in the processing/production of the part designed by the column label. Consider the example illustrated in Table A1. According to the table, part 1 must visit machines A and C, and machine D is used by parts 2, 4, and 7. Note that the machine/part assignment data does not indicate anything about precedent constraints, e.g., part 1 could require sequencing through machine A before it is processed on machine C.

The object of the grouping exercise is to use these processing requirement data to define a grouping of parts and a corresponding cell of machines that can be dedicated to this group of parts. The table can be used to generate a solution that approaches this ideal (and often unreachable) goal. In particular, we seek to reorder the sequence of rows and columns in the table so that it will approach a *block diagonal form* (e.g., where subsets of rows and columns only interact with each other). The following binary algorithm can be used to accomplish this goal.

Binary Ordering Algorithm

1. Order the rows: Let N = the number of columns (parts). Assign value 2^{N-j} to column j. Evaluate each row with these numbers. Sort the rows in nonincreasing order. If the rows were previously ordered and no change just occurred, stop; otherwise, go to 2.
2. Order columns: Let M = the number of rows (machines). Assign value 2^{M-i} to row i. Evaluate each column. If no change, stop; otherwise, go to 1.

Let us apply this algortithm to the example depicted in Figure A1. Column weights are computed and row values are calculated using step 1. This leads to a reordering of the rows to the sequence {C, A, D, B, E} (see Figure A2). The row weights and corresponding column values are then computed according to step 2. The reordered column sequence is {1, 3, 6, 5, 4, 2, 7}, which is illustrated in Figure A3. Step 1 is then applied again to give rise to the row values indicated in Figure A4. Resequencing of the rows to {C, A, B, E, D} is carried out and the final table is generated (Figure A5). No further resequencing is indicated by application of the binary ordering algorithm.

FIGURE A1

Machine/Part assignment matrix
Step 1—Row value calculation

	Part Number							Row Value
Machine	1	2	3	4	5	6	7	
A	1				1			68
B			1		1			20
C	1		1			1		82
D		1		1			1	41
E				1		1		10
Column Weight	64	32	16	8	4	2	1	

In order to define a set of cells for the processing of the parts, we start with machine B, since it has only 2 parts to produce (3 and 5). The complete set of machines needed for these parts is {A, B, C}. We assign these machines to the first cell. Note that the full collection of parts passing through these machines includes part 6, which also requires machine E. If we define the second cell to be comprised of machines E and D, then the corresponding set of parts is {2, 4, 6, 7}. All but part 6 in this group are completely processed by machines E and D. Thus, the proposed floor layout consists of two cells: Cell I = {A, B, C} and Cell II = {D, E}; two dedicated part groups: Group I = {1, 3, 5}, Group II = {2, 4, 7}; and one part, 6, which visits both cells.

FIGURE A2

Step 2—Reordered rows and column value calculation

	Part Number							Row Weight
Machine	1	2	3	4	5	6	7	
C	1		1			1		16
A	1				1			8
D		1		1			1	4
B			1		1			2
E				1		1		1
Column Value	24	4	18	5	10	17	4	

FIGURE A3

Step 3—Reordered columns

	Part Number							Row Weight
Machine	1	3	6	5	4	2	7	
C	1	1	1					16
A	1			1				8
D					1	1	1	4
B		1		1				2
E			1		1			1
Column Value	24	18	17	10	5	4	4	

FIGURE A4

Step 4—Row value calculation

	Part Number							Row Value
Machine	1	3	6	5	4	2	7	
C	1	1	1					112
A	1			1				72
D					1	1		7
B		1		1			1	40
E			1		1			20
Column Weight	64	32	16	8	4	2	1	

Figure A5

Step 5—Reordered rows and termination condition

Machine	Part Number 1	3	6	5	4	2	7	Row Weight	Row Value
C	1	1	1					16	112
A	1			1				8	24
B		1		1				4	84
E			1		1			2	42
D					1	1	1	1	15
Column Value	24	20	18	12	3	1	1		
Column Weight	64	32	16	8	4	2	1		

C H A P T E R

4 Production Automation Technologies

Manufacturing automation is concerned with the application of mechanical, electronic, and computer-based systems to operate and control manufacturing processes. The purpose in automating the function of a manufacturing machine is primarily to replace manual labor and human operator control with automated operations and control, and thereby make manufacturing processes faster, more reliable, more accurate, more flexible, and less expensive. As a background to our discussion of manufacturing automation, we will first review the fundamentals of manufacturing operations and technologies.

4.1 Manufacturing Operations: An Introduction

Manufacturing firms are concerned with conversion of raw materials into finished products. Any discrete-parts manufacturing operation can be conceptualized as a hierarchy of basic parts manufacturing processes, such as cutting and forming, feeding parts to subassembly stations, and subassemblies to a final assembly stage. This scheme is represented in Figure 4.1. The final product is shown to consist of N subassemblies, some of which are used singly, while others are used in multiples. For example, an automobile requires only one chassis and engine, but it also requires four wheels and two headlights. The ith subassembly, in turn, may consist of i parts, some of which again are used singly, while others are used in multiples.[1]

Machine tools are the basic equipment of discrete-parts manufacturing (see Vajpayee [1995]). A machine tool is defined as a machine capable of producing itself (with the help of an operator). For example, a milling machine can be called a machine tool, since it can produce another milling machine, while a sewing machine is not a machine tool, since it cannot make another sewing machine. Machine tools perform two types of basic operations in shaping and forming: *metal cutting* and *metal forming*. The

[1]The "recipe" of a product, which specifies which parts and subassemblies are used, their usage quantities and their linkages (what goes into what), is summarized in a data file called the *bill of materials*.

Figure 4.1

Conceptual structure of manufacturing operations

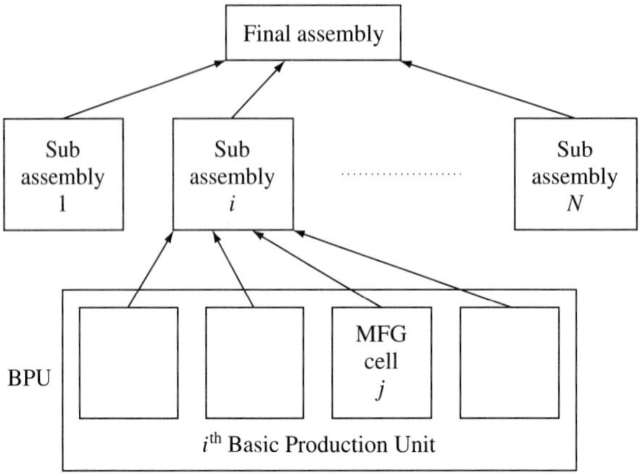

Source: Derived from "Computer Integrated Manufacturing, Vol. I: Revolution in Progress," R. Ayres, 1991, Chapman & Hall, London.

former involves removal of material to create the final shape, while the latter involves little or no loss of material in arriving at the final shape. Examples of metal cutting operations include: *turning* (using a *lathe*), *milling, grinding, drilling,* and *shaping.* All of these operations involve use of a hardened, sharp-edged cutting tool, as shown in Figure 4.2A.

Most metal forming operations use a die (or a mold) to define the final shape. There are four types of forming operations: *forging, shearing, casting* (or *molding*), and *powder forming;* the first two are illustrated in Figure 4.2B. In these operations, the final shape is obtained by applying pressure or heat or both. Forging includes *rolling, bending, punching, extrusion,* and *drawing.* In shearing, the metal is cut by deforming it beyond its breaking point. Casting involves pouring of molten metal (or plastic) into a mold, where the metal (or plastic) hardens to take on the form of the mold. In powder forming, metal or ceramic powder is compressed and hardened by sintering (by the application of heat and pressure).

The most commonly used machine tools are milling machines, lathes (or turning machines), and drills. Milling machines are versatile, as they can generate virtually any surface shape. Lathes can produce shapes with external rotational symmetry. Drills are for making round holes. The most common metal-forming machines are punch presses. A *machining center* is a multifunction machine that combines the capabilities of milling, drilling, tapping, and other machine tools. It also offers automation options such as an *automatic tool changer (ATC).* The main advantage of a machining center is that with one setup it can perform several operations on a variety of parts. This saves the cost and time in setting up work-pieces on different machines and in transporting them from one machine to the next.

The choice of machines to produce a given part depends primarily on matching the machine's capabilities to the part's processing requirements, which are determined by the part's complexity, shape, precision, and production volume. At a more managerial level, the choice of manual or automated mode of operation is determined by the economics of the situation and the above factors. Figure 4.3 provides some general guide-

FIGURE 4.2

Basic operations in metal cutting and forming

A. Basic operations in metal cutting.

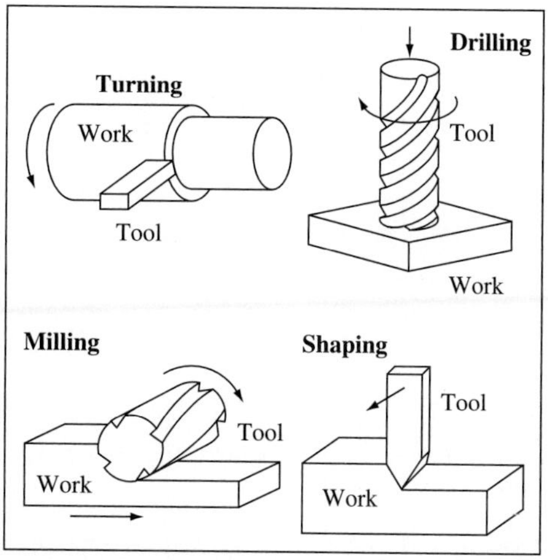

(a) The four basic machining processes can, between them, theoretically produce any contour on a work piece.

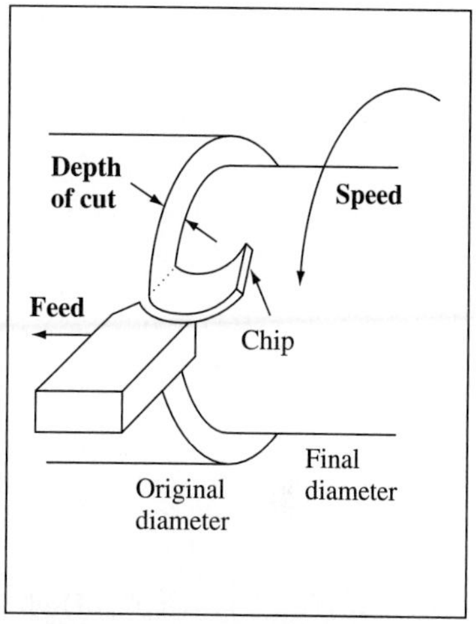

(b) In any of the basic machining processes, speed, and depth of cut determine productivity. The three variables are shown here for turning on a lathe.

B. Basic operations in metal forming.

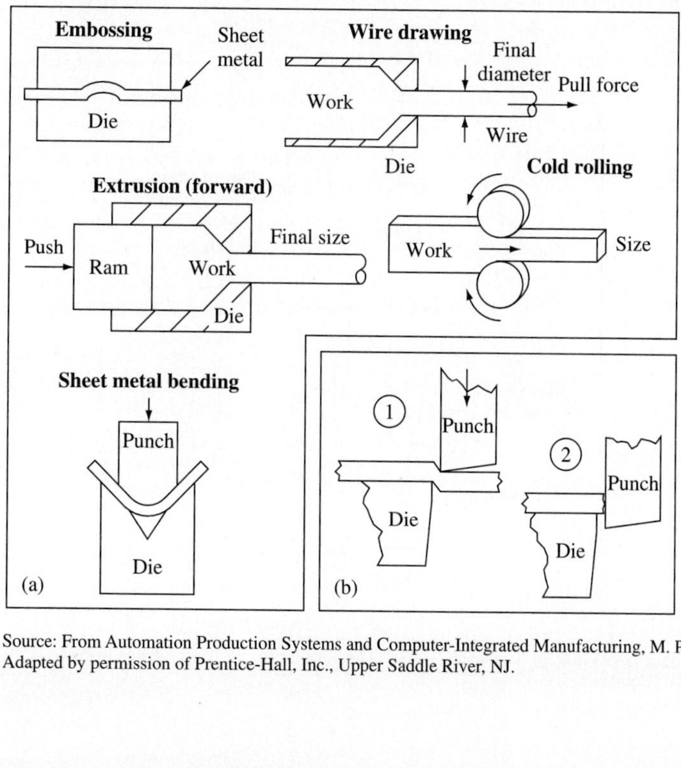

(a) Forging operations bend, squeeze or stretch metal, imparting new sizes or shapes or both.

(b) Sheering deforms metal beyond its breaking point, thereby separating one portion of a metal sheet from another.

Source: From Automation Production Systems and Computer-Integrated Manufacturing, M. P. Groover. Copyright © 1987. Adapted by permission of Prentice-Hall, Inc., Upper Saddle River, NJ.

FIGURE 4.3

Choosing Appropriate Technology

A. Low Complexity Shapes

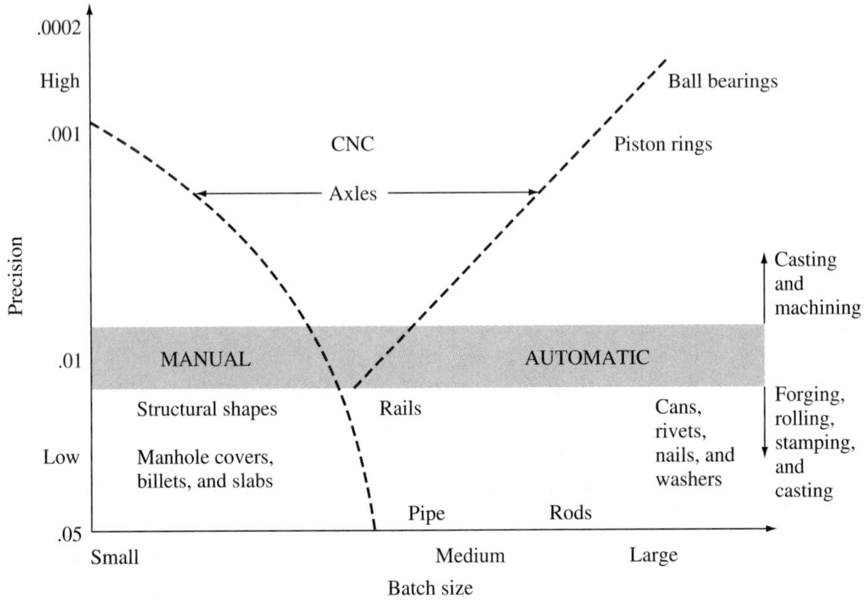

B. Very Complex Shapes

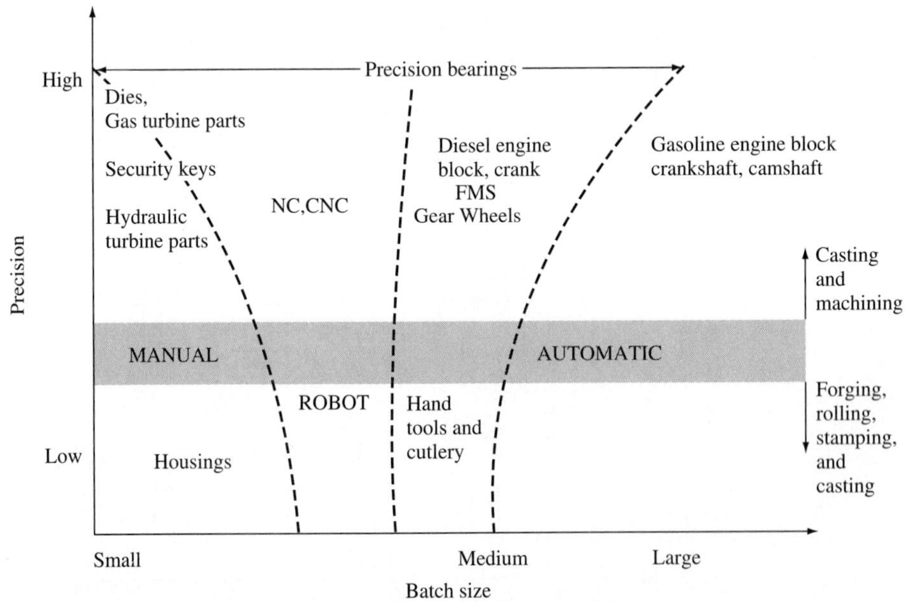

Source: Derived from "Computer Integrated Manufacturing, Vol. I: Revolution in Progress," R. Ayres, 1991, Chapman & Hall, London.

lines regarding the choice of machines. For small batch sizes, manual operation is generally preferred over an automated operation, since it is simply not worthwhile to mechanize tasks that are not carried out at large volumes. Metal cutting or grinding operations are preferred for high-precision parts, while for low-precision simple shapes, forming operations (casting, rolling, or stamping) are considered adequate. Among *automated machine tools* (discussed at length later in this chapter), the preferred choices range from use of stand-alone numerical control (NC) or computerized numerical control (CNC) machines for low-volume production to flexible manufacturing systems (FMS) for midrange volumes, and transfer lines for production of high-volume products.

4.2 Manufacturing Automation

Today's manufacturing automation consists of a collection of technologies, that can be divided into two, interrelated categories: *production automation technologies* and *infrastructure technologies*. Production automation technologies involve processes which directly add value in manufacturing. Infrastructure technologies act to enable the manufacturing process by supporting activities associated with information and materials management. Figure 4.4 provides some examples of both production automation and infrastructure technologies. This chapter will focus on the first class of technologies, and Chapter 5 will review the latter class.

Production automation technologies are concerned with the processes utilized in transforming material inputs into outputs. For example, a numerically controlled (NC) machine transforms a piece of metal into a machined part, while an automated assembly system creates a finished product by assembling the set of required parts. Both of these activities directly add value by transforming materials. The technologies used to automate these activities, therefore, belong to the production automation technology category.

Infrastructure technologies involve activities that deal with the acquisition, storage, processing or communication of information, and those activities that assist in the movement and storage of materials. Infrastructure technologies add value in an indirect

FIGURE 4.4

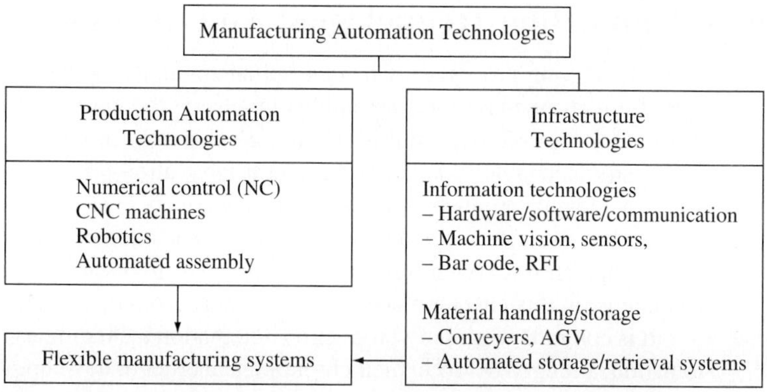

manner. Their role is to facilitate the flow of information and materials which are required in any manufacturing environment. They perform support functions that enable and enhance the value-adding role of the production automation technologies. Examples of infrastructure technologies related to information management include computer and communication hardware and software. Infrastructure technologies supporting handling and storage of materials include conveyers, automated guided vehicles, and automated storage and retrieval systems.

As indicated in Figure 4.4, a *flexible manufacturing system* is formed by combining infrastructure technologies, such as an automated materials handling system and communication/processing systems, with production automation technologies such as numerically controlled machines and robots. This flexible system has a higher potential for adding value than the combined potential of its components operating independently (i.e., the synergistic effect where the whole is greater than the sum of its parts).

An important characteristic of both production automation and infrastructure technologies is the manner in which they contribute to a firm's competitive advantage. Both technologies are available from suppliers on the open market. It is evident, therefore, that no sustainable competitive advantage can be gained from simply purchasing such equipment. The application and integration of both production automation and infrastructure technologies, however, are idiosyncratic to the firm. For example, the software programs that control the NC machines and the automated materials handling system of a flexible manufacturing system are unique to the product line of a firm. Moreover, the compatibility of a firm's product designs to the firm's automated manufacturing process may be high as a result of extensive investment in DFM activities. Therefore, a firm can gain competitive advantage by *selecting* appropriate technology for its environment and *integrating* the components in the context of its overall technology strategy.

We believe, moreover, that the effective utilization of production automation technologies, in particular, is a key success factor for generating competitive advantage. This follows, since these technologies are more directly involved in value creation. On the other hand, we view infrastructure technologies as a strategic necessity, or enabler, since their effective deployment will enhance the value of the production automation technologies. Indeed, the determination of the value of both classes of technologies is the central managerial question which motivates our discussions throughout this book.

4.3 Production Automation Technologies: An Overview

A manufacturing process may be conceptualized in terms of two stages: fabrication of parts from raw materials and assembly of parts to produce a finished good. An assembly process, as we note later in this chapter, is inherently more complex, and therefore is less easily amenable to automation than a fabrication process. This perhaps explains why, in today's production automation technology, we see that the automated approaches to fabrication are far better developed and adopted, as compared to that for assembly operations. Given this current status, our discussion of production automation technology in this chapter places greater emphasis on the automation of fabrication.

It is common to classify production automation technologies into three types of automation, arranged below in their chronological order of development in Figure 4.5A, B:

FIGURE 4.5

Manufacturing Automation Technologies: Volume-Variety Trade-off

A.

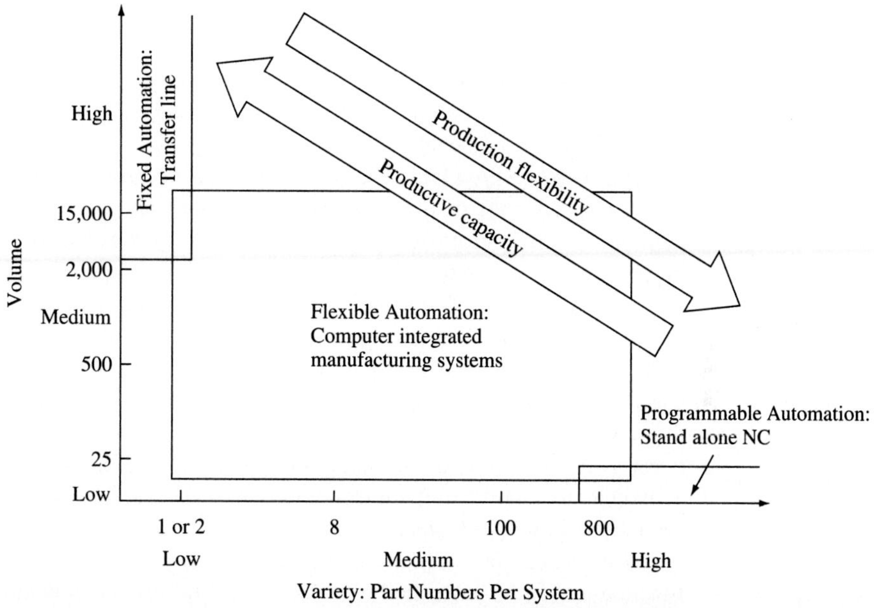

B.

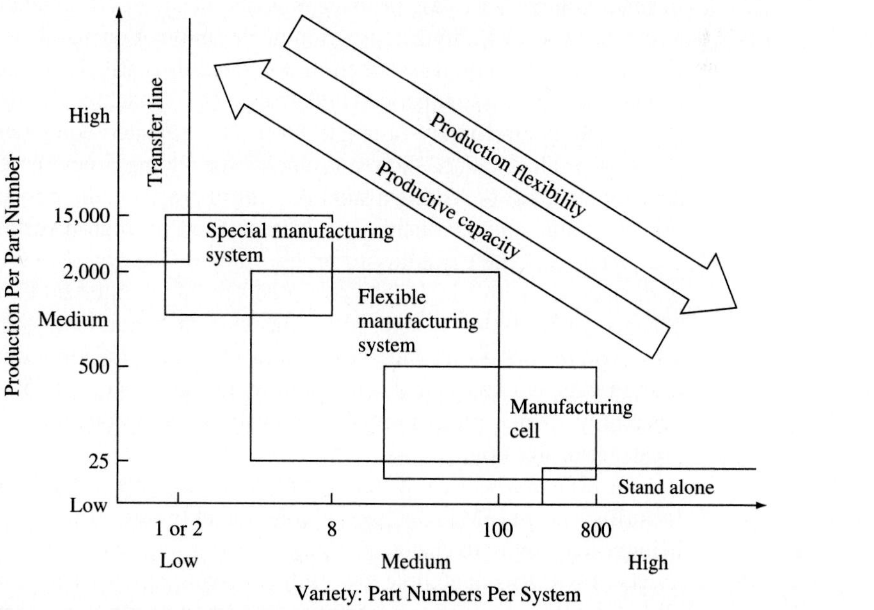

Source: Adapted from Teichoiz and Orr, eds. (1987)

1. *Fixed automation* (ca. 1900): used for high-volume, low-variety products.
2. *Programmable automation* (ca. 1950): used for low-volume, high-variety products.
3. *Flexible automation* (ca. 1960): used for medium-volume, medium-variety products.

Figure 4.5A illustrates the range of applicability (from the perspective of the equipment vendors—who claim to know) of the alternative types of automation with respect to product/part volume and variety.

Fixed automation is characterized by high-investment, special purpose equipment where the sequence of manufacturing operations and the equipment configuration is kept fixed. Fixed automation, also called Detroit-type automation, has been the centerpiece of the mass manufacturing strategy (noted in Chapter 1) since its evolution at the Ford Motor Company in the early 1900s. A prototypical example of fixed automation is the transfer line used in fabrication and assembly in automobile manufacturing. Fixed automation systems are highly efficient in producing the product for which they were specifically designed and configured. But these systems are also highly inflexible in that they cannot accommodate a major product change (i.e., with respect to lost time and expenses of retooling and line setup change). Fixed automation systems are, therefore, economically justified only in situations where the product volume is very large and the product variety is very small.

Programmable automation involves general-purpose production equipment whose functioning can be controlled by a (computer) program that details a set of instructions on how each operation is to be carried out as well as the overall sequence of the operations. Inflexibility of fixed automation (i.e., its inability to produce a wide range of parts in small volumes) was the primary impetus for the development of programmable automation. The evolution of programmable automation parallels the development of computer technology. It has been in use since the early 1950s and forms the basis of technologies such as numerically controlled (NC) machines and industrial robots. Programmable control allows a single piece of automated equipment to produce a wide range of products, since the requirements for adding a new product to the machine's repertoire consist of writing a suitable control program and possibly adding some tooling. Programmable automation is characterized by high-investment, general-purpose equipment, and the capability to produce a wide variety of low-volume products economically and with a high degree of precision. It is important to note that a production changeover from one product to another in a programmable automation environment also involves short, yet measurable loss of production time due to reprogramming, changing the workpiece and equipment setup, and process validation. Accordingly, it is customary to use programmable automation for production in small batches (e.g., greater than size one).

Flexible automation, developed in the mid-1960s in the form of a flexible manufacturing system (FMS), builds on programmable automation, with the main difference being its capability to change programs, and tooling setups, with almost *no loss of production time.* This negligible downtime in changing over from one product to the next allows flexible automation systems to engage in continuous production of a variety of products without requiring that each product be necessarily produced in a batch. Flexible automation is characterized by high-investment, custom-engineered equipment that typically includes computerized control and automated materials handling. Key features of flexible automation are its flexibility in accommodating a medium level of variety in product designs and its efficiency in producing medium volume production of individual products (see Figure 4.5A). Technological advances (which are discussed later in this book) such as numerical control, automated materials handling, and automated tool management have made flexible automation feasible and economically justifiable in the last three decades.

The term flexible automation as used above indicates a solution for a large range of mid-volume, mid-variety production environments (see Tempelmeier and Kuhn [1993]). In practice, flexible automation has been implemented with varying levels of flexibility, depending on the particular needs of the application and the capabilities of the manufacturing personnel to adjust to "flexibility" without causing excessive downtime for changeover. As shown in Figure 4.5B, three categories of flexible automation have been identified:

- *Manufacturing cells:* This category has the most flexibility. A manufacturing cell produces a large variety of parts, and it consists of one or a very small group of NC machines that may or may not be linked together with a material handling mechanisms.
- *Special manufacturing systems:* Also called a *flexible transfer line.* This is the least flexible category suitable for relatively higher production rates. The system is called special, since the operation sequence is specialized (i.e., unique and fixed for a narrow range of parts) and the machines are linked using a "fixed path" type of material handling system. Any change from one set of parts to another requires a complete shutdown of the system, causing severe interruptions.
- *Flexible manufacturing systems:* This category lies between the above two categories in terms of its volume and variety capabilities. Under the control of a computer, a material handling system is used to move parts, either sequentially or at random, to appropriate NC machines.

The above three categories overlap each other in terms of their levels of flexibility and domains of applicability. Our goal here is to understand the capabilities and limitations of flexible automation in general, rather than to concentrate on the specific differences between these categories. Hence, our discussion of flexible automation will focus on flexible manufacturing systems as a representative of flexible automation technology.

Fixed automation (i.e., *transfer lines*) remains an important element of today's manufacturing technology. Development of fixed automation can be traced to the automobile industry in the early 1900s. The concept was implemented and perfected with great benefits at the Ford Motor Company, and it has served as the cornerstone of mass manufacturing since then. Transfer lines consist of several machines, or workstations, linked together by work-handling devices that transfer parts among the stations. As shown in Figure 4.6, a workpart enters one end of the line and the processing steps are

FIGURE 4.6

Four station transfer line

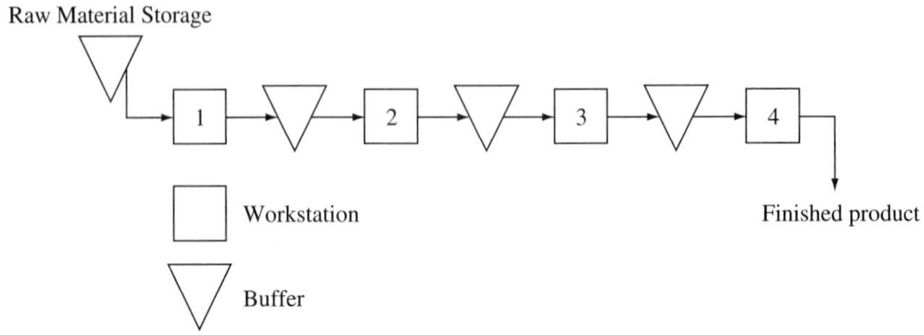

performed sequentially as the part moves from one station to the next. It also is possible to include buffer storage or inspection stations in the flow line. Transfer lines provide the most appropriate production method in cases of relatively narrow and stable products with high production volumes. The advantages of a transfer line include low labor cost, reduced in-process inventory, high production rates, and reduced travel of parts among stations. The flow pattern of transfer lines can take two forms: *in-line* (i.e., straight line) or *rotary*, with the latter used mainly in cases of smaller workparts requiring a few workstations. *Balancing flow* and *synchronization of transfer movement* are two key concepts useful in analyzing and designing a transfer line. Accordingly, the principal design questions involve allocation of work to each workstation on the line, selection of appropriate equipment and personnel for each station, and selection of a line speed.

As stated earlier, the inflexibility of a transfer line to accommodate changes in product design is one of its primary disadvantages. This factor is also the *raison d'etre* for the emergence of flexible automation. Accordingly, our emphasis in this chapter will be on programmable automation and flexible automation. Specifically, we will review technologies such as numerical control machines, computerized numerical control machines, programmable logic control systems, industrial robots, automated assembly, and flexible manufacturing systems. These automation technologies will be discussed in terms of the basic concepts underlying their operation, their advantages and disadvantages, and the requirements for their implementation. A managerial perspective will be adopted throughout.

4.4 Programmable Automation Technologies for Fabrication

A fabrication operation involves transformation of raw material into products according to a predefined specification (in terms of geometry and other physical attributes). An automated system accomplishes this transformation by generating a series of commands to (1) direct the supporting material handling system to transfer the right material and tooling to the machines at the right time and (2) control the movement of machine tools to transform material into the required shape.

Programmable automation systems today range from simple numerical control (NC) and computer numerical control (CNC) machines used primarily for machining and fabrication application, to programmable logic controllers (PLC) used for controlling a multistep manufacturing process involving a number of different machines. We consider each in turn in this section.

4.4.1 Numerical Control Machines

One of the early developments in manufacturing automation was automated tool control, a concept whose origin can be traced to the early 1800s when Joseph Jacquard developed a card-controlled loom for the production of weaving patterns. These looms used a series of cards with punched holes that contained a set of instructions for controlling the loom and thereby producing the desired pattern. Modern automated machines evolved through the work done on automating milling machines and drill presses during and soon after the World War II. These automated tools are based on the concept of *numerical control,* wherein the movement and speed of machine tools, material feed rates, etc., are directed through the use of numbers or symbols which specify instructions for control. Initially these instructions were stored on a paper tape. NC machines of a wide variety were in use by the mid-1960s. The technology underlying machine control has developed considerably since those days. With the evolution of computer technology, the task of machine control has naturally shifted to computers. The general trend is toward creating a fully automated factory where all the elements, such as machine tools working in machine cells, material handling and storage, assembly lines, inspection, etc., are controlled automatically.

The goal of numerical control is to create a manufacturing environment where a machine tool, through automated control, can perform the same range of activities as can be performed by a skilled human operator in control of that tool. For example, consider a NC drill press. The idea here is to provide instructions for controlling the position and motion of a machine tool. Certain other functions, such as the flow of coolant for metal cutting, can also be controlled. For example, by interpreting and executing a predefined program of instructions, the machine control unit of a NC drill press is able to send control commands to a servo motor for moving the drill head 20 inches along the X-axis and lowering it for drilling a 1/2 inch hole into a plate.

The principal advantages of numerical control include:

- *Increased manufacturing flexibility:* Requiring only a change of program, the NC machines make it easier to change over to a new job.
- *Improved quality:* Reject rates are lowered because NC machines can produce parts with greater accuracy and repeatability. A substantial reduction in scrap rate and inspection requirement is also observed.
- *Reduced nonproductive time:* NC can reduce the nonproductive time required for machine setup, tool changes, etc.

In general, the use of NC machines is considered appropriate in the following cases:

- Parts are processed frequently and are required in small batches.

- Part geometry is complex, requiring many operations and removal of much metal.
- Parts require close tolerances.
- Parts are expensive and mistakes in their processing can be costly.
- Parts requiring frequent changes to engineering design.

The principal disadvantages of NC machines involve the effort of higher skilled labor to write the programs and the inability of the machines to react to unforeseen contingencies (such as tool wear, material variation, vibration, etc.). Those latter difficulties can limit the level of precision attainable by NC methods.

Hard-Wired NC Machines

The first NC machine demonstration was held at MIT in 1952, and the first successful industrial application took place in 1957. These early NC machines were *hard-wired*, i.e., the program of instructions was actually wired permanently into the machine tool itself. This naturally reduced the range of operations that the machine could perform. In many hard-wired NC machines, the operator provided some direction, such as dimension and depth of a drill hole or the points at which a metal sheet is to be bent or cut, through a set of dial or digital settings. The hard-wired NC machine is less expensive as compared to a general purpose, fully programmable modern NC machine, but it also has some limitations. First, it is less flexible in terms of the range of operations it can perform, and, second, its dependence on human direction can also lead to lower conformance to specification in producing standardized parts. This makes it difficult to integrate these machines into a line-oriented manufacturing process that depends on parts standardization.

Tape-Programmed NC Machines

The next generation of NC machines appeared in the early 1960s, and these machines used the concept of stored programs. In these *tape-programmed* NC machines, the medium of punched paper tape was used to store the program of instructions. This eliminated the need for hard wiring, since a different tape with a different program of instructions could be used to let the machine follow a different process plan with new specifications. The tape-programmed NC machines are more flexible in being able to change workheads, and, consequently, allow changes in engineering design. They can also accommodate production schedule changes. These machines have a reduced human input and have an ability to achieve closer tolerances leading to improved product quality. Figure 4.7 shows the functioning of a tape-programmed NC machine. A paper tape containing the program of instructions is fed into a paper tape reader. The reader decodes the program and sends commands to the machine control unit which controls the movement and speed of the machine tool.

The benefit of increased flexibility also comes with a cost of increased difficulty in NC programming. Programming requires a detailed analysis of each manufacturing operation to decompose it into a series of small steps representing machine actions. A control word for each action of the machine tool is chosen, and then the entire NC program is put together. Presently, there exist a number of high-level, easy-to-use NC program-

FIGURE 4.7

Tape-programmed NC machine

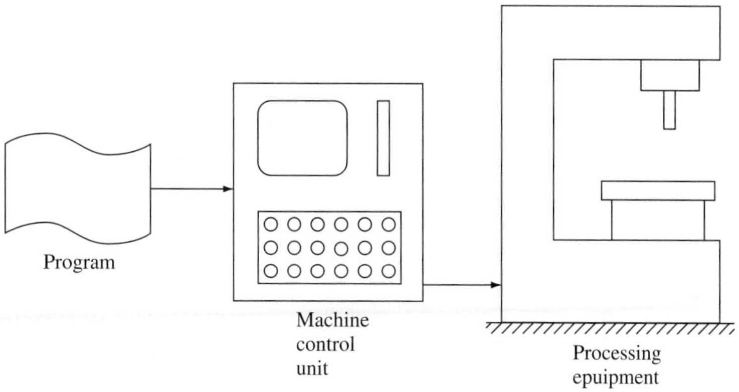

Basic components of an NC system

ming languages, such as *APT (automatically programmed tools)* and *COMPACT II,* that have built-in control words and structures for representing movement and positioning of machine tools. Still, the task of NC programming is too specialized for a typical factory worker to learn. Consequently, the NC machines that lead to a reduction in the number of required machine operators also lead to an increase in the number of skilled NC programmers.[2]

The evolution of NC machines from hard-wired machines to paper-tape machines to computerized NC machines, to be discussed later, is depicted in Figure 4.8. We consider flexibility along two dimensions: *work-head flexibility* and *instruction set flexibility.* Work-head flexibility allows automatic and frequent tool changes, and can accommodate changes in machining operations to be performed on a same job. The flexibility in instruction sets allows a NC machine to work on parts having radically different engineering designs. All that is needed is to change the paper tape (or download a different program in case of CNC machines, to be discussed later) to invoke a different program of instructions.

NC's share of machine tools in the US metal working industry, as of 1989, is given in Table 4.1.

4.4.2 Computer Control of Machine Tools

The evolution of hard-wired and tape-programmed NC machines took place in parallel with the development of computer technology. In the 1950s, computer systems were too costly to permit direct computer control of tools. By the mid-1960s, costs had reduced sufficiently to a level where the use of computers in controlling machine tools was suffi-

[2]It should be noted that with modern technology, it is possible today to train operators to program NC equipment with CAM systems which contain graphical interfaces and object-oriented programming procedures.

FIGURE 4.8
Evolution of NC Machines

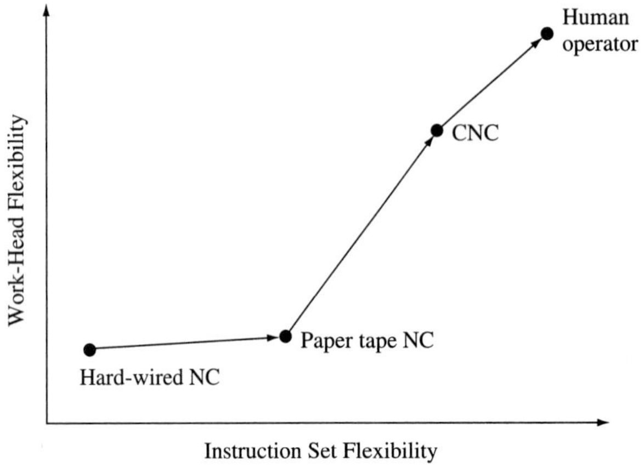

TABLE 4.1 NC Shares of Machine Tools in U.S. Metal Working Industry (1989 Data)

Machine	*New Installations (1984–1987)*	*Installed Base (1989)*
Turning	52.0%	18.3%
Boring	52.0%	22.2%
Drilling	8.6%	3.6%
Milling	26.9%	11.3%
Grinding	8.4%	2.9%
Machining centers	30.9%	9.2%
Total metal cutting	27.1%	10.5%
Total metal forming	14.6%	3.9%
Total machine tools	25.2%	9.6%

Source: American Machinist (1991)

ciently attractive. The first such use was in the form of **direct numerical control (DNC).** DNC was used to control a group of NC machines directly through real-time electronic transmission of numerical control commands from a central computer. A schematic diagram of DNC is given in Figure 4.9. DNC eliminated the need for fragile paper tapes, relatively expensive paper tape readers, and a part of the hard-wired control equipment. DNC took advantage of the storage capability of computers in mass storage of NC programs. Advances in telecommunications technology have allowed the remote location of computers away from hazardous factory floor environments. The main disadvantage of

FIGURE 4.9
DNC machine

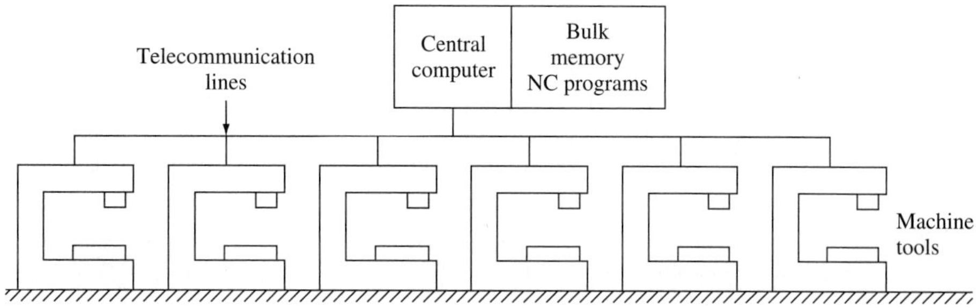

DNC machines was that whenever the central computer went down, all the expensive NC machines being controlled by that computer were also rendered inoperable.

With the introduction of microprocessor technology, the cost of computers had declined by an order of magnitude by the early 1970s. The on-board placement of microprocessors for controlling machine tools therefore became economically feasible and desirable. This use was termed **computer numerical control (CNC).** The essential difference between DNC and CNC is that in DNC one remotely located, large computer controls many machines, while in CNC one small computer is dedicated to one machine and is located very near the machine. With the growth of CNC machines, the use of DNC machines declined. It is interesting to note that today we use the term DNC to mean **distributed numerical control,** in which the NC programs are downloaded to CNC machines from a central computer (server), typically a mainframe. In this configuration, each machine has a dedicated computer control unit, and hence the loss of productivity due to central computer downtime is dramatically reduced.

The CNC machines add several advantages over and above those afforded by NC machines. The overriding benefit here is the increased ability to control machine tools and processes with greater accuracy and repeatability. Other important advantages are:

- *Greater manufacturing flexibility:* CNC allows on-board storage of a large group of programs. These machines are also more reliable and accurate than paper tape NC machines. Programming constructs, such as user defined subroutines, can increase the speed at which certain geometry can be coded (e.g., canned subroutine to make a circular cut with a given radius).
- *Increased potential for integration within CIM:* The existence of an on-board microprocessor makes it potentially easier to integrate CNC machines in CAD/CAM applications and in automated manufacturing systems. Thus, CNC has become an essential building block of computer-integrated manufacturing.
- *Greater flexibility for entry and editing of NC programs:* In CNC machines, the entry and on-line editing of programs are much easier than that in paper tape NC machines. With a simple command it also easy to switch from the English to the metric system, a capability necessary for global manufacturing.

4.4.3 Programmable Logic Controllers (PLC)

In the case of a machine tool process, the control system operates at a low level, i.e., it functions to monitor and control the position and speed of tools. Many other processes, such as fluid handling or assembly-line control, require higher level control for coordination and sequencing of several different machines or subprocesses. Such control is typically provided by technology referred to as *programmable logic controllers (PLC)*. PLCs are solid state devices with programmable memory, designed to control industrial equipment. A PLC is essentially as an electronic switch that can turn a machine on or off. The technology of PLCs is based on microprocessors, and it provides logic functions, such as counting, timing, sequencing, previously performed by electromechanical relays and rotary switches.

Programmable logic controllers were first introduced in 1959, and are now used in almost all automated systems. Applications of PLC include coordination of activities within a manufacturing cell, preheating and mixing of chemicals in a batch reactor, parts assembly in an automated assembly line, distribution of crude oil in a pumping station, and ride control in an amusement park. For instance, consider the example involving the use of PLC in a chemical reactor. A PLC starts the process by setting its step counter to one. A valve is opened and the first chemical starts to flow into the reactor. A weight sensor is active and is being monitored by the PLC. As soon as the required quantity of chemical is collected in the reactor, the PLC increments its step counter to two. As a part of step two, the valve closes and a heater is activated for preheating the first chemical to a required temperature and so on. Thus, the chemical process continues under the overall control of a PLC. Most PLCs are designed to interpret simple binary events, including on/off, high/low, hot/cold. Some advanced PLCs can also interpret analog information, such as voltage, temperature, pressure, etc.

The major advantages of a PLC are:

- Flexibility.
- Ability to coordinate and sequence activities of factory floor equipment in a real-time, multitasking environment.
- High degree of reliability, low maintenance costs and high process uptime.
- Ability to operate in hazardous conditions.
- Relative ease of programming and troubleshooting.
- Small size and low cost.

With the rapid evolution and acceptance of personal computers, it is reasonable to ask if these general purpose computers will replace PLCs on the factory floor. Factors that distinguish a PLC from a general purpose computer include: (1) PLCs can be directly wired to machine level input/output devices such as motor starters, limit switches, digital decoders and other sensors and (2) PLCs are optimized for one task only, i.e., to scan and interpret data from input devices and, based on programmed instructions, to send appropriate control signals to output devices.

Trends in PLC technology are mainly driven by advances in microprocessor and computer technology. For example, the current trends in PLCs are toward the use of custom chips, larger memory, fault tolerant designs, and more sophisticated programming languages.

4.5 Robotics

Robots provide another example of programmable automation. Unlike the image of a human-like robot, popularized by the entertainment industry, a modern industrial robot can be better described as a mechanical arm capable of performing a variety of material handling, metal working and other tasks. Both robots and NC machines are similar in their generic capabilities in that they are programmable and can move and accurately position tools. The main difference, however, is their emphasis on end uses. Robots excel in the grasping and moving of material or positioning of tools, while NC machines are designed mainly to perform metal cutting and shaping tasks. A definition of *industrial robot* offered by the Robotics Institute is:

> An industrial robot is a *reprogrammable, multifunctional manipulator* designed to move materials parts, tools, or special devices through variable programmed motions for the performance of a variety of tasks.

A robot is generally preferred over a human operator if the work environment is hazardous or unpleasant, and/or if the part or tool being handled is too heavy for human beings. Other, perhaps more appropriate, roles for a robot (Paul [1992]) include working under the control of and providing feedback to a human operator in performing a task in a location where the operator may not be present (e.g., on the surface of the moon or underseas). Today, we typically find robots in use in a variety of more mundane industrial applications, including material transfer, machine loading, spray painting, spot welding, arc welding, inspection, and assembly. Robots are now increasingly used as components in flexible manufacturing systems.

The principal components of a robot system include the robot (including the manipulator and joints) and end-effectors, the controller (computer), and the drive system (electric motor or hydraulic power system). An example of a robot that can mimic the movement and rotation capabilities of a human arm and hand is shown in Figure 4.10.

To ensure that a robot is suitable for a given task, a manager should evaluate it on the following performance criteria:

- *Range and speed:* The size of the workspace that a robot can access and the speed with which a task can be performed.
- *Payload:* The amount of weight (including the weight of the end-effector) that can be carried by the manipulator while maintaining the rated precision.
- *Precision:* This refers to a combination of three capabilities—resolution (i.e., the smallest possible incremental motion), accuracy (i.e., the capability to

Figure 4.10
Industrial robot

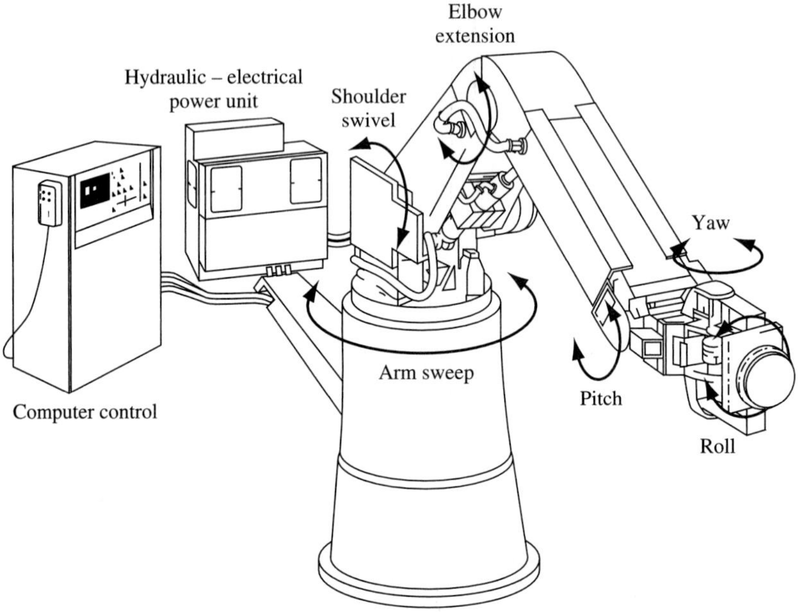

position end-effector at a desired position), and repeatability (i.e., the ability to repeatedly visit the same position).

- *Adaptability:* The ability to be programmed to respond to a variety of conditions.

In implementing a robotic system, the first task is to program the robot. There are essentially two alternatives: (1) use of a device, such as a teach pendant, to lead a robot through positions it needs to visit and tasks it needs to perform. These positions and tasks get recorded in the robot's memory, which then can be played back as needed in operating the robot, and (2) use of a special-purpose robot programming language. The second approach is more flexible and powerful, and necessary in cases where a robot needs to interface with other automation elements, or respond according to the information provided by various sensors.

Cost justification of robots (as well as for many other automation systems) has rested on their use as a labor replacement technology. While this is an important consideration, myopic emphasis on labor replacement usually results in a less successful system. A robot, to be fully effective, must interface with other elements in the manufacturing system, such as an NC machine. Thus, the collective impact of a robot and related automation elements on the performance of the whole system ideally must be evaluated together. In implementing a robotic system, it should be kept in mind that the ultimate cost of the system can be three times as high as the cost of a robot itself (due to fixtures and software support).

The population density of robots in eight industrialized countries is given in Figure 4.11 and Table 4.2.

FIGURE 4.11

Industrial robots: population and population density

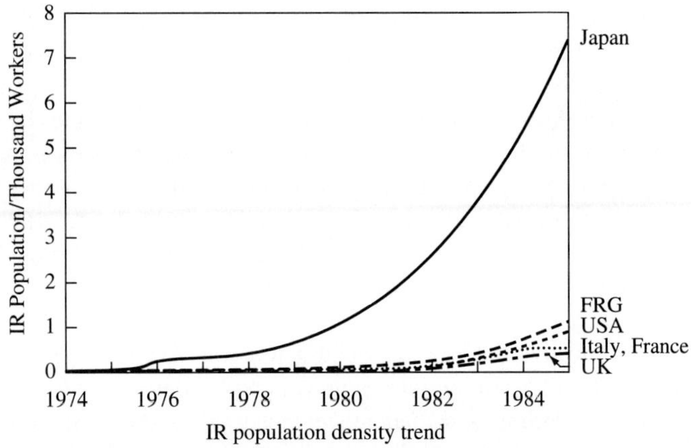

TABLE 4.2

As of 1984 *Country*	*Industrial Robot Population*	*Manufacturing Employment (000)*	*Robot Population Density (Per 000 employees)*
Japan	67,300	12,120	5.55
USA	14,550	19,378	0.75
UK	2,623	5,506	0.48
Germany	6,600	7,516	0.88
France	2,750	4,742	0.58
Italy	2,585	4,205	0.62
Belgium	860	764	1.13
Sweden	1,900	533	3.57

4.6 Flexible Manufacturing Systems

Flexible manufacturing systems (FMS) can be best described as totally automated miniature factories capable of producing a wide variety of components with almost no human intervention. A key feature here is the tight integration of information and decision technologies with the material processing and handling technologies.

A large number of alternate FMS definitions can be found in industry. This variety in definitions simply reflects the fact that the components present in an FMS depend on the ultimate user's viewpoint and on the way the FMS is to be used. All of the definitions, however, contain certain common elements and concepts—NC machine tools linked together with an automated materials handling system, centralized computer control, and randomized loading of parts.

An FMS is made up of various hardware and software components. Hardware components are the tangible elements, such as the NC machine tools, the automated materials handling system (consisting of robots, automated guided vehicles, etc.), the pallets and fixtures, the system for central chip removal and the distribution of coolant, the tooling systems, the part cleaning system, the measurement devices and the central computer system. Software elements are invisible and intangible. In many respects, however, they are more critical for the successful implementation of an FMS, since they determine the throughput of the system and provide the flexibility to respond to contingencies. The software elements include NC programs, material handling and traffic management routines, systems for tool control, part control programs, a resource database, and the higher level business systems for production control, scheduling, costing and invoicing.

Theoretically, an FMS can manufacture a wide variety of dissimilar parts, produce them one at a time, or simultaneously, in any order as needed by the customers, with negligible downtime and changeover costs. To accomplish this efficiently, an FMS needs several types of flexibility:

- *Variety of product mix* that the system can simultaneously manufacture.
- System's *adaptability to design, production and routing changes* (but not necessarily to volume changes).
- Ease with which the system can *change over* from making one part to another.

Since the introduction of the first FMS in the mid-1960s, the number of FMSs installed worldwide has grown to roughly 370 (Reza [1991]). About half of them are in the United States and Japan (both having roughly an equal number), and most of the remaining FMSs are present in Europe. Given the experience with these FMS installations, four benefits of FMS have typically been noted (Luggen [1991]):

- Inventory reduction of 50 percent to 60 percent.
- Direct labor savings of 30 percent to 50 percent.
- Increased asset utilization approaching 80 percent to 90 percent.
- Floor space reduction of 40 percent to 50 percent.

Analysis of 1989 data on FMS installations in 17 countries (Tchijov [1992]) shows that more than half of all FMSs cost less than \$3 million, while one-quarter cost more than \$7 million. The frequency distribution of payback times for FMSs was found to be close to normal, with a mean of about three years. The most typical operational mode of FMS is as follows: three shifts a day, including one unmanned night shift. Of the 276 machining FMSs, 66 percent are used for three shifts a day, 30 percent between one and three shifts, and only nine are used for one shift a day. Of 131 FMSs, 72 percent are autonomous enough to be used during one shift in an unmanned mode, 21 percent are used automatically for less than one shift, and 5 percent are used automatically for longer than two shifts.

FMS flexibility is one of the most important features and advantages of a system. It is usually measured by two indicators: *average batch size* and the *number of product variants supported.* The distribution of the batch sizes shows that the most FMS technologies are really used for small-batch production. One-third of the systems produce parts by batches of 10 or less units, and another third by batches of 11 to 50 units. However, 12 FMSs (or 5 percent) are used for large-batch production, producing parts in batches of more than 1,000 units. Another indicator of flexibility—the number of product variants—reveals a rather moderate flexibility for 30 percent of the FMSs which produce no more than 10 different parts. About 48 percent of the systems produce from 11 to 100 parts, and the "super flexible" FMSs (3 percent of the total) are used for the production of more than 1,000 types of parts.

FMS is not without its inherent limitations and problems in implementation. Many of these result from unrealistic expectations as to what an FMS is and what it can do, and a misunderstanding about the nature of the desired flexibility. As reported by Jaikumar (1986), American manufacturers, by and large, have used FMS in an "inappropriate manner"—they are used for producing a small variety of parts at a larger volume each. A manager, therefore, must critically examine the assumption underlying FMS and its limitations before embarking upon its installation:

- FMS is not a cure-all for all productivity problems, and, in particular, the problems of a poorly managed and organized plant. Information flow must be efficient before computers are introduced, and material flow must be efficient before automated materials handling is introduced.
- FMS can be prohibitively expensive, many costing in excess of $10 million. More modest approaches, such as advanced CNC machines with automatic work changeover, should be considered first before a firm invests in an FMS. An incremental, building-block approach to introducing production automation technology generally works the best. (Refer to Chapter 8 for our discussion of justification and implementation.)
- An FMS is not a standard, commodity-type of machine tool. Hence, the purchaser should see the installation of an FMS as a joint venture with the supplier, which can require up to two years in order to make the system error free and fully operational.
- The hardware components of an FMS are available to anyone in the open market, and, therefore, long-term competitive advantage will not be obtained through their acquisition. Competitive advantage through an FMS, however, is possible by developing efficient and effective software components that are unique to the firm. A manager must pay closer attention to the creation of FMS control software and to the management of the firm's intellectual assets which are embodied in the skills and knowledge of employee teams that design, install, and manage the system.
- Acquisition of an FMS should be viewed as an investment in human capital. Indeed, the most successful users of FMSs, the Japanese manufacturers of FMS

products (such as Yamazaki and Hitachi Seike) understand that the key to successful implementation is learning to organize their manufacturing personnel into teams that both install and operate the systems.

4.7 Automated Assembly

The task of assembly can be conceptualized as a series of steps:

1. Collecting the necessary parts at the assembly station.
2. Positioning an appropriately selected part to serve as the "foundation" for assembly build-up.
3. Repetition of assembly steps, wherein parts are successively positioned and attached to the foundation part, or partial assembly, until such a time that all remaining parts are exhausted and the assembly is complete.

The design of an automated assembly operation involves finding answers to nontrivial questions such as: What part should be chosen as the foundation part? In what order should the parts be assembled? In attaching a part to the partial assembly, how should the part be positioned? Should the part, or the partial assembly, be moved and rotated? How should the system respond to failures (e.g., jammed part, material stockout, or machine breakdown)? These issues make the machines that automate the assembly process extremely complex, requiring highly intricate drives and sensors. The nature and the scope of contingencies that automated assembly must deal with are also enormous when compared to metal cutting. As a result, computerized control and automation of assembly operations are generally much more difficult to accomplish. In manual assembly operations, in contrast, there is no need to answer these questions explicitly, since human judgment can intuitively find satisfactory solutions on a real-time basis. Automated machinery possesses no such intuition, and the dexterity and flexibility of machines is far less than a human hand. Therefore, a minutely detailed specification is needed for defining an automated assembly process. In addition to the above discussion on the inherent difficulties of accomplishing automated assembly, it would be instructive to compare it with the automated operation in the fabrication of parts:

- Parts fabrication involves a greater use of general purpose machinery. Turning, milling, grinding, or drilling machines can fabricate a very large number of parts of varying sizes, shapes, and functions. Thus, by automating the operation of these general purpose machines, a wide range of parts can be automatically fabricated. The assembly machines, on the other hand, are more special purpose and can deal with only a limited range of applications. For example, a machine that can assemble small diesel engines cannot assemble large locomotive engines, and vice versa. This idiosyncratic nature of assembly operations is also a hindrance to the development of automated assembly machines.
- In fabrication, a part can be inspected and its quality can be assured with relative ease. What is normally needed for quality assurance is the use of some

> sensor technology for measurement and verification of a part's dimensions. In contrast, a mere dimensional check of assembled parts is usually insufficient to assure quality at this stage. What is needed is higher level functional and performance testing, which is much more difficult to accomplish automatically, is slower, and is more expensive (e.g., function tests and "burn-in" of a PC). This difficulty in inspection and quality assurance has also slowed the development of the automated assembly operation.[3]

In designing the automated assembly process, it is tempting to mimic the motion of human hands and keep the assembly position stationary, while bringing all parts to the assembly. This approach, however, is usually infeasible and uneconomical in automated environments. Instead, the preferred approach is to combine linear and rotary actions at a series of assembly stations, while the partial assembly moves on a conveyer from one station to the next. Each step is seen as an operation involving the action of a "tool" against a "piece," where the piece is the partial assembly and the tool consists of the part to be attached and the mechanism that performs the attachment. As we have seen in our discussion of DFM, the design of a product should be adapted to the requirements of the manufacturing and assembly processes. Indeed, as illustrated by products such as the IBM Proprinter and the GE dishwasher, design changes required for automated assembly and resulting improvements in process performance can be very dramatic.

In an automated assembly operation, part positioning is usually accomplished through mechanical means, such as a hopper or a feed system (although automated vision systems have also been used in some cases). The assembly tool typically consists of a motion arm and an end effector (i.e., a gripper or a tool). The motion arm positions the tool and the piece for subsequent attachment of a part to the piece. The operational control at each station, involving synchronization and sequencing of operations, is accomplished through either a mechanical or an electro-mechanical mechanism. PLCs are also increasingly used for providing this operational control and sequencing. The programmability of a PLC is found to be attractive for making the individual assembly stations flexible. When the required level of flexibility and control is significantly high, the option of using a robot to accomplish assembly operation becomes highly attractive. In recent years, the use of robots in assembly systems has increased at a rapid pace (especially in the consumer electronics and communications industries).

Recently, a hybrid approach to automated assembly has appeared (Bylinsky [1994]). Under this so-called "soft approach" to automation, an automated system brings kits of parts (via automated material handling) to human operators at assembly stations. These operators are then prompted with digitally displayed instructions, and assemble (or otherwise add value to) the product by using the parts in the kit. Upon completion of the assembly, the worker initiates its transfer (via a conveyor) to the next

[3]It is interesting to note, however, that automated testing equipment is used routinely in electronics manufacturing at both the device and circuit board levels. These testers are themselves examples of sophisticated, programmable machinery.

work station. Such systems, used by IBM, Motorola, and others, combine the best features of human operators (to position parts and adapt to change) with the capabilities of automation (to move material and information).

4.8 Managerial Considerations

A direct consequence of computer-based process control is a reduced need for direct labor. As Ayres (1991) reports, a start-of-the-art FMS can cut direct labor requirements by 80 percent to 90 percent as compared with a conventional machine shop, and by 50 percent as compared with a group of stand-alone NC machines. As CIM technology is implemented at the plant level, direct labor requirements are expected to fall in similar proportion in other production activities, including assembly.

Higher machine utilization and reduced inventory will lead to higher capital productivity. As shown in Figure 4.12, as of 1980, a stand-alone NC machine was being utilized at a far higher rate (20 percent) than a conventional machine. Utilization of FMS was even higher (30 percent). It is expected that this figure will double to 60 percent utilization by the year 2000.

It is important to recognize the broad implications of production automation technologies. A complete change in the thinking of management and labor is required for production automation technologies to succeed, since they have implications at both the micro level (the organization of work) and at the macro level (the efficient size of a firm, product mix, and competition and industry structure).

The evolving production automation technologies have several important implications:

- Increasing the level of *information intensity* of processes.

FIGURE 4.12

Impact of computer control on machine utilization

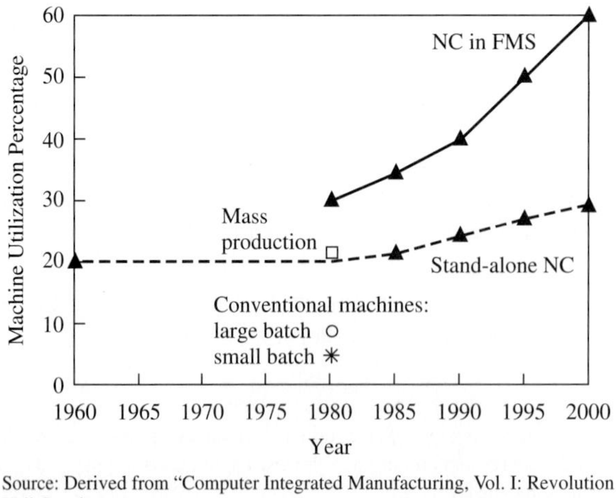

Source: Derived from "Computer Integrated Manufacturing, Vol. I: Revolution in Progress," R. Ayres, 1991, Chapman & Hall, London.

- Increasing the integration of information and material processing activities.
- Evolution from a machinist's art to an engineer's science wherein the conception and execution of an operation can be separated, with increased emphasis on conception (e.g., CNC software development).
- Increased repeatability: if an operation can be correctly captured in a program of instruction for a collection of processing machines, it can be executed correctly every time thereafter (with the exception of system breakdowns).
- The criticality of developing high quality programs of instruction requires that employee skills in this intellectually challenging task be carefully managed and enhanced. In particular, factory workers are becoming knowledge workers.

These implications suggest that the competitive edge of a firm is based on its ability to manage its intellectual assets efficiently and effectively (i.e., the programs of instructions and the knowledge and skills of employees of the firm). Hence, employee selection and training become very important responsibilities of a manager. Training can be done either in a formal manner through training programs or, more advantageously, through indirect measures such as selecting a portfolio of projects to implement and assigning employees to these projects, depending on the skills they possess and their individual developmental needs. (See Chapter 8 for further discussion on the human resource implications of manufacturing automation.)

The increasing information intensity requires that employees possess increasing levels of skill and knowledge. Therefore, the role of an engineer is likely to be dramatically different in this new era of production automation. As opposed to a traditional staff role, an engineer will likely be needed to take on a line responsibility for both the manning of automated equipment and for fixing problems as they arise.

Study Questions

1. What are the principal production automation and infrastructure technologies? How do they relate to each other in a CIM environment?
2. How do product variety and volume affect the choice of manufacturing automation technologies?
3. What are the principal advantages of numerical control of production equipment?
4. Under what conditions is a robot preferred over a human operator in the work environment?
5. *Flexible manufacturing systems (FMS)* have been described as providing the flexibility of a job shop with the efficiency of hard automation (e.g., a transfer line). Explain. Give an example.
6. What are the principal components of a FMS? What are some of the major disadvantages and implementation barriers for FMS?
7. What are the costs and benefits of improving equipment utilization in a FMS environment?

References

Ayres, R. (1991), *Computer Integrated Manufacturing*, Vol. 1: Revolution in Progress, Chapman & Hall, London.

Ayres, R. and S. Miller (1983), *Robotics: Applications and Social Implications*, Ballinger, Cambridge, MA.

Bernhardt, R., R. Dillman, K. Hormann, and K. Tierney, eds., (1992), *Integration of Robots into Computer Integrated Manufacturing*, Chapman & Hall, London.

Bylinsky, G. (1994), "The Digital Factory," *Fortune,* November 14, pp. 93–110.

Cohen, P. H. (1990), "Design for Robotic Assembly," in D. Cleland and B. Bidanda (1990), eds., *The Automated Factory Handbook*, TAB Professional and Reference Books, Blue Ridge, PA.

Delchambre, A. (1992), *Computer-Aided Assembly Planning*, Chapman & Hall, London.

Foston, A., C. Smith, and T. Au (1991), *Fundamentals of Computer Integrated Manufacturing*, Prentice-Hall, Englewood Cliffs, NJ.

Groover, M. P. (1987), *Automation, Production Systems and Computer-Integrated Manufacturing*, 2nd Edition, Prentice-Hall, Englewood, NJ.

Jaikumar, R. (1986), "Postindustrial Manufacturing," *Harvard Business Review*. Nov-Dec, pp. 69–76.

Luggen, W. W. (1991), *Flexible Manufacturing Cells and Systems*, Prentice-Hall, Englewood Cliffs, NJ.

Maleki, R. A. (1990), *Flexible Manufacturing Systems: The Technology and Management*, Prentice-Hall, Englewood Cliffs, NJ.

Melnyk, S. and R. Narasimhan (1992), *Computer Integrated Manufacturing: Guidelines and Applications from Industrial Leaders,* Business-One Irwin, Burr Ridge, IL.

Noori, H. (1990), *Managing the Dynamics of New Technology: Issues in Manufacturing Management*, Prentice-Hall, Englewood Cliffs, NJ.

Paul, L. (1992), Private communication.

Ranky, P. G. (1986), *Computer Integrated Manufacturing*, Prentice-Hall International, Englewood Cliffs, NJ.

Rembold, U., B. Nnaji, and A. Storr (1993), *Computer Integrated Manufacturing and Engineering*, Addison-Wesley, NY.

Tchijov, I. (1992), "The Diffusion of Flexible Manufacturing Systems," in R. Ayres, W. Haywood, and I. Tchijow, eds., *Computer Integrated Manufacturing*, Vol. III: Models, Case Studies and Forecasts of Diffusion, Chapman & Hall, London.

Templemeier, H. and H. Kuhn (1993), *Flexible Manufacturing Systems: Decision Support for Design and Operation*, John Wiley & Sons, NY.

Vajpayee, K. (1995), *Principles of Computer Integrated Manufacturing*, Prentice-Hall, Englewood Cliffs, NJ.

Williams, D. and P. Rogers (1991), eds., *Manufacturing Cells: Control Programs and Integration*, Butterworth-Heinmann, Oxford, UK.

C H A P T E R

5 INFRASTRUCTURE TECHNOLOGIES

5.1 Introduction

Infrastructure technologies perform support functions that enable production automation technologies discussed in the previous chapter to engage in value adding material transformation activities. Such activities can only take place if material, tools, equipment, and process instructions are all brought together in a coordinated fashion. Infrastructure provides such coordination and support through facilitation of the storage and flow of both information and materials between production automation equipment. In addition, infrastructure technologies link production automation equipment with the management staff in charge of planning and control of manufacturing operations. The goal in managing information storage and flow is to ensure that the required information for decision making is made available, in the right format, to the appropriate entity (either a machine or a human being), at the right time, so value-adding transformation activities can actually take place. The goal in managing material storage and flow is analogous—placement of the right type and quantity of material and tooling, at the right time and place.

By providing the information and material linkages, infrastructure technologies support flexibility and allow computer integrated manufacturing to achieve *economies of scope*[1] by reducing or eliminating the switching costs associated with producing different product types in the same facility. The synergistic effect of infrastructure technologies are also significant in that the benefits of CIM are greater than the sum of benefits of the individual technologies that make up the CIM system.[2] Thus, effective infrastructure becomes both an enabler and an incentive for the installation of large scale, multifunction automated manufacturing systems.

[1]Economies of scope are said to occur when the average cost of producing a unit decreases as the number of product types (variety) increases in the factory.

[2]See Milgrom and Roberts (1992) for an economic model which demonstrates this effect.

FIGURE 5.1

Infrastructure technologies

Information Technology	Material Handling/ Storage Technology
• Computer Hardware	• Automated Guide Vehicles (AGV)
• Software Systems	• Conveyers
• Communications Network	• Automated Storage/ Retrieval Systems (AS/RS)
• Database Management Systems	• Robotics
• Artficial Intelligence/Expert Systems	
• Machine Vision/Sensors	
• Data Collection	

In this chapter we provide an overview of the two main categories of infrastructure technologies: manufacturing information technology, and material handling and storage technology. Manufacturing information technology deals with the acquisition, storage, processing and communication of information, whereas the materials handling and storage system deals with identification, handling, storage and retrieval of material. Figure 5.1 lists some examples of these two types of infrastructure technologies.

In general, a manager of an automated production/service facility need not know the technical level details of how infrastructure technologies work in order to use them effectively. A manager, however, must have a good understanding of the performance capabilities of each technology. These include the costs and benefits of its use, and its relationship with other processing and infrastructure technologies. Such knowledge is needed in order to make sound decisions concerning selection, justification and implementation of automated manufacturing systems. In discussing infrastructure technologies, we maintain the managerial perspective of this book. Specifically, we cover the underlying concepts and the functionality of selected infrastructure technologies. We also discuss the costs and benefits of their use, and their role in the successful implementation of CIM.

5.2 Manufacturing Information Technology

Information technology (IT) is a term that loosely covers computer hardware, software, and communications systems. Here the term *hardware* refers to all the physical components, including the computer and other devices connected to it, while the term *software* refers to the computer programs that instruct the hardware to perform desired functions. In general, software is categorized into two groups: *systems software* and *application programs.* Systems software consists of programs that make the computer work and are typically developed and supplied by the computer vendors. Application programs perform user-specific tasks and are typically developed by users or software specialists. Certain standard application programs, however, are developed and sold by software vendors.

Computer technology has generally been slow to adapt to change in manufacturing environments, and was initially developed for and used in commercial applications to support financial and/or administrative tasks. The results of a survey conducted by Gettelman et al. (1990) concerning the use of computers in manufacturing are shown in Figure 5.2. Not surprisingly, larger companies are more frequent users of computers as compared with smaller companies.

Consider the communication needs of three fundamental entities present in a manufacturing environment: computers, machines, and personnel. We can identify a three-

FIGURE 5.2

Percentage of manufacturers using computers

	NC users			Non-NC users		
Functions	*Small*	*Medium*	*Large*	*Small*	*Medium*	*Large*
Accounting	74	98	98	64	91	100
Bill of Materials	39	74	96	37	76	75
Capacity Requirements Planning	22	62	78	21	62	75
Database Management	37	74	87	28	67	75
Design/Drafting	28	74	91	17	48	50
Digitizing and/or Scanning	11	17	44	5	15	25
Drawing/Engineering Retrieval	19	67	70	17	33	25
Finite Element Analysis	11	24	44	4	19	0
FMS/FMC Control	9	17	26	1	0	25
Geometric Modeling	14	26	30	7	14	0
Group Technology	1	26	26	1	5	25
Inspection/Quality Control	19	57	57	19	33	50
Inventory/Purchasing Control	43	93	87	41	90	100
Job Costing/Estimation	37	67	70	32	67	50
Machine Shop	28	36	43	13	19	50
Machine Database Use	19	29	35	8	14	25
Management Data Reporting	39	74	74	31	76	100
Material Requirements Planning	18	79	87	11	71	75
NC Programming (< three axes)	48	60	87	0	0	0
NC Programming (> three axes)	31	45	61	0	0	0
NC Postprocessor Generation	24	45	60	9	0	0
Sheet/Plate Nesting	10	10	30	4	10	0
Order Entry, Scheduling	35	90	74	40	95	100
Plant Layout/Organization	12	43	44	4	24	50
Process Planning	10	43	30	5	43	50
Production Scheduling/Control	31	83	83	25	80	100
Proposal Generation	19	26	44	16	14	25
Shopfloor Control	14	43	52	6	24	25
Simulation Exercises	8	12	30	0	10	25
Tool Crib/Tooling Control	11	33	39	3	14	50
Work Measurement/Standards	14	52	61	12	38	0
Word Processing	61	71	83	55	95	100

Source: Gettelman et al. (1990)

by-three matrix, consisting of the nine possible interaction combinations for these entities. They are person-to-computer, person-to-machine, computer-to-machine, and so on. The communications needs of each interaction pair are different, and yet meeting these needs of each and every pair is important for the successful implementation of CIM. Indeed, it is this diversity in communication needs that makes successful implementation of computer integrated manufacturing so challenging.

Manufacturing automation has evolved considerably from the preinformation technology days of "hard automation," when mechanical constraints were primarily used to guide the path of a machine to accomplish a desired material transformation (Shirley and Jaikumar [1988]). Today's automation technologies, in contrast, are based on information technology, where a microprocessor performs functions similar to those of mechanical constraints as well as of the human brain, in controlling the manufacturing system. Programmability of a microprocessor in the selection of manufacturing tasks and in sequence control, as well as in its ability to respond rapidly to changes in the work environment, are the principal distinguishing features of today's microprocessor-based automation technologies.

Essentially, the information processing and decision-making tasks that humans perform in manual systems must be performed in an *automated fashion* in IT-based automated manufacturing systems. It is, therefore, evident that the operation of an IT-based, automated manufacturing system will be highly information intensive. For example, the following functions, each involving information processing, storage and/or access, are carried out in a typical IT-based automated system:

1. Collect data pertaining to operational *status* of the system, and/or sense environmental conditions relevant to the performance of the system.
2. Make this information available to a microprocessor-based control unit, so an appropriate programmed *response* can be automatically generated.
3. Create *software* programs for use by microprocessors in controlling the material movement and transformation processes.
4. Maintain a *database* containing product and manufacturing process information, system status, and business environment conditions.
5. Install a *communication network* that can link all of the processing machines, the database, and the material handling and storage systems, so they operate in an integrated fashion.

As discussed in Chapter 1, firms that offer a large number of models of complex products with a high level of customer service have created a difficult problem for their organizations' information management and material handling systems. For example, IBM offered (ca. 1975) 55,000 models of its Selectric typewriter, Westinghouse Electric (ca. 1983) manufactured over 50,000 different turbine wheel shapes for its steam turbine, and Caterpillar (ca. 1985) had over 25,000 subcontractors making various parts for its earth-moving equipment. Major manufacturers in industries such as automobiles, airplanes, and earth-moving equipment have become, in essence, "systems integrators," producing only a few key parts and accomplishing the final assembly of several subsystems produced by a network of suppliers. Indeed, direct manufacturing constitutes a small fraction of total cost and employment in these firms. Their main economic role and value added are in functions such as product and process design, financing and co-ordination of production, marketing, and after sales service, and not production *per se.* Thus, as a consequence of increasing complexity and variety, information-intensive activities now constitute a large and growing fraction of all activities within individual manufacturing firms and in the economy as a whole.

Our interest in information technology in this chapter is primarily concerned with how it supports production automation technologies. We discuss, in particular, the cost/performance trend of information technology, alternate computer control architectures, the critical role of databases in achieving true integration, the emerging communications protocols necessary for improving connectivity, and factory floor data collection systems.

5.2.1 Evolution of Computers and Communications Technology

Computer and communications technologies are undergoing revolutionary change. This change has been going on for the last two decades, and the pace of change is expected to continue, or even accelerate, in the years to come. An important feature of this change, as shown in Figure 5.3, is the significant improvement in price/performance ratios of communications and computer hardware technologies. The emergence of "faster, cheaper, smaller, better" information technologies has made CIM both feasible and cost effective. The growth of microcomputer technology is particularly noteworthy in this regard. A microcomputer today gives processing performance superior to both mini-computers and mainframes of even 10 years ago. Evolving workstations are rapidly approaching one BIPS (billion instructions per second) capability. As a result, micro-computers and workstations have become the platforms of choice in automated

FIGURE 5.3

Economic trends in information technology

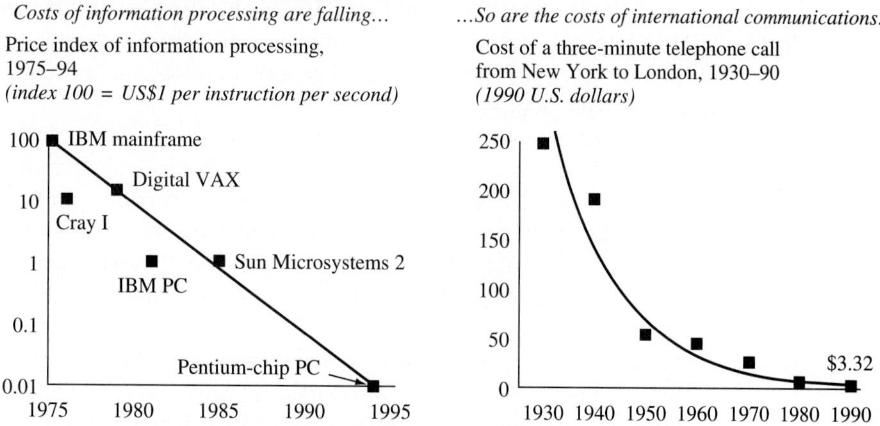

Source: The World Bank (1995), *Global Economic Prospects and the Developing Countries*

manufacturing environments. Computer storage devices are also experiencing similar, strong improvements in terms of both price and performance.

The costs of information processing and long-distance telecommunications also have declined dramatically. The real price of microcomputers, for example, fell by an average of 28 percent a year between 1982 and 1988 (Berndt and Griliches [1993]). At the same time, the quality of services delivered by information technology applications has improved significantly. In the United States, a voice-image hookup for a one-hour, coast-to-coast teleconference call cost approximately the same in 1994 as a coast-to-coast telephone call of four minutes cost in 1915 (Mayo [1994]).

This improvement process is expected to accelerate as the cost of communication bandwidth (the capacity to transmit more bits of information simultaneously) falls by a factor of two per year (Baruch [1994]). Moreover, the cost of communication is also becoming independent of distance, and networks (such as the Internet) are becoming international and pervasive.

In contrast to this spectacular improvement in computer hardware and communications technologies, the slow productivity improvement in the development of software technology (estimated to be about 5 percent per year) is proving to be the main bottleneck in the deployment of advanced manufacturing information systems. Indeed, software has become a major stumbling block for the implementation of CIM. In general, all capital goods, including machine tools and computer hardware, are becoming more and more standardized and available off-the-shelf in the marketplace. As a result, a manufacturing firm's competitiveness is increasingly dependent on the quality of its know-how, as embodied in applications software. As a consequence, software engineering has become a core competency for world-class manufacturers. We believe that software development has become far too important a task to be left in the hands of an outside consulting or a software developing firm. At a minimum, all of the critical elements of a software system must be developed in-house or with a strong level of in-house expert involvement. The importance of this factor has apparently been well understood by Japanese and German firms, but not American firms. It is estimated (Ayres [1991]) that about 10 percent of Japanese FMSs were purchased from outside firms, while the comparable figure for United States firms was 70 percent. Given the knowledge intensity and high skill requirements of software development, the selection and training of employees have also become a critical success factor for computer integrated manufacturing.

In implementing manufacturing automation, it is important to choose an appropriate control architecture for the variety of computers and equipment that are present on an automated factory floor. Having an information architecture blueprint is essential. As various pieces of processing equipment and computers are purchased, it is necessary to place each at an appropriate position in the overall system and to assign it appropriate operational and control tasks. Architectural blueprints make this integration task much easier.

Today's computer control architectures (see Figure 5.4) are based on some variation of a *four-level hierarchy of computers* (which is closely linked to the CIM pyramid introduced in Chapter 1). The top level in the hierarchy is the factory-level resource planning system, based typically on a mainframe or a minicomputer. The next level

FIGURE 5.4

Hierarchy of computers in manufacturing

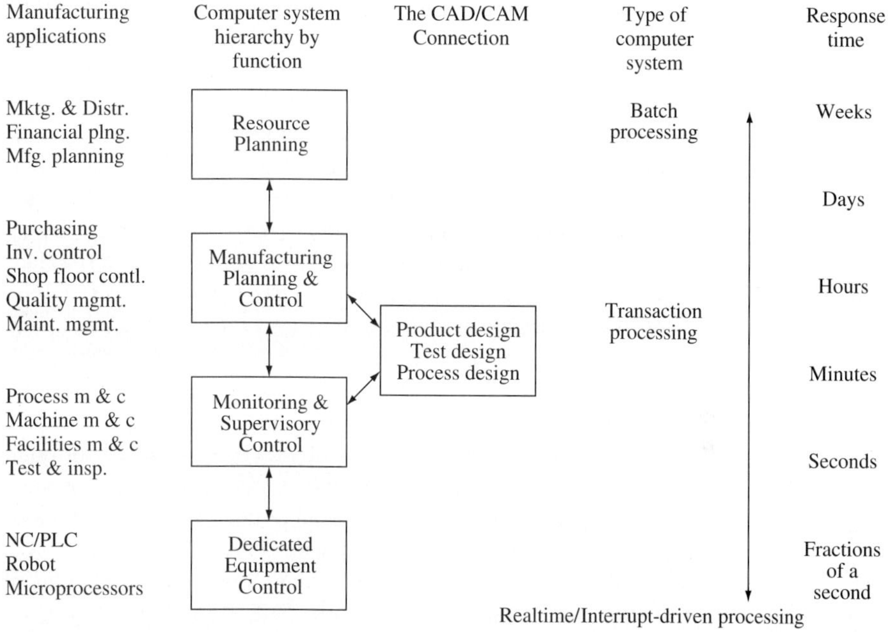

Source: Adapted from Gross (1984)

down is a manufacturing control system assigned to such tasks as inventory control, purchasing, and production scheduling. Below this level is a supervisory control system that monitors and controls work cells, where each work cell consists of several machines and other automated systems, such as a material handling system. The lowest level is typically a microprocessor, a microcomputer, or a programmable logic controller, that is dedicated to the control of an individual piece of equipment, such as a numerically controlled machine or a robot load/unload station. One of the major benefits of a hierarchical control system is its ability to grow gradually by adding components. Such hierarchies must also be designed to be robust to withstand isolated failures of equipment, computers, and network elements through redeploying control to other appropriate computers in the hierarchy.

5.2.2 Database Management Systems (DBMS): An Integrative Mechanism for Manufacturing Information

Manufacturing information, as it relates to both the products and the processes that are used to add value through design, production, and support activities, has become highly data intensive. Management of manufacturing data has always been a challenge, but several factors in today's competitive environment are making it even more difficult. One factor is the increasing variety and customization of products. While in the past a product was typically associated with only one set of product specification data, today's increased variety and customization of products require that multiple sets of product definition data be maintained for each product. For example, information will have to be maintained for specific product configurations used for different customers and for all the options and assemblies required to manufacture a product in different configurations.

The next factor adding to the complexity of manufacturing information management is the increasing tendency of manufacturers to focus only on that part of the process where they believe they add the most value, and to purchase the remaining parts and capabilities from outside vendors. This trend towards increased outsourcing has resulted in the increased flow of data among companies and suppliers along the entire supply chain. The increased adoption of just-in-time systems and the recent emergence of time-based competition require increased speed and accuracy of manufacturing information flow among all participants in the firm's supply chain.

Finally, modern business practices, such as quality management and concurrent engineering, are adding to the increasing importance of manufacturing information. Consider, for example, the emergence of ISO 9000 standard for quality management. ISO 9000 requires that companies maintain proper documentation and audit trails of all their product designs and manufacturing activities so that the root cause of a product's problem can be identified as and when necessary.

With the advent of CAD, more and more companies are using computers to create and store manufacturing information. But, even here, a significant majority of companies have their manufacturing data scattered in many locations and in many media, e.g., on paper, microfilm, microfiche, magnetic and optical disks, etc. Having grown up in days when computers were not easily available, many engineers tend to keep a paper record as a backup to the computer files. Often different people responsible for different aspects of manufacturing a product end up with different versions of what should be the same data. The scope of manufacturing information is wide. The data can also exist in various forms. There is textual data (specifications, schedules, process plans), numeric data (geometry of shape, formulae, physical properties of products), graphic data (drawings, photographs), and voice data. In such an environment, maintaining the integrity of the manufacturing data has become extremely complex.

In a company without properly defined and managed manufacturing information, systems can face a number of serious problems. Table 5.1 gives a list of such problems.

5.2.3 Database Management Systems

For a small company with a limited set of products, it is possible to store manufacturing data using the directory/file-based data storage capabilities of a computer's operating system. A file is nothing but a collection of records that may contain one or more data items of interest to the company. Thus, in the case of a file-based data storage system, each application program needs to develop its own file(s) of data, possibly with different structures of records and data fields. For example, an invoicing system may have one record per customer, while a manufacturing related system may choose to keep one record per product. Some information, such as a customer's address, may be needed in several programs (invoicing, mailing lists, delivery routing, etc.), and hence there may

TABLE 5.1 Problems of Improper Management of Manufacturing Information

Data is lost. Data is not secure.
Data cannot be found. Time is wasted searching for data.
Data does not correspond to the product.
Unnecessary copies of data are maintained. Data is not shared.
Copies of data conflict. Data ownership is not clear.
Interfunctional communication of data is slow.
Transfer of data on paper is slow.
Review and approval procedures take too long.
Engineering change management is too slow.
Engineering changes are not controlled or coordinated.
Project managers lack up-to-date status.
Projects overrun. Project costs are unnecessarily high.
Lead times are unnecessarily long.
Versions are not properly managed.
Configuration management breaks down.
Configuration information does not correspond to reality.
Technical manual development is labor-intensive and slow.
Technical manuals are out of date.
Maintenance information is inaccurate.
Computer programs are not documented properly.

Source: Stark (1989)

be redundant data resident in the system. At the time of any system change, it is always possible that not all instances of information are consistently updated. For example, a customer's address may be updated in the invoicing system, but not in the delivery routing system, leading to obvious delivery problems. Thus, data *redundancy* and *integrity* become major problems in managing the business affairs of the company. The other disadvantage of a file-based data management approach is the difficulty of sharing data among different application systems. As the data storage is not independent of programs, any change in the program is likely to affect the physical and logical structure of the data. This makes any change to an application program a difficult and time-consuming task. *Database management systems (DBMS)* were developed to overcome such problems for a file-based data storage approach.

A *database* is a collection of information required for managing a business. A *DBMS* is software that manages the database. In a DBMS, all applications programs and users are allowed access only through the DBMS. Implementation of DBMS originated in commercial environments, but is now considered a very useful approach for manufacturing situations as well. The main benefits of the DBMS approach are to provide for:

- Reduced data redundancy.
- Physical and logical data independence.
- Data validation standards.
- Controlled access, security, and recovery of data.
- Data to be shared by multiple application programs.

Among the major elements and concepts of the DBMS approach are the *data model,* the *data dictionary (DD),* the *data definition language (DDL),* the *data manipulation language (DML),* and the *data query capability* (see Date [1982] for details on database systems). The three main types of *data models* are the *hierarchical,* the *network,* and the *relational* data model. In database terminology, an *entity* is a thing, such as a product, a part, a machine or a customer, for which the company desires to maintain information. Each entity can have one or more "children" at the level below it. The hierarchical data model, one of the earliest data models implemented, is extremely efficient in accessing data for an application that is designed to take advantage of the hierarchical organization of data.

The network data model is similar to the hierarchical data model, except that it allows entities to have more than one parent at the level above, while still having one or more children at the level below. Although the network data models are somewhat less efficient than the hierarchical model, they are more flexible in being able to better capture the relationships between entities. Both the hierarchical and network models require application designs that explicitly take into account the database structure. In this sense, the data are physically, but not logically, independent of application programs. If the database structure changes, the applications will also have to be changed.

In the relational data model, the data are represented as two-dimensional tables, known as relations. The principal advantage of a relational data model is that application programs become logically, as well as physically, independent of data. Entities can be added or deleted to change the database structure without changing the applications programs. The relational data model offers great flexibility, but it suffers from poor performance when large quantities of data are to be stored.

Keeping pace with computer hardware and communications technology developments, DBMS technology has also evolved rapidly in terms of both storage capacity and data access speed (see Figure 5.5). Acquiring a database machine with a storage capability of hundreds of gigabytes was a dream just 10 years ago. Today, such machines are readily available. A manufacturer also has a choice among the alternate database management technologies, such as a hierarchical, network or relational DBMS, and various hardware platforms to run these DBMS. The main issues of concern to a manager in selecting among alternate DBMS approaches are their efficiency and flexibility in storing, retrieving, and updating information. In general, the hierarchical and network approaches give higher efficiency (e.g., quicker access and ability to handle a larger number of transactions per second) as compared to the relational approach. But the former approaches are less flexible than the latter one in that they can handle only predefined, well structured transactions in information storage and retrieval. Given the increasing and strategic need for manufacturing flexibility, the advantages of using a relational DBMS may prove to be critical.

The DBMS approach was developed originally for commercial applications such as an insurance policy processing system, a saving system in a bank, an airline reservation system, and so forth. Although DBMSs are now implemented in manufacturing applications, the approach has certain limitations due to major differences between the characteristics of commercial and manufacturing data and applications.

FIGURE 5.5

Evolution of database models

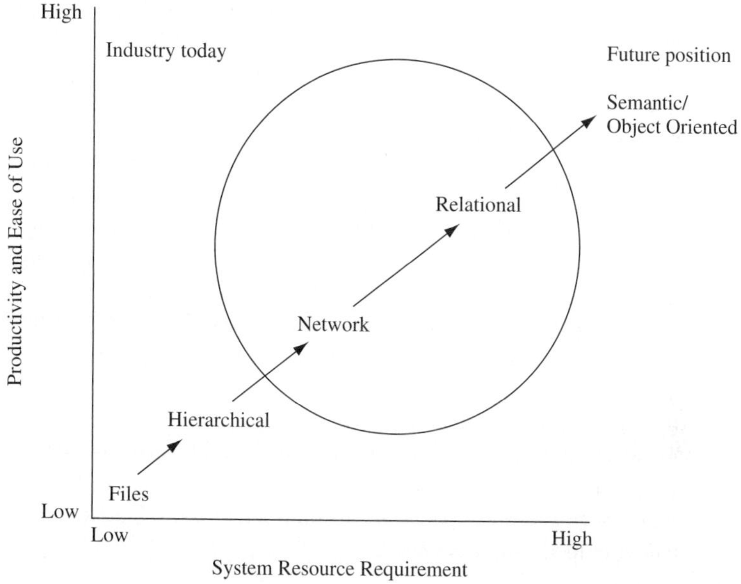

Commercial applications typically handle only alphanumeric data, while the mixed data types and graphics data are common in manufacturing applications. In commercial databases records pertaining to a given entity, such as a customer, have similar lengths, while in manufacturing situations an entity, such as a product, can have a record of dramatically different lengths, depending on the unique product structure. There is also a difference between the time required to complete transactions. Commercial transactions, such as a seat reservation, last for several seconds or minutes, but manufacturing transactions, such as product design, may take hours to days.

Most commercial applications are built around a DBMS, and most data in that environment reside in the DBMS. However, in a manufacturing department, a significant part of the data resides outside the DBMS. For example, today most CAD systems typically do not feed design data into a DBMS, but into a file-based system. In commercial systems the relationships between entities are generally static, well defined, and simple. In manufacturing applications they are dynamic, unclear and complex. For example, as products evolve through various versions and changes, the relationships between parts used can undergo significant change.

In view of the above differences, few manufacturing organizations can put all their data in a single DBMS, or can totally rely on the types of DBMSs that are commonly used in commercial applications. At the same time, however, the disadvantages of file-based systems and the potential advantages of a DBMS approach are as equally valid in manufacturing as in commercial environments. In recognition of this fact, manufacturing-oriented DBMS solutions are becoming available (Stark [1989]). One such approach is the *metadata* approach.

The metadata approach lies between the two extremes of a file-based system and a single DBMS implementation. The approach involves leaving the data in files or other media where it usually belongs and keeping the "data about data" in a DBMS. For example, the information stored in the metadata can include the name, number, version and the release date of a part, the name and location of the file that contains the design specification of that part, the manufacturing attributes of that part, such as the type of material used, and the name of the NC program to be used in manufacturing the part, and so on. The files containing CAD design specs and NC programs can reside in their usual format and locations. The system using the metadata simply keeps track of the underlying files and relieves the users of this work. A user can request a search of specific information on a given entity, and the system will point to a specific file (or even a paper drawing stored in a certain file cabinet). The metadata record is similar to a record in commercial application. It contains only alphanumeric data, and is of normal and standard length. A commercial DBMS can, therefore, be used for storing the metadata. The metadata approach can thus become the central hub of a manufacturing information system. It should be noted, however, that this approach is still evolving, and the full range of software products necessary to implement the metadata approach is not yet readily available in the marketplace.

5.2.4 Connectivity, Manufacturing Automation Protocol

Connectivity, the ability of each system element to communicate with every other system element is considered the most fundamental prerequisite for achieving computer integrated manufacturing. The diversity of computers and communication equipment, automation technologies and OEM suppliers is a fact of life on today's factory floor. But this diversity is a principal obstacle to achieving connectivity.

Local area networks (LAN) have proved to be the most basic solution to the connectivity needs of an automated system. Simply put, a LAN is a collection of communications hardware and software that connects individual PCs and workstations together for messaging services and resource sharing. The technology underlying LAN is conceptually simple. It can be explained in terms of three technological issues, the answers to which define the configuration of a LAN: (1) *Topology:* the methods by which components (e.g., machines and PCs) are connected to form a network, (2) *Media:* the cable or wiring connecting network components, and (3) *Access methods:* the methods by which only one network component at a time sends messages to other components on the network.

Experience at General Motors (Kaminski [1982]) showed that among the 40,000 programmable tools, instruments, controls, and systems, only 15 percent were able to communicate with each other. It was found, moreover, that up to 50 percent of total expenditure on automation was for special wiring and hardware and software interfaces necessary for achieving connectivity among the machines and systems supplied by different vendors.[3] A solution to the connectivity problem was urgently needed since conducting business as usual, while the level of automation was ever increasing, was becoming prohibitively expensive. The option of buying equipment from a single vendor so there would be no connectivity problem was ruled out since no vendor existed that could supply all of GM's automation needs on a factory floor. Hence, development of a standardized approach for communications that all vendors agree to follow became the only viable solution. This reasoning persuaded GM to spearhead a movement for developing a standard manufacturing communications protocol, the well-known *manufacturing automation protocol (MAP).*

The manufacturing automation protocol (MAP) is a specific implementation of the seven-layer open systems interconnection (OSI) standard, sponsored by the International Standards Organization (ISO) (see Figure 5.6). Having a common standard that enables easy communication among all manufacturing automation equipments makes it possible to avoid the expense and difficulties of developing a customized connectivity solution. The vision guiding the development of MAP is very broad. In its ultimate form, the MAP is envisioned to serve the communications needs of a factory at all levels of the four-level control hierarchy—from controlling individual devices in real time at the process level, to coordinating individual devices within a station and individual stations within a work cell, and to providing factory-wide information to manufacturing management.

FIGURE 5.6

Manufacturing automation protocol (MAP) uses OSI model

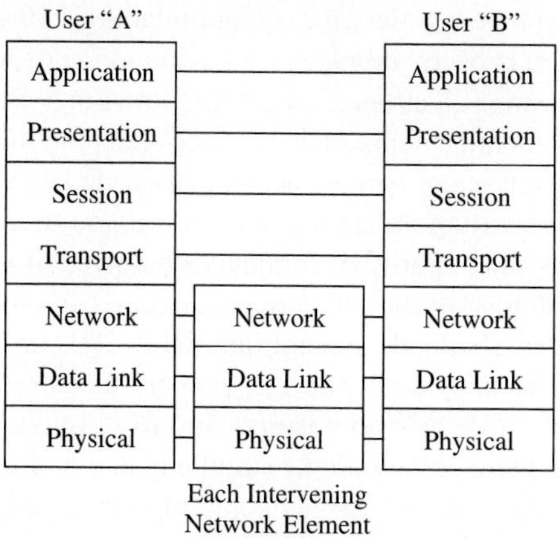

Each layer is its own communicating process, using the services of the layers below and serving the layers above.

OSI model layers and functions

Layer	Function
Application	Specialized functions such as file transfer, virtual terminal, and electronic mail
Presentation	Data formatting and character code conversion
Session	Negotiation and establishment of a connection with another node
Transport	Provision for end-to-end delivery
Network	Routing of packets of information across multiple networks
Data Link	Transfer of units of information, frames, and error checking
Physical	Transmission of binary data over a communications channel

[3] Although more than a decade has passed since this study, the connectivity problems are still prevalent in the industry today.

In spite of the conceptual elegance of MAP, it must be noted that it does add a significant operational overhead and complexity in its implementation. Moreover, as in the case of any standard, a key to MAP's success is its acceptance and adoption by most of the key players in the situation—the manufacturing companies, as well as the vendors of computers, communications, and factory floor automation equipment. The experience with MAP has been disappointing thus far. Contrary to the hopes of its architects, experience suggests that MAP has not yet been adopted as a standard by major players. In fact, in recent years, several start-up companies have begun to offer unique software solutions to the problems of factory floor connectivity (see HBS FASTech Case 9-694-028).

5.2.5 Factory Floor Data Collection

Data related to the status of work and the worker's productivity has always been collected on factory floors, whether through job routing cards, employee time sheets, operations tear strips, or pre-punched cards. In today's increasingly automated manufacturing environments, this data needs to be collected in a timely and accurate manner, since the systems that control the individual machines, or the entire factory for that matter, need such data for planning as well as for real-time control. Hence, there is a need for modern data collection technologies such as *bar codes* and *radio frequency identification.* In general, technology for data collection systems is available; it may cost only a few thousand dollars, but it may be the key to effective utilization of multimillion dollar machinery and other elaborate software systems such as material requirements planning (discussed in Chapter 6).

Bar coding technology was developed in the grocery and retailing industries as a productivity aid. Today, however, it has been embraced by a wide range of industries, including automotive, metal processing, electronic, paper, and trucking. Bar coding technology is essentially a means to identify items such as a stock keeping unit (SKU), a part, or a parcel being shipped. It consists of two elements, a bar code label and a bar code reader. Bar code symbols, representing an identification code, are useful only if they are agreed upon as a standard by the majority of companies involved in a given industry. Over the last two decades, almost all the industries mentioned above have developed their own bar code standards. A bar code symbol familiar to all is the universal product code (UPC), a ten-digit pattern adopted by the U.S. grocery industry that identifies both the product (first five digits) and the manufacturer (last five digits). The main reasons for the popularity of bar code technology are its simplicity and low cost. A bar code label can be printed on almost any kind of surface, including corrugated and plain paper, metal, plastic, ceramic, and glass. The bar code labels are read by a bar code reader or a scanner, and the code is communicated to a computer for operational control, storage, and creation of necessary reports (e.g., retrieve pricing data, update transactions, maintain inventory records, and initiate stock replenishment orders).

Bar code technology has uses far beyond its common role as an identification technology. It can also be used to identify transactions by recognizing transaction elements such as who (the operator), what (the process), how many (the item), when (processing time), and where (location). The transaction identification can then be further used by the central computer for both operational control and productivity analysis.

Radio frequency identification (RFI) is a recent data collection technology that has been found to be very effective in hostile environments where bar coding technology cannot be used. It is based on a small radio transmitter and receiver embedded in a tag which produces a radio frequency signal unique to the item being identified. This signal is recorded by a polling reader under the control of a central computer. RFI systems claim extraordinary accuracy with error rates of one in a trillion trials. Some systems also contain microprocessors that are capable of storing large amounts of data. In such systems, the tag can announce at each station what operations are to be performed and then the operator, in turn, can record the start time and the completion time for each operation on the chip. This chip can be read during or at completion of the job to access its operational history.

A successful implementation of data collection technology ultimately depends on the attention given to the various system issues including personnel training, operational procedures, and organization structure. It should also be recognized that the effectiveness of data collection technology ultimately depends on the design of the software that makes use of the recorded information.

5.2.6 Machine Vision and Other Nonvision Sensors

Manufacturing automation requires that the control units of systems, such as NC machines, robots, automated material handling systems, etc., be aware of their work-space conditions and operational progress. Having a knowledge of these conditions is necessary for deciding what action to take next in continuing and completing a given process. For example, in performing a machining operation, the NC machine's control unit requires that the position of the machine tool relative to the work-piece be continuously monitored. A position sensor is useful for this purpose.

Industrial sensors are specialized devices designed to monitor particular conditions, to measure certain quantities, and to relay that information to a machine control unit for interpretation and action. **Machine vision** is a particularly promising modern technology that is currently being used in some state-of-the-art applications such as robotic assembly and inspection. A range of nonvision sensors has also been developed over the last several decades to monitor and measure quantities such as position, displacement, proximity, force, pressure, and temperature.

Machine vision technology is helping machines perform tasks that were previously not amenable to automation. The technology is currently being used in automotive and electronic industries to find and orient objects so the assembly process can be made more accurate and automated. It is also being used to perform repetitive, yet critical, inspection tasks necessary for assuring product quality. Experience in these industries shows that the use of machine vision in automating assembly and inspection can significantly improve productivity and lower costs, while simultaneously increasing the product quality. For example, using machine vision for inspection allows real-time detection of defects, which makes it possible for a machine control unit to take corrective actions immediately.

A machine vision system typically contains the following components: (1) a source of illumination, (2) a video camera, or some other form of image sensor, (3) a computer and an artificial intelligence (AI) based software that is capable of analyzing image information and/or providing positional movement commands to a control drive, and (4) a monitor to display the product image and the process status. The most critical component of this system is the AI software that imparts to a vision system the capability to recognize predefined shapes, structural patterns, or object classes. The tools and techniques of AI software, however, are still in their early developmental stage. Consequently, any improvement in AI-based vision software is expected to lead to substantial increase in the capabilities and applications of machine vision technology.

Some interesting applications of machine vision technology include installation of windshields in a car, inspection of printed circuit boards to detect minute defects, and inspection of submarine blades for the U.S. Navy. The windshield system contains a robotic guidance system with two video cameras. The system uses stereo images for determining the position and orientation of the car, and the edges of an opening for glass installation. This system is capable of installing windshields with a positional accuracy of up to one millimeter. The printed circuit inspection system can accurately determine the position of a surface mounted device in less than two seconds and can process up to 64 items in the field of view. In inspecting the surface of a 25-foot, 50-ton propeller blade, the Navy's vision system takes about 5000 measurements and compares each to a database of ideal design specifications. Instead of spending hundreds of manual hours inspecting a blade, the vision system can perform the task with higher accuracy in less than eight hours.

Nonvision sensor technology, a well-developed field, is equally useful in manufacturing automation. As an example of a nonvision sensor we will briefly consider alternate configurations of *position sensors:*

- *Potentiometer:* this is essentially a resistor whose resistance is varied by sliding a movable contact in response to a change in the position. The measurement of resistance is then used to compute the precise position of the work-piece. These simple, economical sensors can measure either linear or rotational positions.
- *Optical Encoder:* this sensor converts a linear or a rotational position into a digital signal by interrupting a light signal directed onto a photodiode. Optical encoders give precise measurement, but they tend to be relatively expensive.

Proximity sensors measure the distance to some object without making a physical contact. These sensors are typically based on ultrasonic or other echo systems. Proximity sensors are used in automated guided vehicles for collision avoidance or for vehicle docking. Many automated processes require the use of a *force sensor* to ensure that the force being applied is appropriate for the job at hand. Force sensors generally involve use of an elastic element such as a spring and a displacement sensor that can measure displacement resulting from application of force. Certain force sensors can also measure torque or strain. *Touch sensors* are needed in many automation systems. For example, the gripper mechanism in a sophisticated robot requires a touch sensor. Simple touch sensors are based on a micro-switch mechanism which closes the switch under a small pressure caused by contact with an object.

5.3 Automated Materials Handling and Storage Systems

The physical processes performed on materials can be grouped into three general classes: conversion, transport, and storage (Ammons and McGinnis [1986]). The technologies related to conversion processes were discussed in the previous chapter on production automation technologies. The other two processes of transport (moving from one location to another) and storage (absence of either conversion or transport) are collectively referred as *material handling and storage (MHS)* functions. The MHS system is considered as an infrastructure technology, since its role is primarily to provide the necessary support to the conversion function. It also provides the physical basis for integrating operations on different pieces of (automated) equipment. The fundamental issue underlying the design of a MHS system is providing the required level of service (i.e., achieving certain levels of response time, accuracy, safety, quality/damage rates, etc.), while minimizing the cost (of distance traveled, handling effort, and equipment utilization) of operating the MHS. The key decisions here are how a MHS is designed and how its operations are controlled. These issues are discussed below within the context of two major MHS technologies, the automated materials handling system and the automated storage and retrieval systems.

5.3.1 Automated Materials Handling System

The main purpose in automating a fabrication or assembly system is to improve accuracy and productivity, while also achieving higher production rates in a flexible manner. This investment in automation can pay off, however, only if the material handling function is as efficient in delivering the necessary production inputs at the required rates and in removing the finished goods from the production system as are the conversion processes in transforming the material. Otherwise, an automated machine can become starved for raw materials in some cases and get blocked with unremoved finished items in other cases. As production systems become more automated, extensive, and faster, the role of an *automated materials handling (AMH) system* becomes more critical as an enabling technology in the computer integrated manufacturing environment.

At a first glance, the requirements of an AMH system seem trivial—to move material from point A to point B. But one must consider a number of factors in selecting AMH technology, as it can have a significant impact on the factory layout and on the overall productivity. These considerations are:

- *Capacity:* The AMH system should be capable of moving the material needed for production and storage in the required quantities and at the required rates.
- *Tending:* An important characteristic of an AMH system is its tending capability, i.e., the way in which material gets loaded and unloaded on or off the AMH system and at each workstation. At certain tending points, there may exist a need to move material mechanically to and from the AMH system and the input/output area of the workstation being served. A robot is typically used in these cases.
- *Material flow control:* The system should be capable of identifying all of the items and associating each with an appropriate destination. It should also be able to "steer" the item through its environment. The task of steering an item is easier in a conveyor system where path choices do not exist. This consideration is especially critical for a freely moving automated guided vehicle (discussed below).
- *Integration:* In a fully automated factory, it is important that the AMH system be integrated with the rest of the production system to ensure that the right material, in the right quantity, is made available at the right time at the right place. The computerized control architecture and communications protocols used to achieve this integration should, therefore, be carefully reviewed.
- *Safety:* To prevent worker injuries and to contain rising liability insurance premiums, it is important that the issue of safety be given careful attention in designing an AMH system.

Automated materials handling systems can be broadly classified into conveyor systems or cart systems. The *conveyor systems* represent an old technology that has been in use for the last two hundred years. These systems gained prominence in the era of mass manufacturing when Henry Ford started using conveyors in the transfer lines for automobile production. Conveyor systems are used mainly to transport material on a fixed route, where the stations along the route gain access to the material as it passes next to stations. Conveyor systems generally have a greater capacity for handling a larger volume of material. A major drawback of a conveyor system is that its cost is proportional to the length of the route, and extending its range can involve substantial capital expenditure. Such systems are inherently inflexible as well.

Simple conveyor systems are typically based on the use of rollers or belts. Roller systems consist of tracks in which carriers on fixed rollers are mounted. Goods are moved over the roller system by using gravity, or manual force, or by using a drive system which turns the rollers. A conveyor belt moves over roller assemblies mounted on the floor or suspended from the ceilings. The items to be moved are placed either directly over the belt or are placed in containers. Trolley conveyors are the most commonly found type of conveyors in modern factories. They consist of a series of trolleys supported by a single overhead track. Trolley conveyors are very versatile in that they can follow either level or inclined or curved paths, and can go from one building to another or from one floor to another. The use of overhead tracks also conserves space. Such systems can store and automatically release or pick up loads. Through years of development, the conveyor technology has become fairly mature, but new developments in computerized control are presently underway.

Cart systems have been used in the last three decades. In a cart system, the material moves in a cart or a trolley that may move freely on the shop floor or may move along a predesigned route guided by a wire or some other mechanism. Cart systems are ideally suited for material handling when the relationships between workstations are loose and when the flexibility of dynamically changing routes is deemed important. This routing flexibility implies that extending the range of a cart system is relatively inexpensive. Another advantage of carts is that, aside from being transportation devices, they can also serve as temporary storage devices.

Automated Guided Vehicle (AGV) is a prime example of a cart system that uses advanced technologies, such as vision or nonvision sensors and computerized controls. The major strengths of an AGV are its flexibility in changing its travel path and its low cost as compared to conveyor systems. With these advantages, AGV's use is expected to grow significantly. One of the biggest applications of AGVs is in the automated storage and retrieval systems (AS/RS), where an AGV provides the material handling between the high-reach stacker cranes and the goods shipping and receiving area. Another prominent use of AGVs is in the auto industry for moving engines, transmissions, and auto bodies. A specialized and growing application is in the handling of radioactive and other hazardous waste.

Most AGVs in use have fixed paths for guidance, but free-ranging AGVs are now in the advanced stages of development. Both types require the use of supervisory computers for monitoring and control. A wire-guided AGV is based on a network of buried cables that emits certain frequencies that can be used by the AGV for navigational purposes. In light engineering or office environments, the wire guides are sometimes replaced by painted-line guides that use fluorescent dye. The painted lines have the advantage of being able to change desired paths quickly, but, on the other hand, the painted lines can easily get erased through normal wear and tear. Free-ranging AGVs represent the state-of-the-art in material handling systems research. The paths of these vehicles are software programmed, and hence can be easily altered. The most difficult aspects of a free-ranging AGV are its ability to determine its position accurately and to recognize objects in its environment. The progress of vision and nonvision sensor technologies is critical to the development of free-ranging AGV capabilities.

5.3.2 Automated Storage and Retrieval Systems

The cost of carrying inventory, be it in the form of raw materials, work-in-process, or finished goods, represents a significant portion of the total manufacturing cost. Reducing this cost has been a major goal of manufacturing and purchasing executives since the beginning of this century. *Automated storage and retrieval systems (AS/RS)* work toward this goal. AS/RS is a modern warehousing innovation that combines high-density, high-rise storage racks with automated material handling systems such as AGVs to make the inventory function more efficient and cost effective. It contributes in two complementary ways. First, by significantly increasing the material storage density, it reduces the storage space, with the space cost savings being sometimes as high as 40 to 50 percent. Second, it improves the efficiency and the timeliness of the process of material storage and transportation to and from the production processes and in the receiving and shipping areas.

The main components of AS/RS are high-density, high-rise storage racks; a material picking crane (also called a *storage and retrieval (S/R) machine*); and an interface with AMH systems, such as AGVs or conveyors, all functioning under a centralized computer control. The storage racks can be as high as 150 feet and are typically compartmentalized with openings on one side. The compartments are designed to have a mechanism that makes it possible for an S/R machine, or an AMH system, to place and pick items from compartments. An important task of an AS/RS software system is to associate each compartment with a specific item in a dynamic sense and provide motion control for the S/R machine so that it can properly place and pick items.

AS/RS systems can be classified into *unit-load, carousel, or mini/micro-load* systems. Unit load systems are some of the oldest AS/RSs. Here, each load is treated as a single unit which, in itself, can be a group of products bundled properly for transportation and can weigh between 1,000 to 10,000 pounds (e.g., a pallet or container). These systems have S/R machines that move in a two-dimensional pattern to the correct vertical column and horizontal row for picking and placing individual units. Carousel AS/RS systems involve a rotating set of storage racks and bins. They are mostly used for temporarily storing work-in-process and component assembly kits. In these systems, the operator enters the bin number of the item required, and, in response, the appropriate bin is rotated around to the operator for item picking (a dry cleaner's storage system is an every-day example of a simpler system of this kind). Mini-load systems can inventory items weighing up to 1,000 pounds. They consist of racks about 40 feet high and a variety of S/R systems.

The major drawbacks of AS/RS systems are their large investment requirements, long installation times (sometimes up to two years), and high maintenance costs. Hence, careful design, evaluation, and implementation planning are critical to the success of AS/RS systems. Simulation software packages are increasingly being used as a "what if" analysis tool in the AS/RS design and evaluation process. The purpose of these models is to analyze the throughput rate (capacity and blockage rates) of the system. Such performance is dependent on a variety of factors, transaction arrival rate, configuration of the equipment, buffer capacities at transport interfaces, storage discipline (where to put things), and dispatching rules for selection of the next item (order) to process.

Study Questions

1. What impact does *information technology (IT)* have on information processing tasks within an automated factory?
2. Explain the hierarchy of computers used to manage manufacturing systems.

3. What is a *database management system (DBMS)* and why is it important for CIM?
4. What are the seven layers of the *manufacturing automation protocol (MAP)* systems interconnect model?
5. What are the principal technologies that are used for factory data collection systems? What are their key costs and benefits?
6. What are the principal factors that should be considered in selecting an *automated material handling (AMH)* technology?
7. What is the principal source of uncertainty in material handling systems? How might automation impact this uncertainty?
8. How does automation of infrastructure technology affect factory efficiency, responsiveness and service?

References

Ammons, J. C. and L. F. McGinnis, (1986), "Advanced Materials Handling," *Applied Mechanical Review,* Vol. 39(9), pp. 1350–1355.

Ayres, R. (1991), *Computer Integrated Manufacturing,* Vol. 1: Revolution in Progress, Chapman & Hall, London.

Beeby, W. D. (1983), "The Heart of Integration: A Sound Database," *IEEE Spectrum,* May, pp. 44–48.

Chorafas, D. (1993), *Manufacturing Databases and Computer Integrated Systems,* CRC Press, Boca Raton, FL.

Datapro (1991), "Data Collection Systems: Technology Overview," 2C-002 in *Manufacturing Information Systems,* Datapro Manufacturing Automation Series, McGraw-Hill, Delran, NJ.

Foston, A., C. Smith, and T. Au (1991), *Fundamentals of Computer Integrated Manufacturing,* Prentice-Hall, Englewood Cliffs, NJ.

Gisholt, P. R., R. Jones, and D. Upton (1994), Fastech Integration Inc., HBS Case 9-694-028, Harvard Business School, Boston.

Jones, V. (1987), *MAP/TOP Networking,* McGraw-Hill, NY.

Kaminski, M. A. Jr. (1986), "Protocols for Communicating in the Factory," *IEEE Spectrum,* April, pp. 56–62.

Koenig, D. (1990), *Computer Integrated Manufacturing: Theory and Practice,* Hemisphere Publishing Corp., NY.

Krakauer, J. (1987), ed., *Smart Manufacturing with Artificial Intelligence,* SME Publications, Dearborn, MI.

Krick, J. (1992), "Local Area Networks: Technology Overview," 4C-101 in *Factory Automation Systems,* Datapro Manufacturing Automation Series, McGraw-Hill, Delran, NJ.

Kusiak, A. (1988), ed., *Artificial Intelligence: Implications for CIM,* IFS Publications, Bedford, UK, and Springer-Verlag, NY.

Melkanoff, M. A. (1984), "The CIM Database: Goals, Problems, Case Studies and Proposed Approaches Outlined," *Industrial Engineering,* November, pp. 78–93.

Mitchell, F. H. (1991), "Information Flow and Computer Networks," Chapter 7 in *CIM Systems,* Prentice Hall, Englewood Cliffs, NJ, pp. 339–389.

Rembold, U., B. O. Nnaji, and A. Storr (1993), *Computer Integrated Manufacturing and Engineering,* Addison-Wesley Publishing Co., NY.

Shirley, G. V. and R. Jaikumar, (1988), "Turing Machines and Gutenberg Technologies: The Postindustrial Marriage," *Manufacturing Review,* Vol. 1(1), March, pp. 36–43.

Stark, J. (1988), *Managing CAD/CAM: Implementation, Organization and Integration,* McGraw-Hill, NY.

Stark, J. (1992), *Engineering Information Management Systems,* Van Nostrand Reinhold, NY.

Vail, P. (1988), *Computer Integrated Manufacturing,* PWS-KENT Publishing Company, Boston, MA.

Vajpayee, K. (1995), *Principles of Computer Integrated Manufacturing,* Prentice-Hall, Englewood Cliffs, NJ.

C H A P T E R

6 Production Planning and Scheduling in Automated Manufacturing Environments

The introduction of automated equipment and processes into a factory will affect a wide range of management practices. Perhaps the most significantly impacted areas are the systems and procedures used to manage production. Indeed, if significant reengineering of production planning and scheduling processes does not occur, the investment in automation may not yield its anticipated return. Without such changes, it is possible for the automated environment to be unworkable or for it to perform below the standards of the conventional system it has replaced. This follows, since production management decisions and technology capabilities are closely linked and together are a principal driver of plant performance. Unfortunately, many companies have made the mistake of installing new process technology along with management systems that are more appropriate for conventional processes.

In this chapter, we examine production management from the perspective of an automated, flexible factory. We do so by presenting a framework of the range of production planning and scheduling decisions that are affected by the enhanced intelligence and information handling capacity of automation. Within the context of this framework we focus on the interplay between automation and production management in determining the potential for responsiveness and flexibility. The chapter begins with a discussion of relevant system design decisions. It then goes on to consider methods and procedures for resource planning at the enterprise level, production scheduling for the plant, material requirements planning, detailed scheduling (shop floor control), and real-time (machine level) control.

6.1 Design of the Production System

There are three major classes of design decisions for production systems that have particular relevance to production management in an automated production environment (see Figure 6.1):

FIGURE 6.1

Design methodology

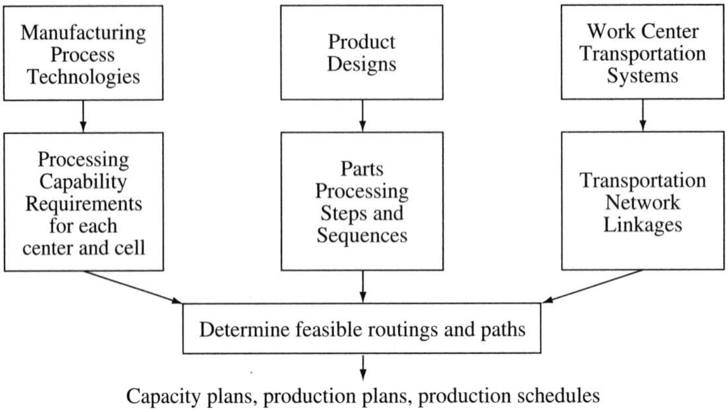

1. Selection of the manufacturing processing technology.
2. Design of the products to be manufactured.
3. Design of the material handling system.

We assume, in this chapter, that these design decisions have already been made, and thus these elements are fixed for the purpose of production planning and scheduling control. Derived from this design is a set of data which include:

1. *The processing capabilities* of the installed manufacturing equipment, i.e., the set of operation steps that can be carried out at each station, processing speeds, setup (changeover) times, yield rates, reliability, etc.,
2. *The processing requirements* for each product and component to be manufactured in the plant, i.e., specific operation steps that must be carried out as well as their tolerances, tool and fixture requirements, precedence relationships, bills of material, process plans, routings, and
3. *The transportation connectivity linkages* of the plant's material handling system, i.e., what machines can be reached from each location in the network; the transport system speeds; the load/unload speeds; pallet, bin, and magazine capacities; etc.

This comprehensive database provides a detailed picture of the firm's manufacturing environment. As indicated in Figure 6.1, these data, in combination, determine the collection of feasible process plans and routings for each product. In fact, one measure of a manufacturing system's flexibility is the number of feasible alternatives for producing products in an environment where dynamic rerouting can take place (see Chatterjee et al. [1987]).

A key requirement for production policy analysis is the development of a model of the factory. Such a model is needed to predict material and information flows under different design and control scenarios. A variety of approaches (e.g., mathematical programming, simulation, optimal control) of varying complexity and detail have been proposed for factory modeling. We will consider a number of simple examples of such models to illustrate how design and production management decisions interact to influence system cost and performance. It is instructive at this point, however, to take a closer look at the manufacturing process in order to gain a better perspective on how process equipment and production management decisions are related.

Consider, first, the nature of production activities in a factory where automated manufacturing equipment and processes have been installed. What are these activities and how should they be controlled? Additionally, what are the requirements for effective system performance? We adopt here the perspective that the manufacturing process can be broken down into a sequence of discrete processing steps (e.g., drill a hole, create a mold, grind a surface, assemble parts). At each step, a transformation of input material into output material (which may become an input to the next step) occurs. In some instances, steps require the presence of more than one item (e.g., in the assembly of components). Both product designs and manufacturing equipment capabilities can be specified in terms of these fundamental processing steps. Production planning and management are thus concerned with ensuring that the required resources are available at the right place and at the right time. Specifically, these decisions must guarantee that each product goes through its required sequence of processing steps by visiting appropriate production equipment with the capability to execute each required step.

The generation of feasible schedules is complicated by the fact that the following resources must be present *simultaneously* to complete a fundamental processing step:

1. The material to be worked on.
2. Material handling capability for loading/unloading of material.
3. The jig or fixture to hold the material in the appropriate orientation.
4. A machine with the appropriate processing capability to execute the step on the material (this could be either a specialized or multifunction machine).
5. Tooling appropriate for the machine, the material, and the processing step to be carried out.
6. Information in the form of an instruction set (i.e., NC program) to control the operation of the machine as it executes the processing step(s) (i.e., instructions to control tool movement, material placement, machine functions).

The requirement for simultaneous availability of these resources makes production planning and scheduling a multiple resource constrained activity. Satisfaction of resource capacity constraints affects scheduling decisions at all levels and for all products. At the same time, it is important to note that the purpose of the schedule is to enable the factory to produce a sequence of product batches which satisfy customer demand requirements. Moreover, as we have noted, both operating costs and performance can be fundamentally affected by the quality of scheduling decisions.

With this perspective in mind, we can state the fundamental objective of a production management system as follows:

> To ensure the availability of all classes of production resources, at the same time and location, in a manner which promotes completion of the overall manufacturing task and the generation of finished products in a timely and efficient manner.

This objective implies that a diverse collection of activities must be successfully coordinated. The necessary activities include the following:

1. Determine next processing destination for material.
2. Transport material to processing location.
3. Transport tools to processing location.
4. Load/unload tools from tool magazine or remote storage location.
5. Load/unload material at processing location.
6. Identify work-piece at the processing location.
7. Transfer instruction set from local or remote data storage to processor.
8. Align tool and material for initiation of processing step(s).
9. Monitor function of equipment as each processing step is executed.
10. Inspect material for conformance to processing step specifications.
11. Record completion of processing step(s) in material file.

As we have noted in Chapter 5, the infrastructure which supports automated processing equipment has the potential to enhance the level of connectivity between machines, both in terms of material and information flow. Such connectivity is essential, since events in a fully automated factory or cell must be managed on a real-time basis, without human intervention, with reduced interstage buffering, and without the occurrence of significant delays. Moreover, contingencies (such as machine failures, tool breakage, material starvation, and queue blockage) that have always been present in manufacturing systems must now be anticipated by the automated factory control system. The control system must therefore contain preprogrammed remedies and procedures for dealing with these contingencies. In general, such events can affect the operation of the entire factory equipment network and require recalculation of the full product schedule. Matters are complicated further by the fact that automated processing equipment is, by design, flexible: it is multifunctional and adaptive. Hence, as noted earlier, there can be an enormous variety of ways to produce a given product (e.g., alternative routings, transport paths, tooling). Selection from this vast choice set must be made either a priori (off-line) or in real time (e.g., on-line) in a fully flexible manner. Taken together, all of these factors increase the complexity of the planning and scheduling task in an automated manufacturing environment.

Some managers of automated manufacturing facilities feel that this increase in scheduling complexity has acted as a barrier to the successful implementation of automated manufacturing technology. It has become necessary, they argue, to create a new paradigm for production planning and scheduling. Others feel, however, that the fundamental issues of planning and scheduling management, noted above, have not been changed by the introduction of automation. Scheduling has always been a complex decision problem requiring the coordination of multiple resources. The introduction of automation, in fact, should serve to improve the performance of production planning and scheduling systems (i.e., produce at a lower cost, with higher quality, with more flexibility).[1] Such improvements would be based on the enhanced information processing, data access capabilities, and precision associated with automated equipment. As we have noted earlier in this book, modern factories fall into a continuum with regard to their use of automation. Consequently, we should expect to see a wide range of production management system implementations. As we shall see, this is indeed the current state of practice in industry.

6.2 A Hierarchy of Production Planning and Scheduling Decisions

The full complement of management decisions which must be made in order to run a factory can be broken down by the time frame, or planning horizon, considered by the relevant management systems and protocols. Figure 6.2 summarizes each stage of this hierarchy. Note that the longest planning horizon is measured in years and the shortest is measured in fractions of a second. In an ideal situation, selection of policies for each

FIGURE 6.2

Production planning decision hierarchy

- Aggregate Planning
  - Long Range (1–2 years)
  - Set planned capacity utilization, # shifts, employment levels
- Master Scheduling
  - Short Range (1–6 months or longest material lead time)
  - Determine future production targets (date, quantity, product)
- Materials Requirements Planning (MRP)
  - Explode the master schedule to obtain NET demand for all components, parts, and subassemblies
- Shop Floor Control
  - Control events from the time jobs are released to completion
  - Determine batch sizes, sequencing, routings
- Real-Time Control
  - Activate and direct the operation of automated equipment as the detailed schedule is implemented
  - Feedback and contingency response

[1] By one definition, a flexible plant should be able to produce a wide variety of products economically, i.e., with low switching costs and minimal delay. Moreover, flexible plants should be able to respond rapidly to contingencies. As discussed in Section 6.6, automated plants do not always achieve such flexibility (see Chapter 1 for a more complete definition of flexibility).

stage is made in a manner which takes into account the linkages across the levels of the hierarchy: the decisions made over the longer term constrain the decisions made for the short term, and the performance at any given level affects the performance at other levels. The complexity of the production management problem, however, often requires that a more separable, disjointed approach be taken.

At the highest level of the hierarchy are the *business planning* decisions. These decisions involve both manufacturing and nonmanufacturing functions (i.e., order acceptance and delivery terms, material sourcing and vendor selection, an aggregate production and factory capacity load plan). Many companies have adopted the goal of coordinating such long-range decisions across multiple business functions, at an enterprise level. The overall objective is to make the most effective use of the investment in plant equipment and infrastructure for production and delivery of value-adding products to the firm's customers.

The second level of decision making is referred to as *master scheduling*. Management, at this level, is concerned with medium-term (master) schedules, which determine the product mix to be made available in each review period to meet expected (firm order and forecast) customer demand. In most cases, it is necessary to consider explicitly the interaction of processing capacity requirements, material availability, and processing capacity constraints in the generation of master schedules (the plant loading problem). Section 6.3 introduces a basic model for analyzing master scheduling decisions.

At the third level of the hierarchy is the *material requirements planning (MRP)* system, which explodes the master schedule to obtain net demand for all components, parts and subassemblies. The timing and quantities of these requirements are computed on the basis of each product's bill of material, expected production/procurement lead times and the availability of both on-hand and on-order inventory. *Detailed scheduling (shop floor) control* constitutes the next (fourth) level in the production planning decision hierarchy. It refers to the management of the specific activities associated with running an actual production process (i.e., batch sizes, product sequencing). The time frame of a detailed schedule is usually quite small (e.g., day, hour). Finally, there is a fifth level of *real-time control* which is typically executed by *programmable logic controllers (PLCs)* to activate and direct the operation of the automated equipment as the schedule is implemented. In this chapter, we focus primarily on the upper and middle levels of production scheduling: aggregate planning, master scheduling, MRP, and detailed scheduling. We do so since these policy areas are especially relevant for determining the impact of factory automation on factory performance.

In order to illustrate the decision hierarchy, consider the production management requirements for automobile assembly. Figure 6.3 illustrates the various stages of production and the break down of the finished product into its constituent subassemblies and principal components. The long-term, aggregate plan is typically formulated for a planning horizon of one year. Facility and technology choices have been made and hence the production planning challenge is to develop the most effective use of the capacity in place. The aggregate plan thus allocates car lines to plants and determines planned levels of output for each assembly, subassembly, and processing line. These output targets, in turn, specify the number of shifts, and the level of employment required at each plant. This aggregate plan is driven by a long-term forecast of shipment volumes for each model and car line.

FIGURE 6.3

Auto assembly

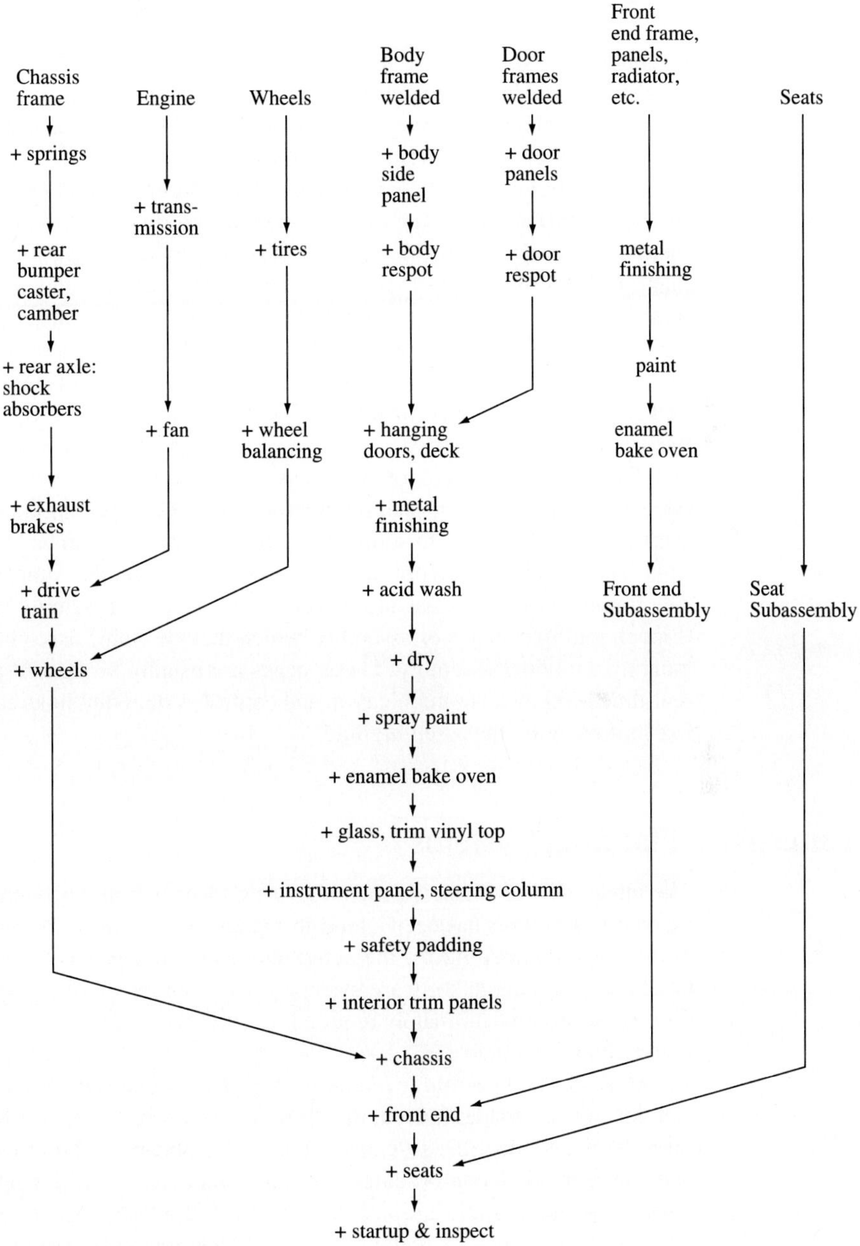

Source: Adapted from Ayres (1990)

The planning horizon for master schedules in auto assembly determines the number of models of each type (color, option mix, etc.) which are to be assembled in the plant over a multiweek planning horizon. The length of this planning horizon has been reduced significantly due to the introduction of lean production practices in the automobile industry. Nonetheless, it is still necessary to plan product mix out through the longest material lead time (which could be several months for an offshore supplier). Co-location of vendors and the use of just-in-time inventory management practices also have reduced these procurement lead times.

The material requirements planning step in auto assembly leads to order releases for each component and subassembly needed to produce a car. As noted in Figure 6.3, there are a small number of major subassemblies which are involved. Certain feeder plants or lines (for engines, the chassis, the frame, or the seats) produce a limited range of products that are used in multiple model choices. Other subassemblies exhibit more variety (e.g., trim, paint, instrument panel). The production schedule, in all cases, is linked to the finished product master schedule. Due dates and net order quantities for all inputs (both internally and externally sourced) are generated by the material requirements planning system.

The detailed schedule for an auto assembly plant determines the sequence of cars to be released to the line on each production day. Each car is unique in terms of its design options and accessories. In some cases, these differences can be extreme (e.g., left- and right-hand-drive cars are produced in random order, on the same line, in some plants). The frame and model class are usually fixed for each assembly line. Automobiles go through multiple stages of assembly, which include frame assembly, engine mounting, painting, and final assembly. These stages are usually sequenced and coordinated at a real-time level by a communication and control system that links each piece of processing equipment on the assembly line.

6.3 Enterprise Planning Systems

The integration of financial, manufacturing, distribution, accounting, purchasing, and marketing functions has been a long-term management goal. The need for such integration is most compelling for the activities involved in production. Accordingly, multifunction management software systems have been developed to support such integration. These systems ultimately require linkage between a wide range of function specific transactions and databases.

Manufacturing resource planning (MRPII) systems have been commercially available for over a decade. Indeed, the effective use of these systems has become a benchmark for excellence (the so-called "Class A" companies). At the same time, the difficulty in achieving conformance and consistency in MRPII environments has been noted, (see the Digital Equipment Corporation Endpoint Model case study (Simpson [1988]), for example). A major stumbling block appears to be the degree of discipline required to enforce the fully linked schedules and plans, generated across the functions, by the MRPII program. Indeed, schedules often must be frozen for weeks at a time. Accordingly, MRPII has been promoted as a means of introducing "rigor" into the chaotic world of manufacturing.

FIGURE 6.4

Graphical overview of the SAP™ R/3™ system

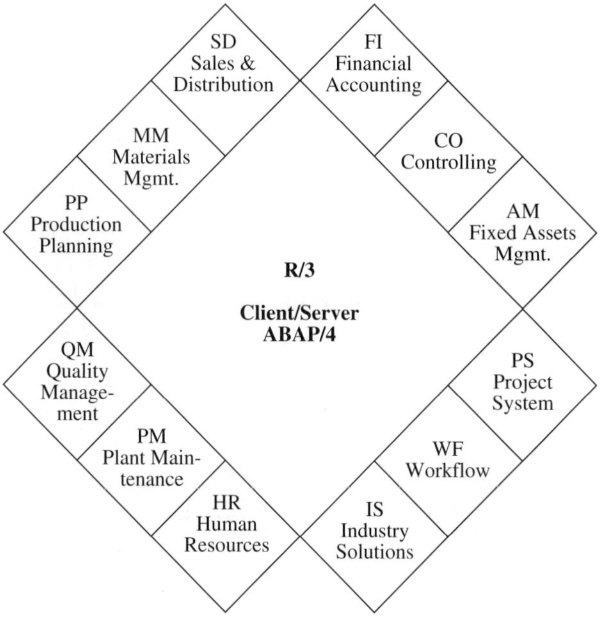

More recently, systems have been developed to support change dynamically in response to changes in business conditions. New generation software and hardware, using distributed processing and client-server approaches, have emerged. These new, *enterprise resource planning (ERP)* systems provide users with the flexibility to change schedules at different levels in the production management hierarchy on a real-time basis. One of the most successful systems is the SAP™ program. Figure 6.4 is a schematic of SAP's R/3™ system architecture. In addition to user friendly interfaces (graphical, mouse driven), the system supports the integration of a wide range of processes. These processes can be linked in responding to customer orders and other changes in the environment.

6.4 The Aggregate Planning Problem

In many industries, demand for finished products exhibits significant variation over time. The source of such variation includes market seasonality (e.g., demand for candies peaks just before Halloween), inherent randomness (e.g., demand for service parts for the maintenance of computer systems), or a combination of these factors (e.g., demand for pesticide chemicals in the agricultural industry). Figure 6.5 illustrates a typical demand pattern for a hypothetical manufacturer of home garden tilling equipment. The pattern in the figure indicates two seasonal surges (one for spring planting and the second for fall cleanup), as well as the pipeline lag associated with getting product from the plant to the retailer for three product models.

FIGURE 6.5

Sales forecasts for rototillers

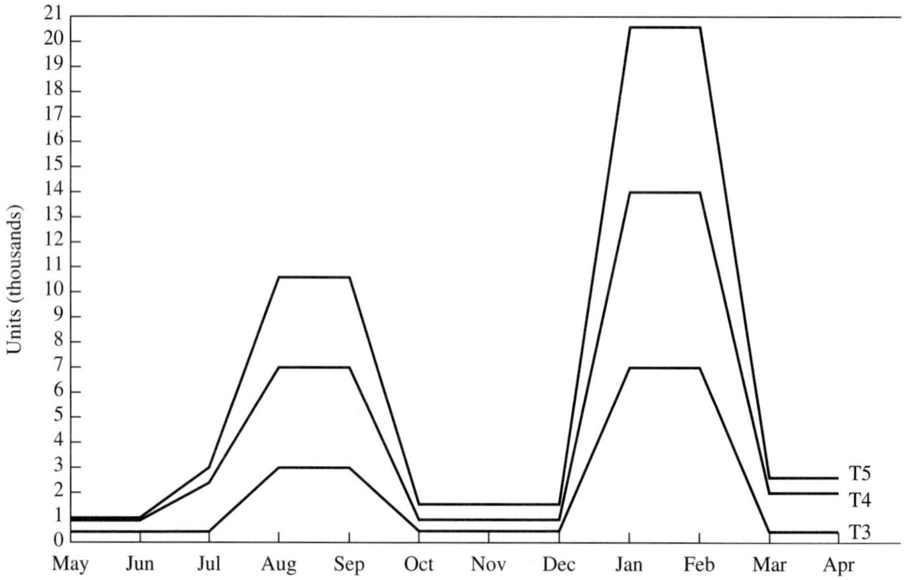

As noted earlier, capacity decisions are typically made on a long-term basis. As a result, firms must devise strategies to absorb demand fluctuations in the context of their existing design solution (i.e., for the given set of facilities, products and processes). Such buffering strategies can be classified into two broad categories (see Galbraith [1969]): *independent* and *coordinated.* The independent strategies are concerned with those actions that the firm can carry out on its own initiative, while the coordinated strategies require joint decision making of multiple firms.

Independent buffering strategies include use of various buffering substitutes that act to absorb demand volatility. The following is a comprehensive list of such tactics:

- Utilization of facilities through adjustment to the rate of production with the current work force level (e.g., use of overtime, length of work week).
- Investment in seasonal inventories.
- Variation in customer service (backorder delay and delivery promises).
- Layoffs and hiring to support changes in the output rate (open second shift or idle lines).
- Subcontracting of bottleneck operations.
- Adjustment to the product mix to smooth production capacity requirements.

Taken together, these tactics define a production smoothing strategy for the plant. Common variants include the "chase" plan, which makes frequent adjustments and the "level load" plan, which attempts to minimize production rate change. The consequences of each strategy option (on cost, performance, reputation, etc.) should be considered explicitly.

The firm, acting independently, can also attempt to influence the timing of its demand through various pricing and promotional activities (e.g., early order discount offers to wholesalers, retail discount coupons, advertising campaigns, peak load pricing). In the long run, the firm may choose to adapt its system design so that it operates in a manner which is more effective in dealing with its demand pattern. The notion of *manufacturing flexibility,* which is central to this book, can be viewed as an adaptation that is achieved through the mechanism of automating the production process. Firms can also select plant locations in communities where the local labor force is willing to undergo repeated cycles of layoffs and rehiring.[2] Finally, a firm can choose to expand its product line to incorporate counter-cyclical markets with similar manufacturing process requirements (e.g., lawn mowers and snow blowers).

Coordinated strategies are reflected in the market structure which emerges in a particular industry. For example, in the coal industry, there are a small number of large mines and a larger number of small, independent operations. The large mines tend to be integrated with steel plants and other major users. They operate at a fixed level and have large fixed cost investments. The small operations, in contrast, are more flexible. They can modify their production rate on a seasonal basis, at a lower adjustment cost. As a result, the smaller operators capture the market for heating and other seasonal uses.

Firms can also form alliances and coalitions to share the costs and risks associated with demand volatility. Customers and suppliers, in particular, can coordinate their activities along the supply chain to develop unified strategies for investing in buffering resources. The current growth in *efficient consumer response* programs in the grocery industry is an example of such behavior. In Chapter 7 we discuss the emerging technology of *electronic data interchange (EDI),* which can facilitate such integrated action.

6.5 The Master Scheduling Problem and Material Requirements Planning

The next (medium-term) level of planning in the decision hierarchy of Figure 6.2 is *master scheduling*. Firms, on a rolling horizon basis, must determine the mix of finished products to be made available by the factory over a planning horizon, which is usually defined by the longest input material lead time. In general, there are multiple products competing for the production capacity which has been made available through the aggregate plan. Demand forecasts (or firm orders) for individual finished product items are

[2] Traditionally, chocolate manufacturers were located in rural areas. They hired farm labor in the high demand winter months when farm workers were less busy. They then laid off the workers in the summer months when demand for chocolate was reduced and when the pace of activity on the farms was higher.

usually available. The key issue here is to develop a schedule which is both feasible and cost-effective. In Appendix A, we introduce a formulation of a standard version of this cost trade-off problem. The key decision variables in this model include: (1) the number of regular and overtime hours a work force of a fixed size will work, (2) the number of units of each product to be produced, and (3) the ending inventory level of each product. These decisions are made in each planning period (e.g., week) over a finite horizon (e.g., several months to a year). The cost structure includes fixed and variable production costs, labor costs, and inventory holding costs. The constraints incorporated in the model account for inventory balance and the relationship among hours worked, hours needed to set up machines due to product changeover, and the effective available production capacity (i.e., the number of hours required to support the product mix of the master schedule).

Many firms solve their master scheduling problem in an iterative fashion, with the aid of a *material requirements planning* system. Figure 6.6 illustrates the logic of such

FIGURE 6.6

Closed-loop material requirements planning

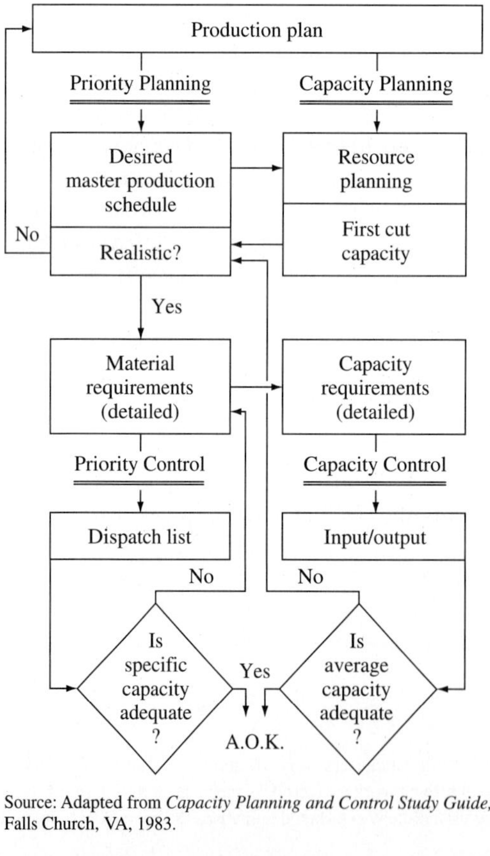

Source: Adapted from *Capacity Planning and Control Study Guide*, American Production and Inventory Control Society, Falls Church, VA, 1983.

"closed loop" MRP systems. The basic idea here is to use the MRP system to simulate the demands on capacity (i.e., capacity loading) and on material associated with a trial master schedule. If the trial schedule is feasible, then the firm can go ahead and implement it by releasing production and purchase orders. If the schedule is infeasible (e.g., a work center is loaded beyond its expected capacity or a component will not be available in time), the firm has two choices—it can proceed as if the schedule is feasible, but be alert to the expected time and location of the bottleneck (infinite loading), or it can adjust the master schedule through the shifting of production quantities over time (finite loading). This iterative procedure can be viewed as a kind of heuristic solution to the constrained optimization problem introduced in the Appendix.

The mechanics of how MRP systems take a master schedule and convert it into a sequence of time-phased requirements for individual components, material inputs and subassemblies are well known (see Nahmias [1993] for a basic reference on this topic). It is important to note that all MRP systems generate such plans on the basis of deterministic expectations for key process variables, e.g., production lead times and process yields. The details of scheduling released orders are accomplished by the detailed scheduling system, which must account for the inter-relationship among products, production resources, and process variables on the automated factory floor. This topic is the subject of the following section.

6.6 Detailed Scheduling—Shop Floor Control

The scheduling of shop floor activities is at the heart of manufacturing control. The *detailed schedule* specifies the sequence of activities, at the work center level, which is required to implement the firm's master schedule. As a consequence, the detailed schedule usually represents the final opportunity for management to exercise some degree of control on the allocation of scarce manufacturing resources. It is also a key lever that determines the overall flexibility of the production system in responding to change.

A detailed schedule is typically represented by a *Gantt chart,* which identifies start and completion times for each job at each operation or work center (see Figure 6.7). There are a number of decisions that must be made as a prerequisite for the generation of such schedules. In particular, it is necessary to decide which parts (or products) are to be produced by the particular production system being scheduled (e.g., on the FMS). The next step is to determine a loading schedule which specifies start times (for system entry) for each part. It is also necessary to assign parts and tools to specific machines (from the set of machines which possess the required production capability). The final prerequisite for shop floor control involves specification of dispatch rules which are used to control the internal timing of each part's movement through the multiple processing stages in the production network.

These dispatch rules answer the question of what to work on next when a choice exists. As we shall see below, there is a final, real-time level of control in the overall production planning hierarchy that is unique to automated manufacturing systems. These real-time control systems run the equipment and contain algorithms for determining actual system response to the realization of production contingencies. The role of dispatch

FIGURE 6.7

Gantt chart (2 jobs, 3 operations)

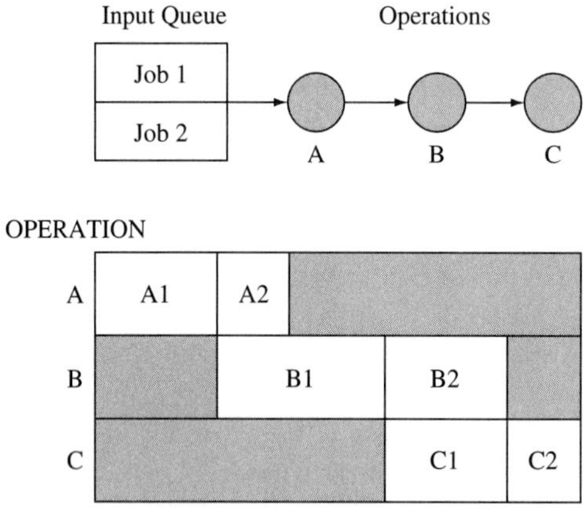

rules is primarily to control input to the automated system. They also should be considered explicitly when management assesses the impact of alternative schedules on factory throughput, WIP, and customer delays. Such assessments are usually carried out with the aid of a detailed model of the manufacturing system.

There are two broad classes of dispatching rules: *static* and *dynamic*. Static rules are order file oriented and are based on part characteristics (i.e., processing times, due dates, unit cost, margin). Dynamic rules are machine-oriented and are based on the load status of all of the machines. Dynamic rules are typically concerned with issues of load balance and equipment utilization. An example of a dynamic rule used in FMS environments is the NEP (next empty pallet) rule, which replaces a completed part with a similar part as soon as an empty pallet has been returned to the input load station. NEP maintains a constant population of part numbers in the system.

In order to implement production schedules, the factory must have the capability to answer specific questions associated with shop floor material flow. Figure 6.8 illustrates how these decisions interact in the context of a FMS in which all parts must be loaded onto pallets prior to launch. The three classes of material flow decisions associated with running an FMS are:

1. *Part Load Plan:* What part should be loaded next at the load station that exists to buffer the FMS from upstream activities (i.e., by using a dispatch rule)? Note that a part can be loaded only if (i) the order for the part has been released, (ii) an appropriate pallet is available, (iii) the raw material or WIP required for process is available, and (iv) there is an open slot in the load buffer queue.
2. *Part Launch Plan:* When should a loaded part be released from the input buffer to the manufacturing system? A rule commonly used in FMSs is ASAP (as soon as possible), in which the part needed to fill an open slot in the routing's first work center input buffer is launched when both the part and the slot become available.

FIGURE 6.8

Shop floor material flow

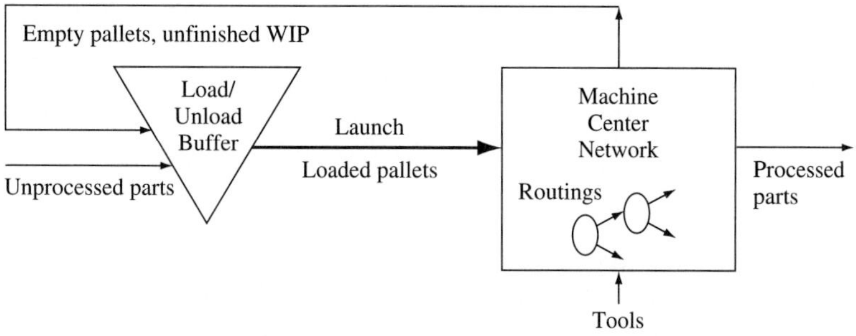

3. *Routing Scheme:* After launch, how should a part be routed from station to station, given the existence of routing alternatives and the occurrence of contingencies such as machine failures, buffer blockage, and machine starvation? A typical routing rule here is to select the machine with the least amount of WIP in its input buffer. More complicated rules could look at both current and projected system status.

As noted earlier, there are a variety of trade-offs and constraints associated with scheduling decisions. A variety of optimization models have been developed to specify algorithms for improving the quality of such decisions. In general, these models conform to the hierarchical decision framework for scheduling decision making, which has been discussed above. They often yield approximate, heuristic solutions since the underlying problem is impossible to solve to optimality.

Recent developments in this area include the *OPT scheduling* system (see Fox [1983]) and *derivative "bottleneck" control* systems, Morton and Pentico [1993]). These systems can schedule a network of machine centers in a manner which accounts for the interaction of each decision with all parts of the system as products contend for limited production resources. In general, bottlenecks shift in an unpredictable fashion in response to even minor schedule adjustments. Black and Decker has recently implemented such a system (known as "Rhythm," supplied by i2 Technologies). B&D's system provides them with the capability to generate schedules "on the fly" for an entire factory: the full shop floor schedule can be regenerated in minutes instead of on a weekly basis (typical of MRP scheduling systems). This capability has enhanced the level of flexibility at the plant level in responding to changes in the production and market environment. Such flexibility, for example, can lead to reduced inventories, since it is no longer necessary to buffer against the weekly schedule regeneration cycle. The capability to schedule at the machine level can lead to a dynamic reallocation of production capacity through the use of "virtual cells." Internally, B&D refers to its system as *MRPIII,* since it represents the next generation of integrated production planning.[3]

[3] This report is based on communication from A. Strumar, former vice president of manufacturing technology, Black and Decker Corporation, and a report on the system in *The Economist* (1994).

An interesting example of an analysis of scheduling in a FMS environment was carried out at Caterpillar (see Flanders and Davis [1995] at the end of this chapter). The system studied consisted of seven milling machines, a fixturing station, a tool station and an automated guided vehicle (AGV) system. It was used to produce a repeated daily ensemble of parts for transmission housings. The simulation model was developed to investigate a shortfall in system throughput: less output was being generated than expected. The simulation model indicated that the root cause was the process for the scheduling of tools for the milling machines. As tool order fill time increased, the overall system became bottlenecked. This is a concrete example of the impact of multiple resources on scheduling effectiveness in an automated manufacturing environment.

By one definition noted earlier, a flexible plant can produce a wide variety of products with a low switching costs and without excessive delay. By another definition, the flexible plant has the ability to adjust to changes in its production environment (such as those caused by a machine breakdown). Experimental results and related empirical observations suggest that there is a paradox in managing flexible systems. In spite of the ability of flexible systems to produce a wide range of products with minimal batch sizes, the effectiveness of such operations declines when there is disruption due to uncertainty. Consequently, flexible systems must be managed with a high degree of precision (i.e., through superior maintenance, high quality inputs, and attainment of high levels of equipment reliability) in order to avoid such disruptions, and with a high level of schedule adaptiveness (e.g., by recomputing the detailed schedule on a rapid or even real-time basis in response to these contingencies). As input variety increases, it is plausible to argue that the likelihood for disruption will increase, as will the computational burden of coming up with even reasonable/feasible production schedules. As Jaikumar (1986) noted in a landmark study, some US flexible manufacturing system managers have responded to this risk by limiting the scope of products to be produced in a flexible system, thereby defeating the purpose of investing in flexible systems in the first place.

6.7 Real-Time Control Systems

We conclude this chapter by noting that automated manufacturing systems must ultimately operate on a real-time basis without human intervention. Consequently, there must be a process for controlling equipment as it executes the production schedule, i.e., to switch equipment on and off. PLCs, which interface with the processor controls of each piece of automated equipment, are used at this level. In general, there are a variety of "disturbances" which can (and do) occur as the system operates. For example, a machine may break down or be restored, or the schedule due date of an order may be changed. In addition, input material may become unavailable or an unexpected priority order may be released to the shop floor.

Figure 6.9 illustrates the hierarchical structure of an on-line shop controller (Odry and Wilson [1987]) that could manage a network of cell controllers. Each controller is connected to a particular manufacturing cell. The management of these controllers can be broken down into two stages: *off-line planning* and *on-line shop control.* The off-line system determines alternative cell routings for different products and allocations of both buffer stock and pallets to each cell. The on-line control system readjusts the off-line

FIGURE 6.9

Hierarchical production planning and control scheme for FMS

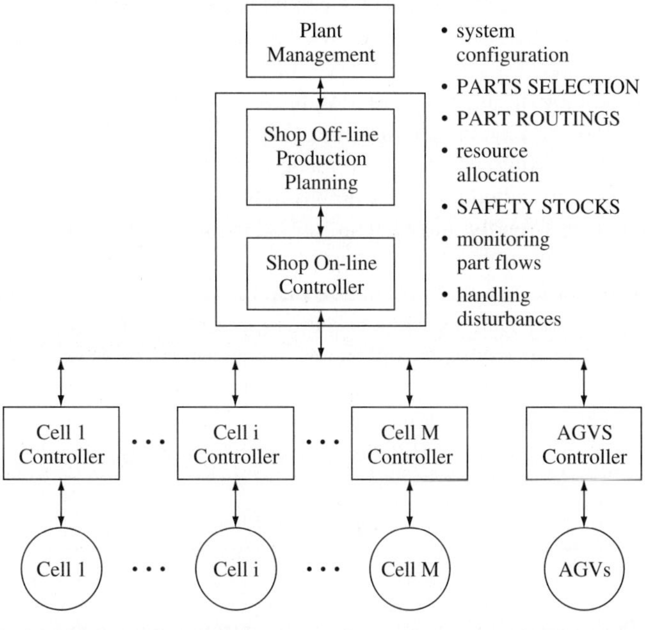

Source: Odry and Wilson (1987)

plan in response to contingencies. These contingencies are based on disturbances which could include a machine going down (failure), a machine going up (restoration), arrival of a priority order, a schedule due date change, and slippage in schedule performance. The control system described by this research adjusts the allocations for each disturbance under the assumption that such adjustment will be completed before the arrival of the next disturbance.

Manufacturers of automated manufacturing systems must include some version of such real-time control systems as a part of their production equipment. Research on improving the design of these control systems is ongoing by equipment manufacturers and manufacturing system scientists.

6.8 Conclusions

This chapter has reviewed the full hierarchy of decisions that must be made by production planning systems in automated environments. As we have seen, a key issue in the design of such systems is *integration.* In particular, there must be integration of decisions over different time horizons, integration of decisions made at different levels of the hierarchy, and coordination of decisions across multiple products, processes, and factory work centers. At the present time, the attainment of complete integration is beyond the capability of most firms.

We observed, in particular, that the attainment of an automated system's flexibility potential is, to a large extent, determined by the performance of its planning and scheduling systems. Such performance can be linked to the responsiveness of these systems to contingencies. At one extreme, we see control systems where many scheduling decisions are made on a real-time basis. Such systems must adopt a myopic planning horizon (i.e., they do not consider long-term, systemic consequences of the local decision being made). At the other extreme are off-line systems where the full range of decisions for a fixed planning horizon are made on a periodic basis (e.g., the master schedule/MRP level of scheduling). In between are the ERP systems, which provide an integrated, user-friendly mechanism for adjusting plans and schedules in response to change.

It is fair to say that production planning and scheduling systems and related integrating support systems have, to date, represented a barrier for the achievement of the full potential of manufacturing technology automation. Recent software system, modeling and algorithm developments hold out the promise for the attainment of true flexibility. Research and development in this area is ongoing. As improved solutions become available, the effective utilization of automation will increase.

Study Questions

1. What components/resources must be present simultaneously to complete production processing steps in a factory?
2. What is the main objective of a production management system, and how does automation affect it?.
3. What are the five levels in the production planning hierarchy, and how are they related?
4. What are the principal trade-offs and constraints that should be considered in solving the master scheduling problem?
5. What makes detailed scheduling/shop floor control so difficult? Does automation mitigate this difficulty? How?
6. What is a dispatching rule? How does it affect the operation of an FMS?
7. Are production planning and scheduling systems barriers for the achievement of the full potential of manufacturing automation technology? Explain.

References

Ayres, R. U. (1990) "Computer Integrated Manufacturing," Vol. I, in Revolution in Progress series, Chapman and Hall, London.

Chatterjee, A., M. Cohen, and W. Maxwell (1987), "A Planning Framework for Flexible Manufacturing Systems," Working Paper, The Wharton School, University of Pennsylvania.

The Economist (1994), "Manufacturing Technology: On the Cutting Edge," March 5.

Flanders S. W. and W. J. Davis (1995), "Scheduling a Flexible Manufacturing System with Tooling Constraints: An Actual Case Study," *Interfaces,* 25, pp. 42–54.

Fox, R. E. (1983), "OPT vs. MRP," *The Magazine for Materials Management,* Vol. 3, no. 6, November/December.

Galbraith, J. R. (1964), "Solving Production Smoothing Problems," *Management Science,* Vol. 15, no. 12, pp. B665–B674.

Hax, A. C. and D. Candea (1984), *Production and Inventory Management,* Prentice-Hall, NY.

Jaikumar, R. (1986), "Postindustrial Manufacturing," *Harvard Business Review,* November–December, pp. 69–76.

Morton, T. E. and D. W. Pentico (1993), *Heuristic Scheduling Systems,* J. Wiley & Sons, NY.

Nahmias, S. (1993), *Production and Operations Analysis,* Irwin, Burr Ridge, IL.

Odry, N. G. and G. R. Wilson (1987), "An On-line Central Structure for Flexible Manufacturing Systems," in Manufacturing Systems Integration and Processes, 15th Conference on Production Research & Technology, NSF Manufacturing System Program, pp. 85–90.

Simpson, J. L., "Digital Equipment Corporation: The Endpoint Model (A)" Harvard Business School Case # 9-688-059, Harvard Business School Publishing, Boston, MA.

Appendix A
Capacitated Lot Size Model[1]

In this Appendix, we present a linear programming model of the master scheduling problem. This model makes the following key assumptions:

- Multiple products compete for limited production capacity.
- Work force size (capacity) is fixed.

There are N products and T time periods in this problem. The model minimizes total costs subject to constraints on inventory balance, production hours capacity and work force size.

$$\min Z = \sum_{i=1}^{N}\sum_{t=1}^{T}[S_{it}\delta(X_{it}) + v_{it}X_{it} + C_{it}I_{it}] + \sum_{t=1}^{T}(r_t W_t + o_t O_t)$$

Subject to

$$X_{it} + I_{i,t-1} - I_{it} = d_{it} \qquad \forall i, t$$

$$\sum_{i=1}^{N}[a_i\delta(X_{it}) + k_i X_{it}] - W_t - O_t \le 0 \qquad \forall t$$

$$0 \le W_t \le \tilde{R}_t \qquad \forall_t$$

$$0 \le O_t \le \tilde{O}_t \qquad \forall_t$$

$$\delta(X_{it}) = \begin{cases} 0 & \text{if } X_{it} = 0 \\ 1 & \text{if } X_{it} > 0 \end{cases}$$

Decision Variables:

X_{it} = production volume of product i in period t
I_{it} = ending inventory level of product i in t
W_t = manhours of regular labor available in t
O_t = manhours of overtime labor available in t

Model Parameters:

d_{it} = demand (forecast) for product i in t (units)
v_{it} = unit production cost for product i in t (\$/unit)
c_{it} = inventory carrying cost in t (\$/unit)
S_{it} = setup cost in t (\$/setup)
r_t = cost of regular labor hour (\$/hour)
o_t = cost of overtime labor hour (\$/hour)
k_i = manhours required for production of one unit of i
a_i = manhours required for setup of production batch for product i
$\tilde{R}_t$ = total regular hours available
$\tilde{O}_t$ = total overtime hours available

[1] See Chapter 3 in Hax and Candea (1984) for a more detailed discussion of the model formulation.

It is instructive to consider alternative sequences of production. Consider, for example, a case where there are three periods under consideration. There are four production sequences, for each product, which do not involve lot splitting. Tables A1 and A2 indicate the production quantities and total cost of each feasible sequence, respectively.

TABLE A1 Amounts to Be Produced for Each Sequence

Sequence	$t = 1$	$t = 2$	$t = 3$
1	$X_{i11} = d_1 + d_2 + d_3$	$X_{i12} = 0$	$X_{i13} = 0$
2	$X_{i21} = d_1 + d_2$	$X_{i22} = 0$	$X_{i23} = d_3$
3	$X_{i31} = d_1$	$X_{i32} = d_2 + d_3$	$X_{i33} = 0$
4	$X_{i41} = d_1$	$X_{i42} = d_2$	$X_{i43} = d_3$

where X_{ijt} = amount produced of item i in period t by means of sequence j

TABLE A2 Total Cost for Each Sequence

Sequence	*Setup and Holding Costs for Each Sequence* (t_{ij})
1	$t_{i1} = S_{i1} + V_{i1}(d_{i1} + d_{i2} + d_{i3}) + C_{i1}(d_{i2} + d_{i3}) + C_{i2}(d_{i2})$
2	$t_{i2} = S_{i1} + S_{i3} + V_{i1}(d_{i1} + d_{i2}) + V_{i3}(d_{i3}) + C_{i1}(d_{i2})$
3	$t_{i3} = S_{i1} + S_{i2} + V_{i1}(d_{i1}) + V_{i2}(d_{i2} + d_{i3}) + C_{i2}(d_{i3})$
4	$t_{i4} = S_{i1} + S_{i2} + S_{i3} + V_{i1}(d_{i1}) + V_{i2}(d_{i2}) + V_{i3}(d_{i3})$

The costs indicated in Table A2 can be defined (for item i in sequence j) as

$$t_{ij} = \sum_{t=1}^{T} [S_{it}\delta(X_{ijt}) + V_{it}X_{ijt} + C_{it}I_{it}].$$

Similarly, total labor resources consumed by producing X_{ijt} under sequence j are

$$l_{ijt} = a_i\delta(X_{ijt}) + l_iX_{ijt}.$$

Linear Programming Formulation

A simple linear program of the mixed integer problem defined above can be formulated as a problem of selecting the optimal sequence. It is necessary to list every possible production sequence for each time $(j = 1, \ldots, J)$.

Let

Θ_{ij} = fraction of j'th production sequence used for product i

J = total number of dominant (not lot-splitting) production sequences

The LP can be written as

$$\min Z = \sum_{i=1}^{N} \sum_{j=1}^{J} t_{ij} \Theta_{ij}$$

Subject to

$$\sum_{i=1}^{N} \sum_{j=1}^{J} l_{ijt} \Theta_{ij} \leq \tilde{R}_t, \qquad t = 1, \ldots, T \qquad (*)$$

$$\sum_{j=1}^{J} \Theta_{ij} = 1, \qquad i = 1, \ldots, N \qquad (**)$$

$$\Theta_{ij} \leq 0 \qquad \forall i, j$$

Note that there are $T + N$ constraints and $N \times J$ variables, where $J = 2^{T-1}$.

This LP is a "relaxation" of the sequence selection problem, since we do not require Θ_{ij} to be a 0/1 variable. The following observations can be made about this model:

- There are $T + N$ constraints, which implies that at most there will be $T + N$ positive (basic) variables in the optimal solution.
- There is at least one positive variable for each sequence as a result of constraints (**).
- Therefore, at most there will be T instances where more than one Θ_{ij} in a sequence is positive.
- Since, in general, $N \gg T$, there will be few fractional values, and hence the LP will generate a good approximation for the underlying integer program (with $\Theta_{ij} = 0$ or 1).
- It is easy to include other resources with this model.
- Note, for example, that if $T = 12$ and if there are $N = 1000$ items, then the number of variables $N \times J = 1000 \times 2^{12-1} = 2{,}048{,}000$.

In practice, one must use decomposition algorithms to solve this problem.

Reading 6.1

Scheduling a Flexible Manufacturing System with Tooling Constraints: An Actual Case Study

Interfaces

Seth W. Flanders
Lehigh University

Wayne J. Davis
University of Illinois at Urbana-Champaign

Material handling and the provision of supporting resources, particularly tools, in flexible manufacturing systems (FMSs) can constrain production. To improve the productivity for an existing FMS, we first developed a simulation model that included constraints arising from material handling and tool management. Detailed simulations of the FMS showed that a simple set of heuristic rules could produce several alternative schedules of nearly optimum performance. They also showed that when the flow of tooling was considered the FMS was operating near or over capacity. The goal of the study—to generate slack production time for the manufacture of additional parts was unachievable. Clearly companies must consider the constraints arising from material handling and the flow of supporting resources in scheduling, yet these constraints have seldom been considered in simulations of FMSs.

Scheduling flexible manufacturing systems (FMSs) has been addressed by hundreds of technical articles, yet scheduling remains a major concern for most FMS operators. Numerous approaches to solving the problem have been attempted, including mathematical programming, heuristic searches, genetic algorithms, and simulated annealing, but all methods have limitations. Currently, the goal of developing a robust solution methodology to address the wide breadth of scheduling problems arising in a flexible manufacturing environment remains unfulfilled.

Most published scheduling approaches focus upon job flow only. That is, given a set of jobs to be done, the scheduler seeks to discover the order in which to assign each manufacturing task to the available processes. Seldom do schedulers consider the assumed secondary concerns pertaining to material handling and the provision of supporting resources, such as tooling. These constraints are nearly impossible to express mathematically, and their inclusion in a simulation model is difficult to accomplish.

In 1990, Caterpillar, Inc. came to us with a scenario that appeared to present an ideal scheduling problem. The FMS under consideration was used to manufacture a repeated ensemble of parts each day. The manufacturing engineers had optimized the processing plans for the required part types so that every part type of a given material type (either steel or aluminum) could be produced using a single set of tools positioned at each machine dedicated to processing that material type. The goal was to determine the schedule that would maximize the availability of the FMS for the production of additional parts while insuring that the ensemble of required parts would be produced each day. While we initially believed that we could achieve this goal, our efforts to generate this schedule demonstrated otherwise.

Source: Reprinted by permission, Seth W. Flanders and Wayne J. Davis, *Interfaces*, Vol. 25, no. 2, 1995, pp. 42-54.

Description of the FMS

Figure 1 shows the FMS. The system consists of seven Scharmann numerically controlled milling and drilling machines, one fixturing station where parts are introduced into and removed from the system, one tooling station where tools are introduced into and removed from the system, and four automated guided vehicles (AGVs) that deliver both jobs and tools to the stations.

The FMS produces several different transmission bell housings made of either iron or aluminum. The processing of these two material types must be segregated on different machines because different tools are used to machine the two materials and there is a desire to segregate the cuttings for recycling. At present, two of the seven milling machines are used to process aluminum parts while the rest are used for iron parts.

Orders for replacement tools eventually become too frequent.

A typical part requires three distinct fixturing steps, each corresponding to a different orientation of the part. For each fixturing step, the milling machine will execute several machining instructions. The process engineer has designed the machining instructions so that any fixturing step for any part of a given material type (iron or aluminum) can be executed using the standard ensemble of tools provided at each machine dedicated to that material type.

FIGURE 1

The modeled FMS includes seven milling machines, a fixturing station, and a tool station. Four AGVs travel along the one-way paths.

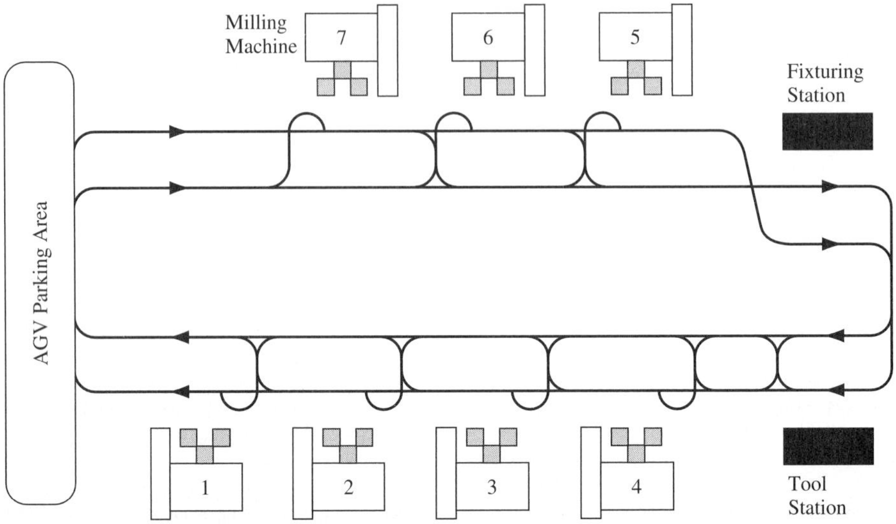

The AGVs deliver tools using the same network they use for parts. To accomplish a tool delivery, the tool setter first places the replacement tool on a tool carrier that can hold up to 16 tools. After the AGV delivers the tool carrier to the milling machines, the machine's worktable positions the tool carrier so that the tool can be placed into the machine's collet. The carrier is next withdrawn, and the tool exchanger unloads the tool from the collet and places it in the machine's tool magazine. This process is repeated for each replacement tool on the carrier. The procedure is reversed to remove worn tools from the tool magazine and to place them on the tool carrier. The AGV finally returns the tool carrier with the worn tools to the tool station. The machine's tool magazine can store up to 80 tools, but an oversized tool may require several positions in the magazine.

The tool station is responsible for refurbishing and storing tools. Each tool has a different maximum life, which is defined as the total machining time the tool should be able to withstand without becoming so fatigued that it breaks or so worn that it will not cut properly.

Because the AGV tracks are one-way only, travel distances between points within the system are direction dependent. For example, the travel distance from machine 1 to machine 2 is 374 feet, while the travel distance from machine 2 to machine 1 is only 52 feet. Clearly, the scheduler must consider travel distances in determining which AGV will respond to a given transportation request. Finally, the AGVs are capable of carrying only one part pallet or tool carrier at a time.

At each machine are two on-line queuing positions for parts that are waiting for processing, parts that have finished processing, or tool carriers. Each machine also has two off-line buffer positions where parts waiting for processing can be stored. These off-line areas do not interact with the work area; the AGV must move a stored entity from an off-line queue into an on-line queue. For this reason, the off-line positions are typically used for storing only pallets that are not in use, not for storing fixtured parts or tools needed in production.

There are 24 pallets available to the system. The pallets are metal platforms that have been specifically designed to be handled by loading systems for the AGVs and the machining centers. Each pallet employs a custom fixture to hold a part during the machining process. Some pallets can carry only one specific fixture of a given part type, but others can carry multiple fixtures, allowing more than one part at a time. For the multiple fixtured pallet, usually all three fixturing steps for a given part type are processed on the same pallet.

The pallets enter the system through the fixturing station. Following a predefined production schedule, the fixturing station takes parts from bins behind this station, mounts the part(s) on the pallet, and places the pallet in a queue awaiting introduction into the system. If a processing machine cannot currently accept a pallet in its on-line queue, the AGV can take the pallet to one of the 14 off-line buffer stations within the system or the pallet can wait at the fixturing station until an on-line queue position at the assigned milling machine becomes available. After the parts have completed the processing required by a given fixturing step, they return to the fixturing station where they are either refixtured or exit the system as finished parts.

Problem Statement

Currently, the company uses the FMS to produce a standard ensemble of bell housings each day. After reviewing the performance statistics for the FMS and the previous simulation studies, the manufacturer had concluded that the FMS was being underutilized. The company asked us to define an improved production schedule that would permit it to produce the daily requirement of parts in the most efficient manner so that the resulting slack (unscheduled) production time could be reassigned to produce additional parts. Because the company wanted to produce the same ensemble of parts each day, we expected that we could define a schedule that could be repeated daily.

Our first task was to define an objective function for the scheduling problem. We chose minimizing the makespan (the total time required to produce the daily ensemble of parts) as the performance criteria for the study. From the outset, we knew that the two milling machines dedicated to producing aluminum parts were underutilized and would not affect this performance criterion. Our major concern was scheduling the five milling machines dedicated to producing iron parts. By minimizing the makespan, we thought that we could maximize the utilization of the iron machines. Furthermore, since the milling machines were completely interchangeable (every machine could process every fixturing step for every iron part), we also believed that minimizing the makespan would balance the utilizations among the milling machines dedicated to the production of iron parts to the greatest extent possible. The difference between the determined minimum daily makespan and the total available production time for each day would be the maximum possible slack time that could be assigned to the production of parts beyond those contained in the required daily ensemble.

Developing a Detailed Simulation Model

To determine the makespan required to implement a given production schedule, we developed a detailed simulation model for the FMS using SLAM [Pritsker 1986], which was requested by the company sponsor. In developing the SLAM model, we relied upon experience that we had gained developing a similar simulation model using SIMAN [Pegden, Shannon, and Sadowski 1990] to consider the tooling constraints for an FMS operated by the US Army Rock Island Arsenal [Dullum and Davis 1992; Hedlund, Davis, and Webster 1990]. In both FMS models, detailed consideration of the processing plans for each part type was essential. We also made a complete accounting of the remaining tool life for each tool residing at a given milling machine as each processing task was implemented. Whenever tools became worn, the model generated orders for replacement tools and incorporated the steps needed to ready the replacement tools at the tool station and deliver them to the machine via the AGV system. We also included the detailed mechanics of loading replacement tools from the tool carrier into the machine and removing worn tools.

Using SLAM, we considered the dynamics of the AGV system including the control logic for AGV assignment. However, we had only limited documentation for the AGV control logic because it was contained in proprietary computer code, which was not available for analysis. Perhaps our most significant deviation from real-world operating conditions in the model was omitting both machine breakdowns and the occurrence of system deadlock. Watching the operation of the FMS, we observed several modes of system deadlock. For example, an AGV with a low battery might wait to be recharged in the middle of a primary track and block passage of the other AGVs.

That there are only two on-line buffer positions at each machine also caused problems. In more than one instance, the AGV deilvered a second input part to a milling machine while another part was being processed. When processing was completed, the completed part could not be unloaded from the work area because no on-line buffer positions were available. In this situation, the FMS operator had to direct the controller to bring an AGV to the deadlocked machine to remove one of the waiting parts to an off-line buffer position, allowing the finished part to be removed from the work area. During this intervention, most other material handling was delayed. In some instances, it took over 30 minutes to restore a normal operating condition. Since we did not model breakdowns or system deadlock, we realized that the detailed model would overestimate the actual performance of the system.

The detailed simulation included a dispatching mechanism that insured that parts were assigned to machines in the exact order dictated by the schedule. The development of this mechanism was complicated by the fact that sometimes two or three different parts had to be mounted upon a fixture before it was dispatched to the machine.

In addition to maintaining the scheduled processing order and machine assignments, the dispatching mechanism prevented pallets from being transferred to their assigned milling machines until it ascertained that tools with enough tool life needed to process the fixtured part(s) were available. After dispatching the pallet, the dispatching mechanism computed the life that would remain for each tool after the pallet of fixtured part(s) had been processed. Using this remaining life, it reviewed the processing plans to determine which fixturing steps would exceed the maximum tool-life of any tool currently positioned at the assigned milling machine and ordered the necessary replacement tools. The dispatching mechanism then prevented any part that violated the maximum tool-life constraint for any tool from being assigned to that machine until the essential replacement tools were replaced.

The FMS operator may have suspected that tool management and handling was the primary constraint.

Since the tool magazines at the machines were nearly full, replacement tools could not be delivered until the currently assigned parts had been processed. Replacing tools before assigned parts were finished would waste available tool life and ultimately result in more tool devliveries. Because the tools are delivered by the

AGV system and introduced into the machine through its work area using the same spindle and tool exchanger used in processing, tool deliveries decrease both the availability of a given machine for production and the availability of the AGV system to transfer parts. Veeramani, Upton, and Barash [1992] discuss the many issues associated with tooling management in the modern manufacturing environment and survey the related literature.

Selection of an Optimization Approach

From the outset, it was clear that we needed a detailed simulation to accurately assess the internal dynamics associated with operating the FMS. As we constructed the simulation model, we realized that employing a mathematical programming approach for scheduling the FMS would require too many simplifying assumptions. We would have to ignore constraints pertaining to material handling, tool management, and the details of the fixturing process to use this approach because it is nearly impossible to express these constraints in the form of mathematical equalities and inequalities. Even if we could formulate the mathematical programming problem, we would still face the task of finding the optimal solution.

Since simulation models provided the only viable means to express these constraints, we sought an optimization procedure that would permit direct consideration of the simulation model. We eliminated experimental factorial design-based methodologies because we would need too many variables (factors) to define the production schedule. Davis and Stubitz [1987] have shown the difficulties that can arise in attempting to optimize a performance criteria (response) over a discrete-event decision space. Their example considered only seven factors.

We did briefly attempt to apply genetic optimization [Goldberg 1989] to define the optimal schedule. However, as we simulated the generated populations of schedules, we made a series of observations that indicated that a simple set of heuristic rules could generate a set of alternative schedules with a nearly optimal makespan. First, the simulations confirmed that machining aluminum parts did not constrain the overall makespan. Second, five machines were assigned to producing steel parts. If each machine is utilized at a 75-percent level, the probability that all five machines will be busy at any moment is approximately 24 percent. Even at 80-percent utilization, the probability of all machines being busy at a given time is less than 33 percent. Given that any machine dedicated to producing steel parts can process any fixturing of a steel part, machine availability is not the issue. Whenever any part is fixtured, a machine is likely to be available (or nearly available) to process it.

Intuitively, we could see other obvious policies that should be adopted. Pallets capable of carrying multiple fixtured parts should be filled whenever possible to minimize the number of their trips. Second, we should attempt to keep the machines dedicated to producing steel parts busy by giving priority to fixturing a steel part over fixturing an aluminum part. To minimize the overall makespan, we should keep the process utilizations across the five machines nearly equal. Using these simple heuristics, we could generate many schedules requiring a makespan of approximately 1,270 minutes (88.3 percent of the available daily production time). For this makespan, the average utilization for the iron machines was approximately 85 percent, which is near the feasible upper limits for most machines in operating FMSs.

The Real Production-Scheduling Problem

Given this discovery, we began to ask ourselves why the company was having a problem. The heuristic was so simple that the company must have discovered it during the years it operated the FMS. Still, the company was having problems in meeting the minimum daily requirement, and it often needed overtime production on the weekends to insure parts would be available for the assembly area. We had neglected several considerations in the simulation model, including breakdowns and deadlocks. Also we had not assessed lost production time during the shift changes. Nevertheless, about 12 percent of the available production time could serve as slack time to cover these omissions.

However, we had not tested one major assumption made in our simulation model. We had assumed that each simulation run would begin with a new ensemble of tools at each machine. During the simulation run for schedules with a 1,270 minute makespan, less than 20 tools were replaced. Obviously the number of tool replacements depends upon the remaining life of the tools at each machine at the beginning of the simulation run. Since we had no basis for assigning this initial configuration, we chose to allocate a complete set of new tools for each simulation run considered during the optimization.

To test the effect of this assumption, we ran a 20-day simulation run. Figure 2 shows the slack time left each day after meeting the daily requirement. In performing the 20-day simulation study, we assumed that any slack time generated each day would be used to produce additional parts. Therefore, when the simulation finished a given day's production, we advanced the simulation time to the beginning of the next day. Slack time was available for reassignment during the first nine days. On the tenth day, the FMS needed nearly 1,550 minutes to produce the daily requirement. This implied that the 10th day's production spilled over into the 11th day, which fortunately generated enough slack time to complete the 10th day's requirement and do the 11th day's requirement and leave a small amount of slack for additional production. The system then returns to a situation in which there was slack time each day. By the 14th day, however, it became unstable; the system got behind and never caught up. The only source of variability in this simulated scenario is the remaining life for each tool at the beginning of the day. That is, the simulation time assumes that the processing times are deterministic and that the outcome of each processing task is a successful completion. Because of the level of detail in this simulation model, there are few (if any) random variables.

The above simulation results are conservative in their projection of the consequences arising from the consideration of tooling. First, there is no way that the current simulation study could account for the tool wear resulting from assigning slack production

FIGURE 2

This figure gives the observed slack times during the 20-day simulation run. A negative slack time implies that the simulation took longer than the available time to complete the daily production. The system becomes unstable after the 14th day.

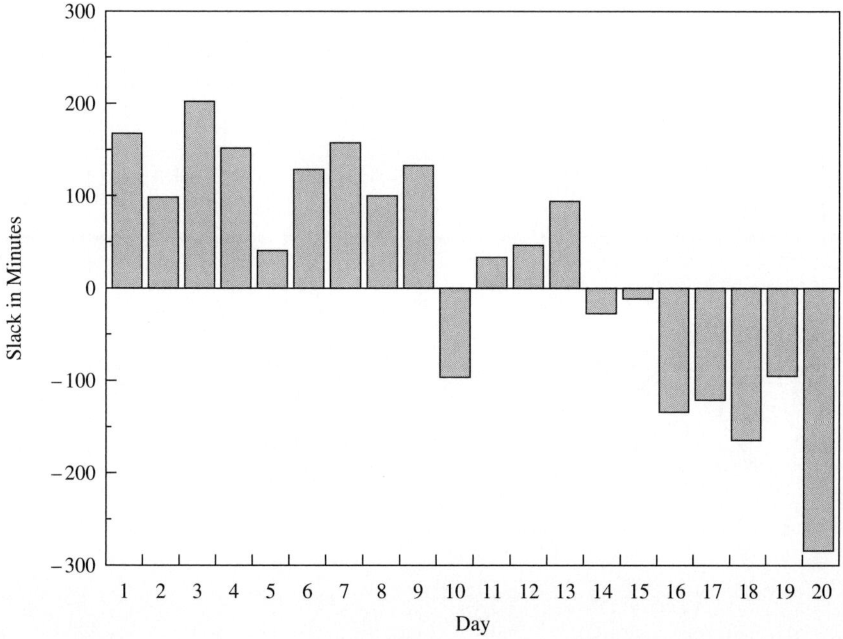

time to the production of other parts. Hence, using the slack production time would wear out tools and require additional orders for replacement tools beyond those observed in the 20-day simulation study. Second, we assumed that only orders requiring more than one replacement tool would use the AGV delivery system. An order for only one tool would be delivered and installed by the operator with no production downtime. The company used this replacement policy under some circumstances, and we decided to adopt it for all single tool orders since the company could not define the conditions under which it performed manual transfers. We conducted several experiments to determine the cause of the declining slack. We collected statistics on machine utility, cart utility, tool-carrier utility, and the number of tools ordered per day. After these experiments were performed, the most significant factor appeared to be the tooling replacements ordered by the machines. The number of replacement tools increases each day over the 20-day period and the number of multiple tool replacements (requiring the use of the AGV delivery system) also increases (Figure 3).

We believe that the increasing number of multiple tool orders is the main cause of the declining slack. When a machine needs a tool for processing, it will be effectively shut down until that tool arrives. Production time is lost every time a tool carrier of replacement tools are loaded into the machine magazine and worn tools are removed. As the initial condition deteriorates and tools wear, the number of requests for tools rises dramatically. Orders for replacement tools eventually become too frequent, and the time it takes to fill a tool request also rises (Figure 4). The number of multiple tool orders per day and the time required to fill tool orders are correlated. The order is generated after a part is fixtured, and the replacement order cannot be put in the machine's magazine until the part that initiated the order is finished. Also, only three tool carriers are available to the system, which becomes a production constraint when several orders for replacement tools arrive in a short period. In addition, tool orders must wait for a single tool setter to install cutters on the replacement tools and put them on the carrier for delivery.

After making these observations, we counted the replacement tool orders for each tool over the 20-day period. We found that eight tools constituted almost 60 percent of the tool orders. We then decided to place two of these tools in the tool magazines to see if this would increase slack time. We ran this configuration for a 20-day period, assuming again that the slack time would be allocated to other production.

Placing redundant tools in the magazine allowed the system to run smoothly for a longer period, but the number of tool orders

FIGURE 3

The total number of tools ordered and the number of orders requesting multiple tools each day increased during the 20-day simulation study.

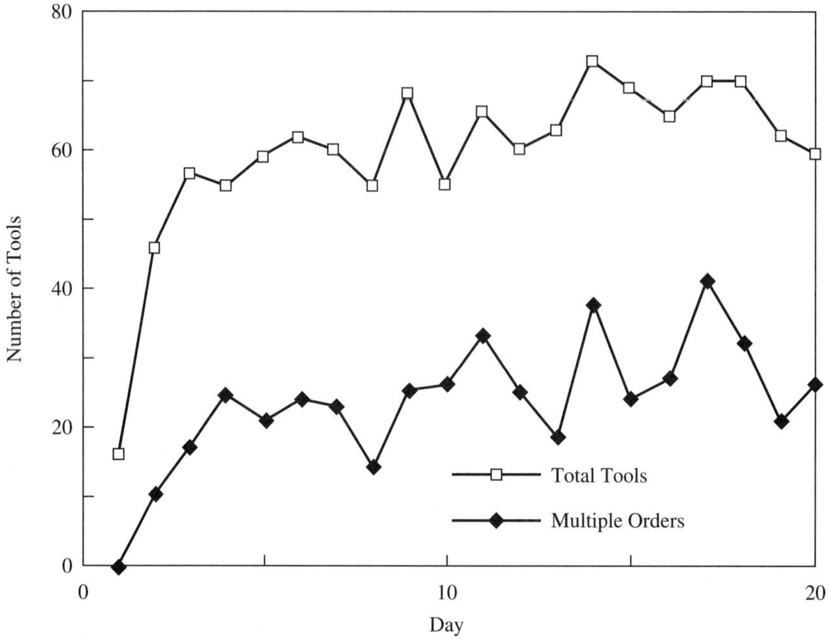

still increased beyond the capacity of the system. The 30-day trial ended three minutes over the due date of 28,800 minutes. The average slack per day was 92 minutes, but four days in the period were still so overburdened with tool orders that they exceeded the time allotted for daily production (1,440 minutes). Thus, placing spare tools on the machines just delays the time when the system becomes saturated and then unstable.

The simulation also tabulated the percentage of downtime on each machine that can be attributed to tool replacement. With no duplicate tools, the five machines average 17.3 percent of the production time waiting for tools to be replaced (a maximum value of 25 percent was observed at one machine). With duplicate tools, the five machines spent an average of 8.6 percent of the production time waiting for tool replacements. This downtime is clearly a significant detriment to producing parts within the specified makespan. For the single-day simulation runs in which we assumed all tools were new, the number of requests for multiple tool orders were minimal (single tool orders incur no loss in production time), and the machines took 88 percent of the available production time to complete the daily production requirement. With average downtime due to tooling replacement of 17.3 percent, the FMS, even with duplicate tooling at each machine, operates near or over saturation. And we have not modeled breakdowns or the occurrence of deadlocks, which further diminish the available production time. It is clear why the company currently needs overtime to keep up with its daily production requirements.

Conclusions and Recommendations

We learned a great deal from our retrospective analysis of this scheduling exercise. Many troubling issues can be raised. An important question is, has this exercise helped the owner of the FMS? The answer is probably no. We did ascertain the true bottleneck to production, but we did not provide any solutions. The FMS operator may have suspected that tool management and handling was the primary production constraint before the study was authorized and hoped that we could prove otherwise.

The true scheduling concern for this FMS lies in the management of tooling. In our simulations, the tool handling constraints eventually made the system unstable, regardless of whether or not the slack production was reassigned to machine additional parts beyond the minimum daily requirement. In the real world, this system does not become significantly unstable because the company suspends production on the weekends and performs system maintenance to replace worn tools. In addition, it can use overtime production on weekends to catch up with production requirements.

FIGURE 4

The line graph gives the number of orders requesting multiple tools for each day of the 20-day simulation study. The bar graphs depict the statistics pertaining to the time required to fill the tool orders on each day, including the minimum observed time, the maximum observed time, and the average time.

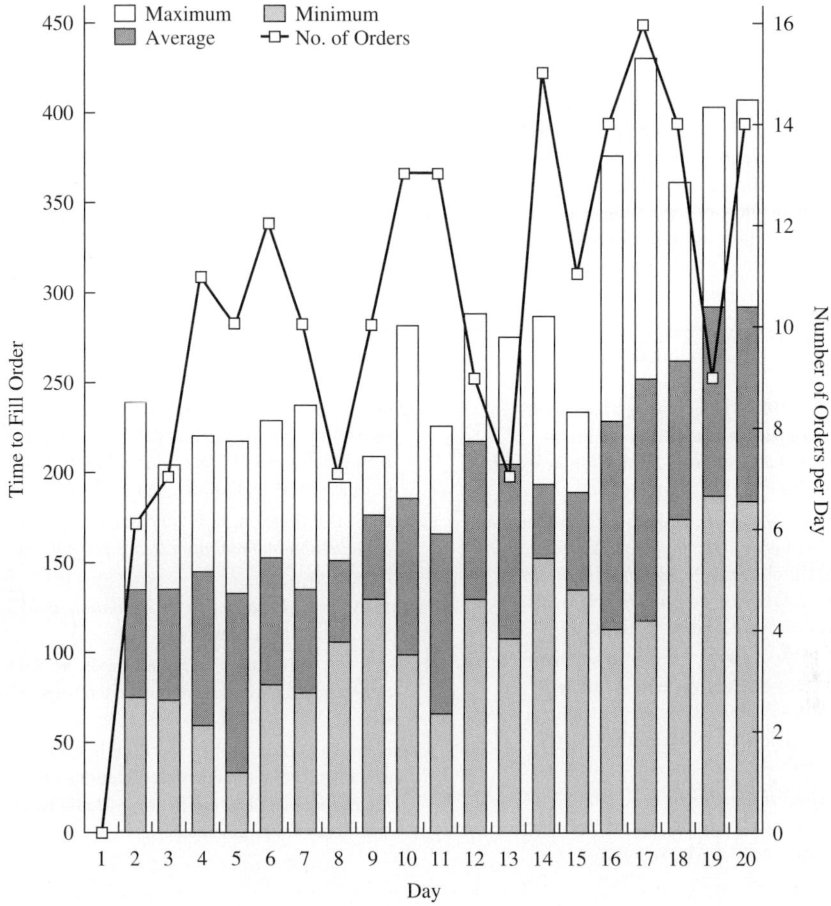

It is very difficult to formulate an optimization problem to assess the optimality of a general tool management policy. We would first specify the set of alternative policies that can be considered and the planning horizon. Considering an extended planning horizon would greatly increase the simulation time required to evaluate each policy.

The tooling should not be the only area of focus, however, because the FMS is not an entity unto itself. The dynamics of part arrival and the demands of the downstream assembly areas must also be taken into consideration when this model is used. The variability of daily demand should be considered. In addition, if the company uses the slack time provided through improved scheduling to produce other parts, then these additional parts must also be specified because their production will likely affect tool replacement. All of these concerns must be addressed to develop a model capable of assessing the true system dynamics.

For the modeler of FMSs, we have an additional observation. Few simulation studies of FMSs consider the flow of such supporting resources as tooling. Had we not considered tooling in this simulation study, we would have overlooked a major production constraint. This is not the first FMS we have modeled in which the flow of supporting resources is the primary bottleneck. In fact, our experience indicates that the flow of jobs is becoming a secondary concern in modeling FMSs. For every FMS we have modeled, the flow of supporting resources is much more complex and is not understood.

Most scheduling approaches focus upon job flow only.

There is a fundamental rule in optimization. Whenever a new constraint is appended to a decision, the ability to optimize the decision is at best unchanged and more likely, it is diminished. This implies that whenever operational constraints are not included in a simulation model, the simulation model will probably overstate the expected performance for the modeled system. In our collaboration with industry, we find a continued concern that FMSs cannot provide the production capability predicted by simulation studies. Invariably the cited simulation studies have ignored the constraints associated with the flow of supporting resources.

Furthermore, most FMS simulation studies assume that there is a steady state for operating the system. This is not a valid assumption. There is simply too much variability in the operation of FMSs to assume that a steady-state operating condition be maintained over an extended horizon. Therefore, we conclude that any detailed schduling must be performed on a near-term basis. To analyze proposed near-term schedules, one must use real-time, discrete-event simulation [Tirpak, Deligiannis, and Davis 1992]. Davis et al. [1993] state additional concerns in the modeling, scheduling, and control of FMSs and then provide an integrated solution approach to address all three tasks.

Acknowledgment

This research was supported by Caterpillar, Inc. with a grant to the University of Illinois Manufacturing Research Center.

References

Davis, W.J. and Stubitz, S. 1987, "Configuring a kanban system using a discrete optimization of multiple stochastic responses," *International Journal of Production Research,* Vol. 25, no. 5, pp. 721–740.

Davis, W. J.; Setterdahl, D.; Macro, J.; Izokaitis, V.; and Bauman, B. 1993, "Recent advances in the modeling, scheduling and control of flexible automation," *Proceedings of the 1993 Winter Simulation Conference,* eds., G. W. Evans, M. Mollaghasemi, E. C. Russell, and W. E. Biles, pp. 143–155.

Dullum, L. M. and Davis, W. J. 1992, "Expanded simulation studies to evaluate tool delivery systems in an FMC," *Proceedings of the 1992 Winter Simulation Conference,* eds., J. J. Swain, D. Goldsman, R. C. Crain, and J. R. Wilson, pp. 978–986.

Goldberg, D. E. 1989, *Genetic Algorithms in Search, Optimization and Machine Learning,* Addison-Wesley Publishing Company, Reading, Massachusetts.

Hedlund, E.; Davis, W.; and Webster, P. 1990, "Using computer simulation to compare tool delivery systems in an FMC," *Proceedings of 1990 Winter Simulation Conference,* eds., O. Balci, R. P. Sadkowski, and R. E. Nance, pp. 641–648.

Pegden, C. D.; Shannon, R. E.; and Sadowski, R. P. 1990, *Introduction to Simulation Using SIMAN,* McGraw-Hill, New York.

Pritsker, A. 1986, *Introduction to Simulation and SLAM I,* John Wiley and Sons, New York.

Tirpak, T. M.; Deligiannis, S. J.; and Davis, W. J. 1992, "Real-time scheduling in flexible manufacturing," *Manufacturing Review,* Vol. 5, no. 3, pp. 193–212.

Veeramani, D.; Upton, D. M.; and Barash, M. M. 1992, "Cutting-tool management in computer-integrated manufacturing," *International Journal of Flexible Manufacturing Systems,* Vol. 3, no. 4, pp. 237–265.

C H A P T E R

7 Interorganizational Coordination: Electronic Data Interchange

7.1 Introduction

As competition has intensified, manufacturing organizations have dramatically changed the nature of both internal and external processes in order to become more responsive and agile. Advances in automation and information technologies—and evolving management practices, such as just-in-time supply and modular design—have become critical enablers for these efforts. Some industries, particularly automobiles and electronics, have wholeheartedly embraced the just-in-time approach for material management during the last decade. As a consequence, companies in these firms have restructured their production and distribution processes and have realized the benefits of reduced inventory and improved availability. They have also redesigned their products to take greater advantage of outsourcing from external vendors.

Big Three automakers are focusing their information system spending to enhance their lines of communications to suppliers via electronic data interchange (EDI) and dedicated networks (LaPlante [1992]). Improved communications have allowed automakers to cut delivery time, reduce excess inventory, and improve efficiency. In recent years, leading consumer goods manufacturers, such as Procter and Gamble and Campbell Soup, have also restructured linkages and forged close partnerships with major retailers such as K Mart and Wal-Mart (Buzzell and Ortmeyer [1995]). The aims of these partnerships are to lower product costs, reduce pipeline inventories, smooth out production schedules, and quickly identify problems in quality and service associated with moving goods from manufacturers to retailers and then to the ultimate customer. The application of information and automation technologies is proving to be a key to the success of these partnership arrangements, since they are predicated on a heightened level of communication and coordination. Examples of such technologies include: (1) EDI for the exchange of information on standardized business transactions, (2) *bar-coding and scanning* technologies for capturing sales data and consumer preferences at the point of sale, and (3) *shipping container marking (SCM)* for tracking of containers in shipments and for identifying their contents and destinations.

This chapter begins with a discussion of EDI. It then moves on to consider the impact that EDI and related information technologies (IT) are having on the economics of interorganizational (firm to firm) coordination and, as a consequence, on the structure of industries. The full range of applications of this technology and their dramatic impact can be best illustrated with the examples of supply chains in the apparel and grocery industries. Hence, deviating from our earlier practice of giving examples from discrete parts manufacturing industries, we will briefly discuss the applications of EDI and other information technologies in the apparel and grocery industries in the fourth section. The chapter ends with some comments on managerial guidelines for implementing EDI. The reading by Apte et al. (1993) provides a more detailed discussion on the use of EDI and quick response strategies in the apparel industry through the reengineering of business transactions.

7.2 Electronic Data Interchange (EDI)

EDI can be defined as intercompany, computer-to-computer communication of standard business transactions supported by a standard format (Sokol [1989]). With EDI, one company's order information feeds directly into its supplier's order processing system. Likewise, invoice data from the supplier feeds directly into the accounts payable system of the purchaser. The communication linkages in EDI cross organizational boundaries and require a high level of cooperation and coordination between companies engaged in a trading relationship. Such information exchange is accomplished through the medium of data communication, taking place between the computers of each company, i.e., without any human intermediation or paperwork. The scope of EDI, however, goes well beyond simple computer-to-computer communication or the transfer of files (which have been in place for years). Rather, EDI enables the implementation of software applications that support coordination of critical business functions.

A prerequisite for EDI is the availability (within both the seller and the buyer organizations) of computer applications that support the relevant business transactions and that have the capability to generate, transmit, and receive transaction data. In order for these computer applications to understand each other, the transaction data being communicated must be transmitted in recognized and previously agreed-to formats. EDI, therefore, requires standard formats for business transactions, such as requests for a quotation, generation of purchase orders, bills of lading, invoices, electronic payments, and so forth.

The EDI standard defines a common syntax and the specific manner by which this syntax is to be interpreted. This standard can be easily explained using the analogy of a letter (i.e., transaction information) enclosed within several layers of envelopes. The outside envelope identifies the communications protocol being used. Inside, each envelope layer provides further clarification. The innermost envelope contains the message, consisting of variable-length transaction data (e.g., an invoice specifying the product identification code, shipment quantity, unit price, payment terms, etc.). The role of EDI software is analogous to *opening* the envelopes, *routing* the letter, and *interpreting* the business transaction inside. The interface file, also called the application-link software, then takes over in *mapping* the transaction data to the appropriate fields within company-specific applications (e.g., an order entry or the MRP system). EDI software is available today as off-the-shelf packages offering support for one or more of such transactions.

Communication with a trading partner using EDI can be accomplished in two ways. First, via a direct telecommunications link or a leased line established by the partners. Assuming that sufficient communication volume exists, this is the least expensive alternative. Communication can also be directed through a third-party, value-added network. The value-added networks typically provide an electronic mailbox facility which holds EDI transactions in the trading partner's mailbox until such a time that the partner requests delivery of the accumulated transactions.

In contrast to the traditional mode of mailing paper-based purchase order forms, EDI allows a supplier to receive orders almost immediately. Unlike the orders placed via a telephone or a fax, the EDI order does not have to be keyed in. Thus, keying errors, delays, and misinterpretation of orders are avoided. This enables the supplier to send the appropriate shipment to the customers quickly, and thus enhance responsiveness and accuracy. The improved quality and speed in distribution resulting from this approach creates the foundation necessary for success in implementing a just-in-time program for material management.

The current state of EDI technology has two drawbacks. First, the batch nature of EDI, especially in working with the third-party networks, introduces a time lag which is contrary to the true philosophy of just-in-time procurement. It can also cause problems when the reengineered transaction is based on the assumption that the EDI transaction is received almost immediately (e.g., the Advanced Shipment Notice [ASN] transaction discussed later). Efforts are currently underway to enable real-time exchange of transactions between trading partners. The second problem with EDI is the lack of a universal EDI standard for electronic communication. Presently, each industry has settled on certain protocols and EDI standards. The full benefits of EDI will only be realized, however, once a standard that is common across all industries and countries is established.

Examples of industry-specific EDI standards include *UCS,* for the grocery industry, and *ORDERNET,* for the pharmaceutical industry. The widespread acceptance of the ANSI (American National Standard Institute) X.12 definition of standard business transactions as a cross-industry standard has led many industries, new to EDI, to support X.12 for use within their industries. Examples of such industries include automotive, electrical, chemical, metal, apparel, and office products. While the X.12 standard defines the data format, individual industries are developing guidelines for usage and interpretation of data fields. We note that these issues concerning the EDI standard are very similar to those raised in Chapter 5, where we discussed the communication protocols (e.g., MAP) for automated manufacturing equipment on the factory floor.

A survey of companies using EDI technology (Hwang et al. [1993]) revealed that 60 percent of the firms using EDI belong to manufacturing sectors, such as automotive and chemical. Nearly 75 percent of the EDI systems were developed after 1987, suggesting a growing technology trend. About a third of the companies surveyed spent between $10,000 and $50,000 to install an EDI system, with the most frequently reported amount being $40,000. Of those surveyed, 83 percent use EDI for sending and receiving purchase orders, while 65 percent use it for sending and receiving invoices.

FIGURE 7.1

EDI purchase order

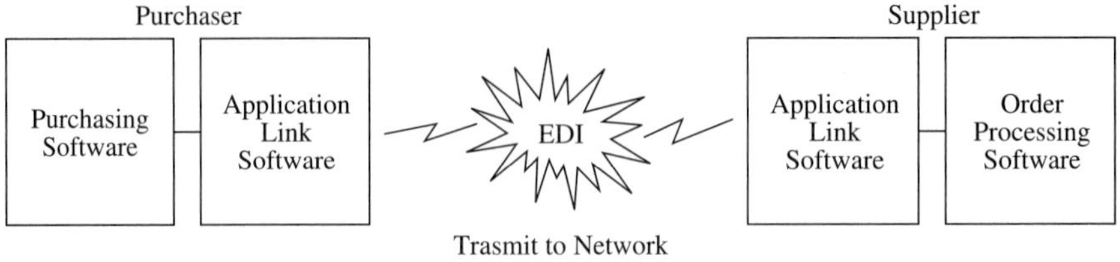

7.2.1 How Does EDI Work?

The concepts underlying EDI are simple. As noted above, EDI involves transmission of machine readable data between trading partners' computers, either via a direct telecommunication link or via a third-party, value-added network. Figure 7.1 illustrates the interchange of data for a purchasing transaction. The buying company generates a purchase order using its current purchasing application software. The main difference under EDI is that, instead of generating a paper-based purchase order, the system passes the purchase order data through "application link" software (i.e., the interface file) which maps the order into a standard EDI transaction data. These data are then transmitted via a third-party, value-added network to the seller's computer, where they pass through the seller's "application link" software, which finally maps them into formats expected by the seller's order entry application software. Thus, compatible application link software packages are required at both the seller's and the buyer's ends, since existing applications are rarely designed to generate or receive purchase order data in a common format.

7.2.2 Benefits of EDI

Interorganizational integration of business processes facilitated through the use of EDI can lead to several benefits:

- *Reduced cost of business transactions.* EDI eliminates not only the keying and rekeying of data and associated data entry errors, but it also eliminates tasks such as mailing, sorting, filing, and reconciliation of paper forms. Thus, the costs of mailings and of materials and labor needed to support paper-based transactions are reduced considerably. According to a recent estimate by an electronics manufacturer, the $50 cost of processing a paper purchase order has been reduced to $4 per purchase order with the use of EDI (Datapro [1991]).
- *Reduced cycle time of business transactions.* Products can be shipped and received sooner, since the time required for mailing an order is eliminated and the time involved in processing an order transaction is reduced. Similarly, an invoice can be sent and payment can be made electronically by using EDI, although this particular use of EDI is not well developed. Reduced cycle times

increase material and cash turnover and thereby reduce the level of safety stock inventory and the cash reserve required to support business activities.

- *Reduced uncertainty and improved accuracy of information on business transactions lead to improved capability for planning and forecasting.* EDI allows for frequent exchange of demand information and rapid receipt of accurate, reliable, and complete information on business transactions. This information can then be used to improve demand forecasting and to enhance the planning of production and the control of inventory. These improvements lead directly to reduced levels of inventory—which is a major measurable benefit of EDI. EDI can also be used to track vendor performance accurately, and to create better audit trails of transactions for management control. Companies, therefore, can plan more effectively for the receipt of goods as well as for the scheduling of manufacturing tasks.
- *Potential for business process reengineering.* The elimination of unnecessary transactions and the introduction of new transactions that did not exist before can lead to the complete reengineering of supply chain business processes. Such reengineering can lead to significant improvement benefits (Buzzell and Ortmeyer [1995]). Examples of new, EDI-enabled transactions are the *advanced shipment notice (ASN)* and *point-of-sale (POS)* data capture systems. In ASN, the supplier sends, in advance of the actual shipment of goods, an EDI transaction that completely specifies the shipment (e.g., the number of containers and the contents of each container) that the customer is about to receive. This allows the customer to avoid several manual processes involved in receiving the shipment, such as checking the contents of containers against the original purchase order. The associated lead time for receiving an order is correspondingly reduced. The customer's cost in receiving is also reduced, and his/her record keeping accuracy is considerably improved. The supplier also stands to gain since reduced cycle time leads to getting paid sooner, which reduces working capital requirements. The problems of the batch nature of EDI can be better understood in the current context. For nearby suppliers, the physical goods can arrive before receiving ASN. In this case, all of the benefits of ASN mentioned earlier are completely lost.

The use of EDI is changing many aspects of the way business is being conducted. For example, consider the warehouse management function that deals with order shipping and receiving, stock control, and other activities related to material handling. With the help of EDI, it is now possible to view the flow of inventory beginning at the vendors' shipping dock and ending at the customers' receiving dock. This has given rise to the warehouse manager's need for EDI software to be seamlessly integrated within the warehouse management software system. A recent survey of warehousing firms (Rogers et al. [1992]) showed that firms that used EDI were demonstrably able to improve customer service by accommodating special requests and events for customers. A study of EDI's usage and adoption within marketing and logistics channels showed that both shippers and carriers are driven to establish EDI to reduce demand and supply uncertainties and to reduce costs (Walton [1994]). It was noted that carriers adopt EDI to differentiate themselves and to improve their competitiveness in the transportation marketplace.

EDI also has been found to be critical in implementing outsourcing and concurrent engineering. As DeFosse and Barr (1992) illustrate through a case study of the Plastics Technology Center of Lexmark International Inc., there must be constant communication among the product designer, manufacturing engineer, model designer, and tool builder for concurrent engineering to work. EDI allowed for the immediate transfer of massive graphic files between CAD systems, and thereby proved to be critical in achieving rapid, efficient, and error-free communication among disparate product design project team members.

As noted above, one of the principal benefits of EDI in manufacturing is that it enables the implementation of a just-in-time supply strategy. For example, General Motors has reportedly saved several million dollars with the use of its European EDI network. Navistar has reduced its inventory from 33-days to 6-days supply, thereby lowering its inventory costs by $167 million. Finally, the use of EDI has enabled LTV to have JIT deliveries made to some customers in only four hours (Datapro [1991]). Kekre and Mukhopadhyay (1992) analyzed the impact of EDI on the suppliers of LTV steel. Their study found that the use of EDI improves supplier performance in terms of higher quality and lower inventory.

In a survey of EDI and MIS managers (Scala and McGrath [1993]) on costs and savings of EDI, improved information accuracy and reduced errors ranked first on the list of 10 advantages, which also included reduced data reentry, reduced time of data transmission, and reduced inventory and inventory carrying costs as other major advantages. The requirements of high level management commitment and difficulty in quantifying a return on EDI investments were listed as top disadvantages. Major (1995) estimates that EDI can eliminate as much as 70 percent of data reentry common in business transactions and reduce the associated cost. Error reduction from using EDI can also significantly add to the bottom line.

The use of EDI in conjunction with JIT supply is becoming increasingly more important as the use of international sourcing increases. Sungchil (1994) analyzed the experience of a Korean consumer electronic firm in supplying its finished products to original equipment manufacturers in the United States. He found that time-based competition can still be implemented with overseas production, and that the use of EDI is a critical element in achieving quick response. In a 1991 keynote address to the Data Interchange Standards Association, the president of TI's information technology group stated, "In the business of the twenty-first century, speed and efficiency—and therefore EDI—will be the keys to survival and prosperity."

7.3 Impact of Information Technology on the Economics of Interfirm Coordination

As discussed in Chapters 4 and 5, manufacturing automation and information technologies (IT) have thoroughly transformed the production function in manufacturing companies. This transformation has benefited manufacturing companies in terms of product cost reduction and improvement in quality, flexibility and responsiveness.

In addition to enabling changes in a firm's production function, information technology has significantly changed the way coordination activities are carried out within and between organizations. This is due to information technology's vital role in the creation, transformation, storage, and movement of information—the essential ingredients for carrying out any coordination task. With the orders-of-magnitude improvement in IT's performance to cost ratio, the nature and the economics of coordination tasks are undergoing dramatic changes.

These changes can be best understood from the vantage point of *transaction cost economics* (Williamson [1985]). As Williamson points out, transaction costs capture a firm's efficiency in coordinating within and across its organization boundaries. The existence of such transaction costs is considered to be a principal factor in explaining the very existence of organizations. As per transactions cost economics theory, the governance structure of a firm is chosen to minimize the total of its production and transaction costs. Here, production costs are defined as the costs associated with producing the goods within the firm, while transaction costs are considered as those costs incurred in coordinating and procuring goods and services. Transaction costs are also defined to include transaction risk related costs, such as the cost of monitoring supplier performance. Due to the relative ease of coordination internal to thc firm, transaction costs are generally lower when production activities are performed within the firm. Production costs, on the other hand, can be lower when goods are procured from an outside supplier who may generally enjoy lower costs due to economies of scale and scope.

As we noted in Chapter 1, Chandler (1962) pointed out that vertical integration became the norm at the end of nineteenth century for a wide range of industries. In contrast, recent advances in information technology and its use have resulted in reduced transaction costs, which, in turn, have led to increased use of marketplace procurement. The net result of all this is that the boundary of the firm is changing (Clemons [1993]). Specifically, the value-added supply chains in a great majority of industries are being disaggregated, with different activities being conducted by different firms. A description of this trend in the U.S. auto industry is discussed by Taylor (1994).

Malone (1990) and Clemons (1993) argue that IT makes a dramatic reduction in transaction cost in coordinating actions of a firm with that of its customers and suppliers. Therefore, IT should be ideally viewed as a *coordination technology* that will result in greater reliance by firms on cooperative efforts, including outsourcing, joint ventures, and strategic partnerships.

Applying transaction cost economics theory, Clemons and Reddi (1992) hypothesize that the primary impact of IT on the degree of outsourcing and on the nature of governance structures used is a "move to the middle," which states that:

- A greater degree of outsourcing will take place (a move away from ownership/hierarchy),
- But firms will rely on fewer suppliers than before, with whom firms will have close and long-term relationships (a move away from the market to intermediate governance structures when outsourcing).

Seidmann (1994) has developed an economic model of buyer-supplier interaction that suggests that only the buyer and the end consumers will ultimately benefit from

EDI. The model also shows that, under certain conditions, it may be in the best interest of the buyer to subsidize suppliers to encourage them to adopt EDI.

With regard to the impact of IT, Scott Morton (1991) suggests that IT's impact on the economics and functionality of coordination stems from the improvements it makes in the flow of information in supply chains. Specifically, two aspects are important: (1) that the concepts of location and distance are becoming increasingly irrelevant for the flow of information[1] and (2) that the amount of time involved in coordination tasks is being dramatically reduced. Thus, EDI, which is an important type of interorganizational system (Johnston and Vitale [1988]), may be seen as the first step in a long-term move towards the emergence of electronic markets and hierarchies.

Under this scenario, a parts supplier receives and acts on orders instantaneously with the aid of EDI technology. Without the benefit of EDI, these tasks may require several days or even weeks to complete. Such changes in coordination, brought about by electronic integration, have affected all of the tasks that an organization carries out to produce and to deliver its products to its customers. Consequently, the evolution of IT, especially communications networks, has had an impact on both the procedural aspects of production and coordination and on the organizational structure of firms in a wide range of industries.

IT-based integration of processes is showing up in multiple forms within and across organizations:

- *Value chain within an organization:* Xerox has used LANs to connect its design, engineering, and manufacturing personnel for creating teams that are focused on individual products. These teams have implemented the principles of simultaneous engineering, previously discussed in Chapter 2, to accomplish design-related tasks in less time, with higher quality and with greater morale, as compared to the performance afforded by previous organization structures.
- *Value chain that spreads across multiple organizations:* A supplier's shipping department can be electronically connected using EDI to a buyer's purchasing department to speed the flow of goods and to eliminate unnecessary paperwork and non-value-added activities. EDI has, therefore, created a "virtual" organization that extends the boundary of an organization to include activities contained in other organizations.
- *Substitution of a stage in the value chain with a subcontract or alliance:* A common example involves a firm asking its supplier to design a component needed for the firm's product. CAD/CAM and EDI technologies are very useful in this context, since they allow for joint design involving specialists from both the purchaser and supplier organizations. Examples of such electronic integration include: (1) Ford Motor Company's subcontracting award to Ryder

[1]However, these concepts of location and distance are still very important and relevant for the economics of the flow of physical goods. Hence, these factors must be considered in any decisions related to supply chains.

Truck for handling Ford's new car shipments and (2) Nissan's giving general specification of seats to Johnson Controls and letting Johnson do the detailed design.

High-quality public and private telecommunication networks now provide the infrastructure necessary for coordinating business processes that are spread over multiple firms. With the aid of this infrastructure, a firm can share data and information with its suppliers and customers through electronic connections. This is termed the *interorganizational integration* of business processes. This interorganizational integration, also referred to as *external integration* (Hunt [1989]), is having an impact on manufacturing firms that is equal in scope to the impact of internal integration of manufacturing automation technologies embodied by the CIM framework.

Jaikumar and Upton (1993) provide another perspective on the disaggregation of value-adding chains caused by manufacturing automation and communications technologies. As we discussed in Chapters 2 and 3, product and process specifications now exist in the forms of specific databases and computational procedures generated by systems such as CAD and CAPP. These specifications and manufacturing procedures are transportable, via telecommunication links, to machine controllers governing automated processing machines anywhere in the world. An important advantage here is that of *reproducibility,* i.e., any machine following the recipe provided by the coded manufacturing procedures has the capability of producing an identical part with a high degree of precision (i.e., with the same cost, quality, and timing performance). Such transportability and reproducibility, made possible by the union of manufacturing automation and communications technologies, reduces the need to colocate engineering and design with manufacturing. Indeed, manufacturing firms need not even have their own engineering and design functions. These capabilities can now be procured from an outside vendor specializing in design and engineering.

7.4 EDI and IT in the Apparel and Grocery Industries

To appreciate the impact of information and telecommunications technologies on industry structure through the interorganizational integration of firms, we need to consider the entire supply chain. This perspective includes the channel structure linking the firm to its market, and encompasses activities from raw material acquisition to manufacturing through delivery of final products to consumers.

Figure 7.2 shows the supply chain for the textile and apparel industry. Advances in information and telecommunications technology here have resulted in improved information flow that allows for greater flexibility and increased speed in responding to the everchanging customer needs of this industry. The firms that use IT to link effectively to other firms in the supply chain have developed a commanding competitive advantage. Examples of companies benefiting from this "quick response" strategy in the apparel industry include retailers such as Wal-Mart, J.C. Penney, The Limited, and Home Depot, and manufacturers such as Procter and Gamble, Haggar Apparel, and Levi Strauss.

FIGURE 7.2

Flow of products and information in textile and apparel industry

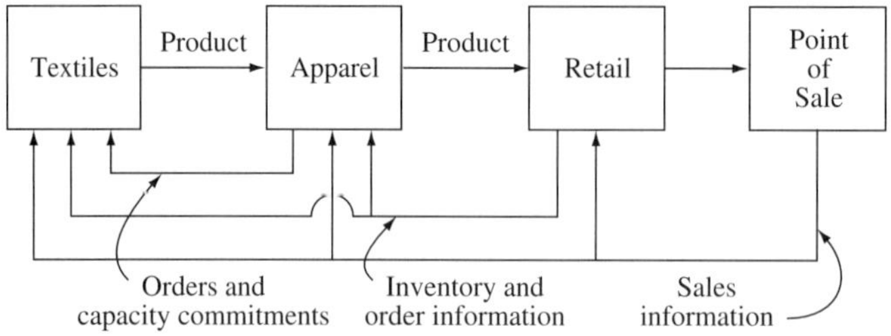

Source: J. D. Blackburn, "The Quick Response Movement in the Apparel Industry," Irwin Professional Publishing, Burr Ridge, IL, 1991.

As documented by Apte et al. (1993) and Hammond (1991, 1993), the *quick response (QR)* strategy in the apparel industry involves several key components:

- An effective, integrated information architecture across the entire channel that combines information technologies such as EDI, bar coding, and point-of-sale scanning.
- Effective consumer testing to improve product selection.
- Shortened product development cycle.
- Accurate forecasting and replenishment systems.
- Short-cycle manufacturing and rapid product fulfillment.
- Transformation of corporate culture including changes in measures and incentives.

Fisher et al. (1994a, 1994b) have developed a system which is based on many of these features (especially accurate forecasting). Their approach is based on gaining information from key customers early in the season on order intentions. The resulting estimates of both the mean and variance of demand are then used to allocate stock optimally throughout the supply chain. The method was successfully applied at a ski-wear manufacturer.

Efficient consumer response (ECR) is the grocery industry's version of the QR strategy, in which suppliers, distributors, and retailers are working closely together to bring better value to grocery consumers. Currently, the grocery supply chain consists of two replenishment loops operating in a relatively uncoordinated manner (see Figure 7.3):

1. Supplier to distributor's warehouse, and
2. Distributor's warehouse to retail store.

In the first loop, inventory is "pushed" into the distributor's warehouse by forward buying and trade deals, while in the second loop, inventory is "pulled" by store replenishment orders (loosely driven by customer purchases). This inventory push and pull meet in distributor warehouses in a manner which causes the buildup of inventories and

FIGURE 7.3

Grocery industry: current supply chain

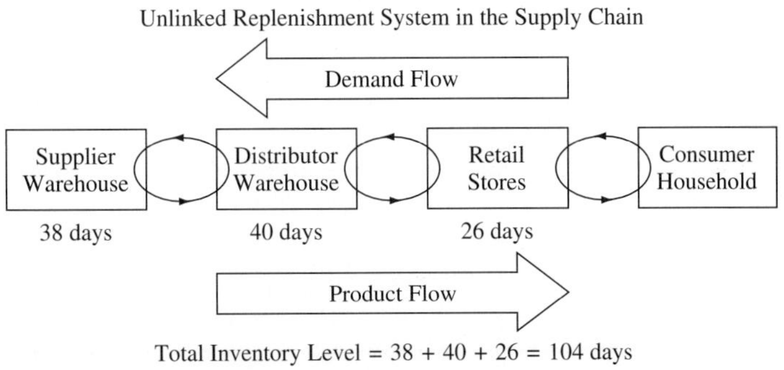

Source: Kurt Salmon, "Efficient Consumer Response: Enhancing Consumer Value in the Grocery Industry," Kurt Salmon Associates, Inc., Atlanta, 1993. Reprinted by permission of Kurt Salmon Associates, retailing and consumer products consulting firm.

promotes inefficiencies across the entire channel due to a lack of synchronization of material related operations.

The goal of ECR is to integrate these disconnected, unsynchronized replenishment loops into a single, unified, synchronized replenishment loop (see Figure 7.4). In this loop, product flows are to be triggered solely by the pull extended by customer demands. The ultimate goal of ECR is to achieve a responsive, consumer-driven supply chain, in which all parties work together in partnership to maximize consumer satisfaction while minimizing system cost.

The guiding principles of ECR include:

- Provide better value to grocery consumers through better product, better quality, better assortment, better in-stock service, and better convenience.
- Replace traditional win/lose trading relationships with mutually profitable win/win business alliances.
- Create and use accurate and timely information to support effective marketing, production, and logistics decisions.
- Manage product flow value-adding processes in the supply chain so the right product is available to consumers at the right time.
- Use common and consistent measurement and reward systems that focus on the effectiveness of the entire system.

The benefits of ECR can be substantial. In a recent report (1993), Kurt Salmon Associates, a consulting firm, estimated total supply chain savings in excess of $30 billion for the grocery industry in the United States. Supply chain inventories are expected to drop significantly as products move more rapidly from production/packing to the consumer's basket. As illustrated in Figures 7.3 and 7.4, supply chain inventory level in dry grocery is expected to drop from the current supply of 104 days to 61 days of supply under ECR. In a recent study of Campbell's successful implementation of ECR, Cachon (1994) validated these kinds of savings.

FIGURE 7.4

Grocery industry: supply chain with ECR

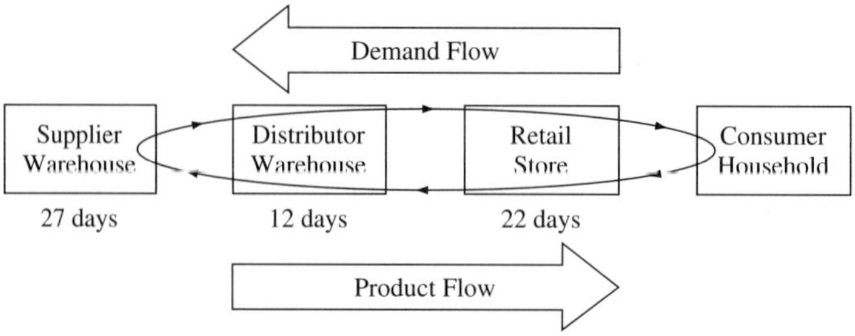

Total Inventory Level = 27 + 12 + 22 = 61 days

Source: Kurt Salmon, "Efficient Consumer Response: Enhancing Consumer Value in the Grocery Industry," Kurt Salmon Associates, Inc., Atlanta, 1993. Reprinted by permission of Kurt Salmon Associates, retailing and consumer products consulting firm.

In shortening cycle time and reducing costs, ECR focuses on efficient implementation of four core processes:

1. *Efficient store assortment* addresses the optimum use of store and shelf space by adopting effective category and space management systems. This can significantly increase sales and gross margin per retail square foot.
2. *Efficient replenishment* links the consumer, retail store, distributor, and supplier into a synchronized supply chain. Accurate information flows quickly through the EDI linkages between trading partners, while product flows with less handling and fewer interruptions and spoilage.
3. *Efficient promotion* refocuses suppliers' promotion activities away from selling-in to the distributor and towards selling-through to the consumer.
4. *Efficient product introduction* improves the process of developing and introducing new products. By working together, suppliers and distributors develop better products at lower costs.

Of these four core processes, *efficient replenishment* is considered to be the fundamental technological platform for implementing the overall ECR strategy. The technologies of EDI, UPC bar code, and POS scanners are critical enablers of efficient replenishment. They allow information on customer sales to be quickly and accurately known to all parties in the supply chain. Product handling can be significantly reduced, and non-value-adding activities, excessive inventories, and their associated costs can be eliminated.

Similar to QR strategy in the apparel industry, ECR focuses on the reengineering of business processes to make them more effective. In addition, UPC bar-coded product identification and electronic data interchange, using industry standards, are an essential part of ECR.

The ECR strategy requires a two-phase implementation. The goal of the first phase includes *automation* of existing replenishment loops by creating a paperless logistics system using EDI, UPC bar-code, and scanning technologies. In this phase, today's best practices are also to be implemented along with automation of each of the two replenishment loops. Examples of best practices to be implemented include every day low pricing by the supplier (EDLP), cross-dock shipments at the distributor, and scanned receiving and POS transactions at the retailer. The implementation of such best practices and installation of modern information technologies will improve and automate the flow of information and the processing that takes place in today's operating environment.

The second phase builds on the foundation laid by the first phase. It requires the *integration* of the two replenishment loops into one. In this phase, information and decision support systems and logistics elements are introduced at both the distributor and the supplier. Such elements include weekly orders and forecast updates by the distributor, and full EDI implementation of invoices, billbacks, coupon redemption, and funds transfer between the supplier and distributor partnership.

7.5 Conclusions and Managerial Implications

In spite of the substantial benefits attained by early users, EDI should be viewed as a basic requirement for doing business rather than as a source of competitive advantage. The reason for this is the relatively low investment required to adopt EDI, which leads to a low entry barrier. Thus, EDI can deliver competitive advantage only when the firm is innovating continuously in the use of EDI to reengineer its value chain, thereby preventing competitors from catching up quickly.

As discussed earlier, EDI implementation requires changes across the entire organization. The impact of EDI on the people directly involved in purchasing and logistics departments can be profound and comes as no surprise. The impact on other functions is often overlooked, however. For example, the accounts payable staff will now have to deal with EDI invoice transactions; the legal department must deal with the lack of signatures on electronic documents; and so forth. Given these sweeping changes, it is important that senior corporate and functional management be involved and committed to the success of quick response and EDI strategies.

The successful adoption of EDI depends largely on the standardization of organizational procedures, and not merely on the standardization of messages and protocols (Brousseau [1994]). This means that EDI should be viewed primarily with respect to the coordination processes linked to it, rather than as a means to transmit and process information automatically.

Given that EDI implementation requires the crossing of organizational boundaries, close cooperation and coordination between companies engaged in an EDI-based trading relationship is very important. In fact, partnerships built on trust are a prerequisite for success in quick response and EDI since only such partnerships allow for the sharing of information credible enough to be used in making important production and stocking decisions by both parties. Partnership means that support in the form of education, problem resolution, advice on technology selection, etc. be provided to trading partners. Long-term commitment and relations with a relatively small number of trading partners are implicit in making these partnerships work.

Study Questions

1. What is the role of electronic data interchange (EDI) software in manufacturing and distribution networks?
2. What are the principal benefits of an EDI system?
3. How does EDI affect transaction costs between buyers and suppliers?
4. Will the Internet reduce the role of mediators in the supply chain? Illustrate with examples from the apparel and grocery industries.
5. How does *efficient consumer response (ECR)* create a win-win situation? Explain.

References

Apte, U. M., W. Lane, T. Sample, and W. Vaughn (1993), "Quick Response and EDI at JC Penney and Haggar Apparel: Using Information Technology for Re-engineering of Business Processes," Working Paper, Cox School of Business, Southern Methodist University, Dallas, TX.

Blackburn, J. D. (1991), "The Quick Response Movement in the Apparel Industry," in J. D. Blackburn, ed., *Time-Based Competition,* Business One Irwin, Burr Ridge, IL.

Brousseau, E. (1994), "EDI and Inter-Firm Relationships: Toward a Standardization of Coordination Processes," *Information Economics and Policy,* Vol. 6, nos. 3–4, pp. 319–347.

Brown, et al. (1991), "Quick Response: Special Report," *Chain Store Age Executive,* March.

Buzzell, R. and G. Ortmeyer (1995), "Channel Partnerships Streamline Distribution," *Sloan Management Review,* pp. 85–96.

Cachon, G. (1995), "An Investigation of Continuous Product Replenishment," Dissertation Proposal, Operations and Information Management Department, The Wharton School, University of Pennsylvania, Philadelphia, PA.

Clemons, E. K. (1993), "Information Technology and the Boundary of the Firm: Who Wins, Who Loses, Who Has to Change," in Bradley, Hausman, and Nolan, eds., *Globalization, Technology and Competition,* Harvard Business School Press, Boston, MA.

Clemons, E. K. and S. P. Reddi (1993), "An Analysis of the Impact of Information Technology on the Organization of Economic Activity," Working Paper, Operations and Information Management Department, The Wharton School, University of Pennsylvania, Philadelphia, PA.

Datapro (1991), *Manufacturing News and Perspectives,* Vol. 5 (8), Datapro Information Services, McGraw-Hill, Delran, NJ.

Davenport, T. H. (1993), *Process Innovation,* Harvard Business School Press, Boston, MA.

Dearing, B. (1990), "The Strategic Benefits of EDI," *The Journal of Business Strategy,* January/February, pp. 4–6.

Defosse, S. F. and T. D. Barr (1992), "Implementing EDI/EGI Will Reduce Time to Market," *Industrial Engineering,* Vol. 24 (8), pp. 30–31.

Dolen, P., R. Grottke and J. Luker (1989), *Quick Response,* Andersen Consulting, Arthur Andersen & Co., NY.

Feder, B. J. (1991), "Moving the Pampers Faster Cuts Everyone's Costs," *The New York Times,* July 14, Section F, p. 5.

Fisher, M., W. Obermeyer, J. Hammond, and A. Raman (1994a), "Accurate Response: The Key to Profiting from QR," *Bobbin,* February, pp. 48–62.

Fisher, M., W. Obermeyer, J. Hammond, and A. Raman (1994b), "Making Supply Meet Demand in an Uncertain World," *Harvard Business Review,* May–June, pp. 83–93.

Hammer, M. (1990), "Re-engineering Work: Don't Automate, Obliterate," *Harvard Business Review,* July–August, pp. 104–112.

Hammond, J. H. (1993), "Quick Response in Retail/Manufacturing Channels," in Bradley, Hausman and Nolan, eds., *Globalization, Technology and Competition: The Fusion of Computers and Telecommunications in the 1990s,* Harvard Business School Press, Boston, MA.

Hammond, J. H. and M. A. Kelly (1991), "Quick Response in the Apparel Industry," *Harvard Business School Publication #9-690-038,* Boston, MA.

Hunt, V. D. (1989), *Computer Integrated Manufacturing Handbook,* Chapman and Hall, NY.

Hwang, K., C. Pegels, R. Rao, and V. Sethi (1993), "Electronic Data Interchange Systems—State of the Art," *Journal of Systems Management,* Vol. 44, no. 12, pp. 12–15.

Jaikumar, R. and D. M. Upton (1993), "The Coordination of Global Technology," in Bradley, Hausman, and Nolan, eds., *Globalization, Technology and Competition,* Harvard Business School Press, Boston, MA.

Kekre, S. and T. Mukhopadhyay (1992), "Impact of Electronic Data Interchange Technology on Quality Improvement and Inventory Reduction Programs: A Field Study," *International Journal of Production Economics,* Vol. 28, pp. 265–282.

Kurt Salmon Associates (1993), "Efficient Consumer Response: Enhancing Consumer Value in the Grocery Industry," Food Marketing Institute, Washington, DC.

LaPlante, A. (1992), "Troubled Automakers Rethink IS: Big Three Abandon Inflated Expectations, Aim Squeezed Funds at EDI, Dedicated Nets," *Computerworld,* Vol. 26, no. 3, pp. 67–68.

Major, T. (1995), "The Economics of EDI," *CIO,* Vol. 8, no. 13, pp. 62–72.

Rogers, D.S., P. J. Dougherty, and T. P. Stank (1992), "Enhancing Service Responsiveness: The Strategic Potential of EDI," *International Journal of Physical Distribution and Logistics Management,* Vol. 22 (8), pp. 15–20.

Scala, S. and R. McGrath (1993), "Advantages and Disadvantages of EDI: An Industry Perspective," *Information and Management,* Vol. 25, no. 2, pp. 85–91.

Seidmann, A. and E. T. G. Wang (1994),"Electronic Data Interchange: Competitive Externalities and Strategic Implementation Policies," Working Paper, Simon School of Business, University of Rochester, Rochester, NY.

Scott Morton, M. S. ed. (1991), *The Corporation of the 1990s: Information Technology and Organizational Transformation,* Oxford University Press, NY.

Sokol, Phyllis K. (1989), *EDI: The Competitive Edge,* Intertext Publications, McGraw-Hill, NY.

Taylor, Alex III (1994), "The Auto Industry Meets the New Economy," *Fortune,* September 5, pp. 52–60.

Walton, L. W. (1994), "EDI: Its Usage and Adoption within Marketing and Logistics Channels," *Transportation Journal,* Vol. 34, no. 2, pp. 37–45.

Williamson, O. E. (1985), *The Economic Institutions of Capitalism,* Free Press, NY.

Reading 7.1
Quick Response and EDI at JCPenney and Haggar Apparel

Using Information Technology for Re-engineering of Business Processes

Uday Apte, William Lane, Tom Sample, and William Vaughan

Cox School of Business

Abstract: The Quick Response movement and the use of information technologies, including electronic data interchange, bar coding and scanning, have emerged over the last decade at the center of business strategies in the apparel and retail industries. Quick response requires improving information linkages, and shortening manufacturing cycle time. The goal is to respond quickly and accurately to the ever-changing market needs so as to contain billions of dollars of losses currently experienced by the industry in overstocking and stockouts. As detailed in the paper, implementation of Quick Response, with information technology as a critical enabler, has resulted in significantly shortened replenishment cycle time, improved sales and in-stock position, and reduced on-hand inventory. The net impact of all these factors is tens of millions of dollars of benefits to JCPenney Company, and equally impressive gains for Haggar Apparel Company.

Quick Response and Electronic Data Interchange cross organization boundaries and, to be successful, require a partnership based on trust. JCPenney and Haggar have developed such a partnership. In addition to describing this partnership experience, we will also look at the resultant strategic benefits, and analyze the lessons that can be drawn from this experience. Specifically, we will discuss two categories of information technology applications: first, the use of EDI for incrementally improving traditional transactions such as purchase orders and invoices; and second, the use of information technology in re-engineered business processes such as advanced shipment notice, and point-of-sale data capture and processing.

1.0 Introduction

For the last decade or so, the apparel and retail trade has noted a steady and disturbing increase in the cost of non-promotional markdowns and lost sales due to stockouts, estimated to be more than twenty-five billion dollars a year (Hammond and Kelley 1991). The Quick Response movement and the use of information technologies, including electronic data interchange, have been since 1985 at the center of efforts to contain these losses.

Mr. James Oesterreicher (1993), President of JCPenney's Stores and Catalog Division, describes the partnership requirements and benefits of Quick Response and EDI,

> ". . . we began developing some true partnerships with key suppliers. . . . One of these is the Haggar Apparel Company. . . . one of the true pioneers in the concept of supplier/retailer partnerships. They were also one of our early test suppliers on Electronic Data Interchange. Open communication was key to the success of our partnership. We electronically shared product forecasts, transmitted point of sale data and interfaced with each other at all levels . . . using all primary Electronic Data Interchange functions.
>
> The success of this Quick Response partnership is apparent in the average replenishment time on Haggar's poly wool pant. It now takes only eighteen days for our stores to receive merchandise from Haggar. Twelve days faster than the other suppliers we track. In 1992, our unit sales were up by sixteen percent, while our on-hand inventory was down by three percent. In-stock percent in stores was the highest of any of the three pant suppliers we track. Truly a win-win program for both Haggar and JCPenney."

In this paper we explore this partnership between JCPenney and Haggar in implementing Quick Response and EDI, and evaluate the lessons learned. In the next section we provide background information on both the companies and their efforts in implementing Quick Response and EDI. In the third section we describe the concepts underlying these efforts, and illustrate their applications of information technology to both traditional transactions and re-engineered business processes. In the fourth section, we evaluate the strategic benefits. The paper concludes in the fifth section with a discussion on the managerial implications of using information technology for re-engineering of business processes.

Source: Working Paper, Edwin L. Cox School of Business, Southern Methodist University, Dallas, TX, September 1993.

2.0 Quick Response and EDI at JCPenney and Haggar

JCPenney Company was founded April 14, 1902 in Kemmerer, Wyoming, by Mr. James Cash Penney. The original store financed with Mr. Penney's life savings, was named the "Golden Rule." Today, approximately 150,000 associates operate company's 1300 stores and 1980 Catalog Units across the United States, Hawaii and Puerto Rico. In 1992, JCPenney Company realized profits of $777 million on total sales of $18 Billion.

Haggar Apparel Company is a leading designer, manufacturer and marketer of men's apparel. Established in 1926, the company coined the term "slacks" and has been the number one seller of men's dress slacks for many years. The company is also a market leader in sales of custom-fit suits comprised of a pair of slacks and a separate jacket that may be matched to form a suit requiring little or no alterations. The company's product lines also include sport coats, casual slacks, shorts, and shirts. Haggar markets its products primarily through department stores and mass market retailers.

JCPenney has been one of the early users of Electronic Data Interchange (EDI) technology which has been defined as the intercompany computer to computer communication of standard business transactions specified in a standard format (Sokol 1989). With EDI, a retailer's purchase order information feeds directly into its supplier's order processing system. Likewise, invoice data from the supplier can feed directly into the accounts payable system of the retailer.

JCPenney began using a proprietary format to send and receive purchase orders and invoices with its largest suppliers in the early 70's. While this was more efficient than paper, it required that each supplier develop and maintain a different computer program for each retailer they electronically conducted business with.

In 1988, JCPenney moved its corporate offices from New York to Dallas. Unexpectedly, this move proved to be a catalyst for replacing the company's proprietary format with the EDI standard.[1] The Catalog Division previously had a staff of about 75 that handled telephone follow-up on purchase orders. Most of these associates chose not to relocate to Dallas. A study was launched to determine if the use of EDI to automate purchase order follow-up could help alleviate this staffing problem. That study, in turn, led to a broader, corporate evaluation of EDI for other business documents exchanged between JCPenney and its suppliers.

In 1993, JCPenney's EDI effort is supported by a staff of thirteen. They work under the direction of a cross-functional EDI Steering Committee chaired by the EDI manager. The primary driver of EDI is the company's commitment to Quick Response for both improved customer service and profitability. Quick Response is intended to improve profitability through increased sales, better in-stock position, reduced on-hand inventory, and cost containment. Customer service is enhanced by having the merchandise on hand that the customer wants, when the customer wants it, and in the right sizes and colors. The development, sourcing, purchasing, movement, sale and replenishment of merchandise under a Quick Response strategy require that fast, accurate information be made available to all parties along the merchandise pipeline. Hence, the need for EDI. Currently, JCPenney processes most of the standard EDI transactions listed in Table 1. In 1992, JCPenney processed ninety-three percent of their purchase orders and seventy-five percent of their invoices through EDI. The total volume of EDI documents exchanged in 1992 was 32.5 million including 14.7 million purchase orders, 13.6 million invoices, and 1.5 million advanced shipment notices.

Haggar Apparel Company began using EDI in 1984 with a request from a department store customer to replace its monthly replenishment orders by weekly replenishment orders using EDI. The project was a success. Sales to the store increased by about 25 percent with a 25 percent reduction in inventories. To capitalize on the benefits EDI, a proprietary purchase order message standard was developed and was offered to all customers.

The emergence of industry standards in early '80s for barcoding and EDI were important to the development of Haggar's EDI and Quick Response programs. To address an internal inventory accuracy problem, Haggar began to automatically capture inventory transactions using the Universal Product Code (UPC). By May 1985, all products were being ticketed using UPC.

In mid '80s, Haggar participated in the development of industry standards for EDI, and the company became one of the early users of the industry-wide EDI standard that emerged. Haggar enhanced its internal systems to conform to the industry EDI standard and encouraged its customers to move to the industry standard formats. In 1993, Haggar performs a majority of the EDI transactions listed in Table 1 with its retail customers and fabric suppliers. In 1992, Haggar processed a total of 1.3 million EDI transactions including 0.5 million purchase orders and 0.6 million invoices.

---

[1]In EDI, the computer applications that support business transactions and that have the capability to generate, transmit and receive transaction data must be present at both the seller and the buyer organizations. For these computer applications to understand each other, the transaction data being communicated must also be in the form of recognized and previously agreed-to standards. EDI, therefore, requires format standards for business transactions such as request for quotation, purchase order, bill of lading, invoices, electronic payments, etc. Since the mid '80s, retailers and apparel and textile manufacturers have developed and used a subset of ANSI X.12 standard known as VICS (voluntary inter-industry communications standard) as the industry standard for EDI.

TABLE 1 STANDARD EDI TRANSACTIONS

Business Transactions		Transportation Transactions	
810 -	Invoice	110 -	Air Freight Invoice
820 -	Remittance Advice	210 -	Motor Carrier Freight Invoice
830 -	Planning Schedule/Material Release	213 -	Motor Carrier Shipment Status Inquiry
832 -	Price/Sales Catalog	214 -	Motor Carrier Shipment Status Message
846 -	Inventory Advice		
850 -	Purchase Order		
852 -	Product Activity (Point-Of-Sale data)		
855 -	Purchase Order Acknowledgment		
856 -	Ship Notice/Manifest (ASN)		
864 -	Text		
869 -	Order Status Inquiry		
870 -	Order Status Report		
997 -	Functional Acknowledgment		

3.0 Information Technology for Improving Business Processes

Quick Response is being increasingly adopted by retailers, apparel manufacturers, and textile mills for speeding the flow of information and merchandise in the textile-apparel-retail distribution chain. This improves their ability to respond quickly to changing market needs. The primary motivation for Quick Response stems from the fact that sales forecast error diminishes in proportion to the time remaining until the event. Hence, forecasts and stocking decisions made at the start of a season are generally much more accurate, and are financially far more rewarding, than decisions made a year earlier—a typical time lag in this distribution chain (Blackburn 1990). Thus, Quick Response is the apparel and retail industry's time-based competitive strategy for responding quickly and accurately to shifting market needs and thereby reducing losses associated with both overstocking and stockouts. Studies of the retail industry suggest that Quick Response and information technology application together can lead to annual savings as high as a few percent points of sales (Dolen, et al. 1989; Brown, et al. 1991).

At a conceptual level, Quick Response can be seen as a strategy to improve information and logistical linkages, and to shorten the cycle time involved in activities in two flows: (a) product flow moving forward in the value added chain from textile to apparel to retail, and (b) information flow consisting of consumer demand, and orders moving backward in the chain (Blackburn 1991).

Quick Response implementation involves a combination of reengineered manufacturing and business practices (Hammer 1990). Information technology is the enabler. The main information technology tools and the role played by each tool in Quick Response are:

- *Electronic Data Interchange* facilitates and speeds communication of business documents among retailers, apparel manufacturers and textile mills. Communications time and cost is substantially reduced while lowering the need for data entry and clerical work in buying and administrative functions. A decision to implement EDI, however, has much broader implications than the establishment of simple computer-to-computer communication, since, the full potential of EDI can only be tapped by reengineering of business process that use EDI transactions.
- *Bar coding with UPC and point-of-sale scanning* allows manufacturers and retailers to track merchandise at the item level and provide detailed sales data for better merchandising and inventory control decision. Furthermore, ticketing efforts are reduced, merchandise flow to the sales floor is improved, and with added price look-up functions the speed and accuracy of processing at the point of sales are significantly improved.
- *Shipping container markings* used in conjunction with the EDI's advanced shipment notice document provide significant savings by eliminating manual verification of a carton's contents and by speeding the flow of merchandise from a manufacturer through a retailer's receiving center to the sales floor.
- A variety of *application systems* including automatic replenishment systems for basic goods, forecasting systems (Evans and Apte 1992), inventory management systems, and computer aided design systems at manufacturers.

The applications of information technology for improving business processes can be conveniently (though somewhat imprecisely) classified into two categories: those applications that result in *incremental improvements* versus those that achieve *dramatic improvements.* In the former applications incremental improvements come from automating certain steps in the overall business processes that, for the most part, remain unchanged. The dramatic improvements of latter applications result from re-engineering (i.e., complete redesign) of business processes using information technology as the enabler (Davenport 1992). Presented below are implementations of these two categories of information technology applications at JCPenney and Haggar. First, the use of EDI for incrementally improving traditional transactions such as purchase orders and invoices. Next, the use of information technology in re-engineered business processes such as advanced shipment notice, and point-of-sale data capture and processing.

3.1 EDI for Incremental Improvements—Purchase Orders and Invoices

EDI implementations generally begin with the purchase order and invoice transactions. There are several reasons: these are critical business documents that must be exchanged in any buy/sell process, they usually have the highest volumes, their business function is well understood, and the benefits of implementing these two transactions are usually clear and measurable. Figures 1(A) and 1(B) respectively present the processing of purchase order transaction before and after the implementation of EDI.

3.1.1 EDI Purchase Orders. Prior to EDI implementation, purchase orders were generated at JCPenney by the order writing system, printed on paper forms, and finally sent to the mail room where they were folded, stuffed and mailed to suppliers via the post or a courier service. All these steps together took about 10–12 days. As these orders reached a supplier, such as Haggar, they were received in the mail room, and then batched and forwarded to data entry. The order transactions were then processed through an editing program. when errors were found in editing, the orders were re-keyed to resolve the errors and, once error free, were added to an order file. At Haggar, it took from 10 to 30 days for an order to get added to the order file and for it to become eligible for shipment.

With EDI, the purchase orders at JCPenney are generated by the order writing system, formatted into the EDI standard formats, and then transmitted to a supplier's computer. EDI purchase orders are now available to suppliers in one business day. The reduced handling and mailing of EDI purchase orders has translated into an annual saving of approximately $37 million at JCPenney. After receiving the EDI purchase orders, Haggar processes all orders directly into the order file within twelve hours of receipt, with

FIGURE 1(A)

Paper purchase orders

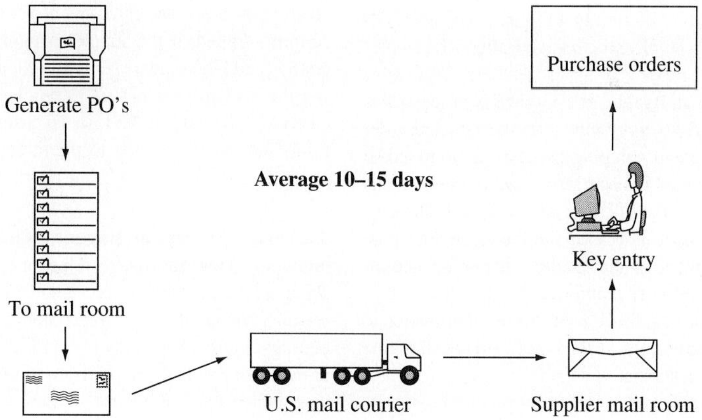

FIGURE 1(B)

EDI purchase orders

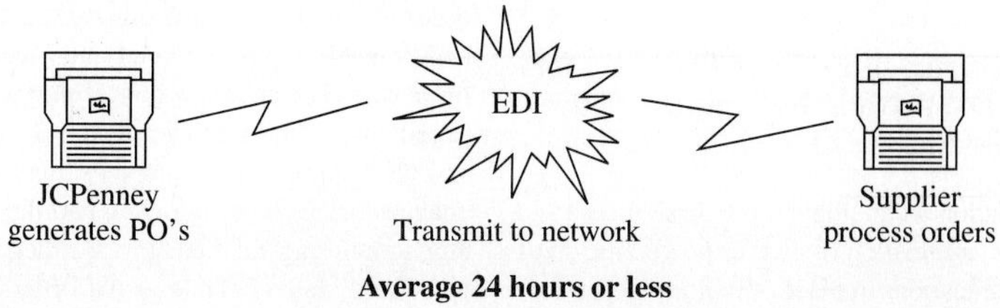

urgent orders being processed immediately. Errors detected in the processing of an EDI order are written to an on-line exception file for further correction. When all errors are corrected, the order is re-processed. As a result of EDI implementation, the error rate in purchase order processing is now less than one-tenth of the previous rate, the processing time for purchase orders is reduced to one day, and total cost savings in processing of EDI purchase orders at Haggar amount to about $380,000 per year.

3.1.2 EDI Invoices. Prior to EDI, Haggar prepared its invoices based on the merchandise included in the orders that were selected for shipment. Deviations, if any, between the ordered merchandise and packed merchandise were recorded in the system, and the invoice amounts were suitably adjusted prior to printing of invoices. This process worked reasonably well but it did lead to a high average invoice error rate (i.e., percent of invoices having one or more errors) of 1.2 percent. Most of the errors occurred when a part of an order could not be shipped due to the unavailability of merchandise. The paper invoices were finally mailed to the retailer.

At JCPenney, each store places its own orders with suppliers. Therefore, a duplicate copy of the invoice was sent to the store that placed the order, while the original invoice was sent to JCPenney's central accounting office where it was keyed into the payment processing system. Each store reviewed and filed all paper invoices for possible future inquiries. At JCPenney, processing an invoice took about 10–12 days on average. Figures 2(A) and 2(B) below show processing of invoices before and after EDI implementation.

In implementing EDI at Haggar, the invoicing process was changed to accurately produce an invoice after the selected order was shipped. The packing and shipping processes, discussed in detail later, were re-engineered to accurately capture the contents and the unique bar coded number of each carton shipped. Thus, all EDI invoices are created and sent to a retailer based on the actual contents of the cartons shipped to the retailer. This delays the invoice by one day, but this is more than offset by the reduction in time, from three days to one day, that it takes for an EDI invoice to reach the retailer. Most importantly, EDI has reduced the error rate in invoicing to less than 0.1 percent.

The EDI invoices are received for payment processing by the central accounting staff at JCPenney in less than 24 hours. The electronically filed invoices are now available to all stores for online inquiries, thus eliminating the distribution and storage of store invoices. At JCPenney, the savings from EDI Invoices is conservatively estimated at about $3.8 million in 1992.

3.2 IT-Based Re-engineering of Business Processes for Dramatic Improvements: Advanced Shipment Notice and Point of Sale Processing

The application of information technology to purchase orders and invoices primarily involve automation of certain processing steps and the use of data communications in place of mailing paper documents. Though the resultant benefits of reduced cost and time are substantial, these applications do not illustrate the dramatic power of information technology applications to improve business performance. Advanced shipment notice (ASN) and point-of-sale (POS) processing are such dramatic examples. Interestingly, these transactions did not even exist in the pre-information technology era. They provide excellent illustrations of re-engineered business processes using information technology as the enabler. Both of these transactions use bar code technology for item-level identification, and scanning technology for recognition of the bar code identification. As an integral part of the Quick Response strategy, these transactions exemplify the significant opportunity to improve cost and time performance of the merchandise pipeline.

The advanced shipment notice (ASN) is sometimes described as an "electronic packing slip." It is much more than that. The major difference is that the advanced shipment notice, consisting of the detailed information on the contents of a shipment, is transmitted and made available to the retailer before the arrival of the merchandise. This allows for the retailer to receive and "check and verify" the shipment without actually opening the cartons. When the merchandise inside the cartons is UPC bar coded by the supplier, opening can be delayed until the merchandise is stacked on the shelves. Advanced shipment notice reduces time and labor required in the merchandise pipeline. But realizing these benefits depends crucially on the supplier's accuracy in ticketing the merchandise, picking and packing it as per the purchase order, and creating the advanced shipment notice to match the contents of the cartons shipped. If the supplier shipments are undependable, the retailer will be forced to verify the contents of all cartons as before and the principle advantage to retailer in automating stock room receiving process will be lost. To ensure a high level of accuracy, a supplier usually resorts to the re-engineering of pick and pack process.

3.2.1 ASN Process at Haggar—Picking, Packing, and Shipment of Merchandise. Haggar's implementation of ASN began with a thorough understanding of the existing process. This review revealed that the process could not guarantee a sufficiently high level of carton content accuracy. Management committed to re-engineer the pick and pack process using information technologies such as bar coding and scanning.

Prior to ASN implementation, Haggar printed a three-part picking ticket that displayed the details of an order. When the products were picked from inventory, unavailable items were noted on the picking ticket. The orders were then sent to a CRT operator who entered these changes into the system. The order was subsequently assigned to a packer who, if necessary, split the order into carton size increments, packed the order in the carton, and applied a shipping label and a copy of the picking ticket to the carton. The accuracy concerning the contents of the carton, as stated on the picking ticket, was only as good as the attentiveness and the legibility of the handwriting of the packer who noted the changes on the picking ticket, and the accuracy of the CRT operator in entering the changes. Further, the information was organized and was available at the order level but not at the carton

FIGURE 2(A)

Paper invoices

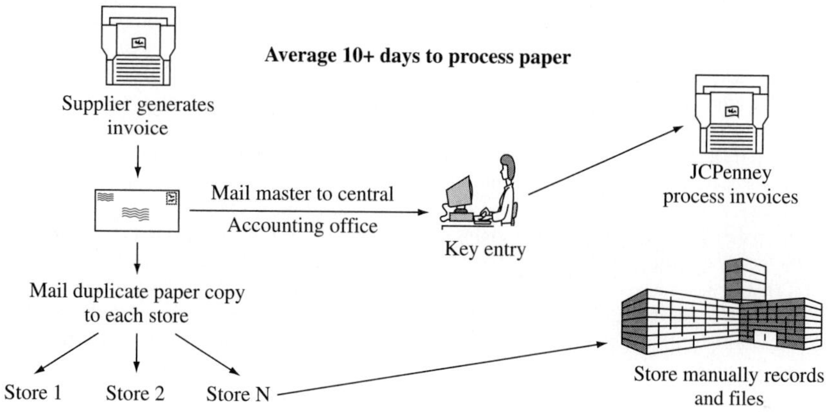

FIGURE 2(B)

EDI invoices

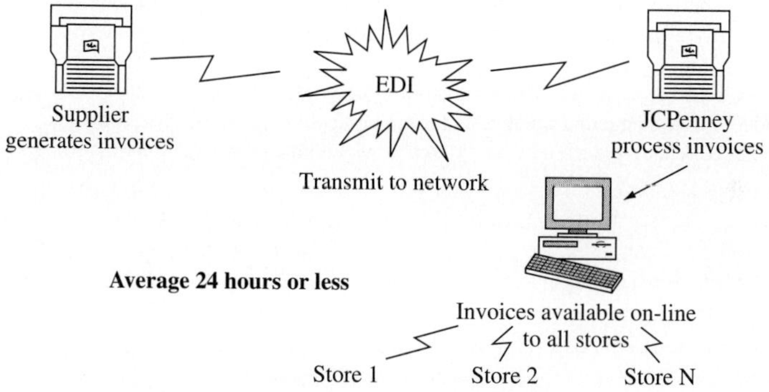

level. Quality control checks indicated that while the total units numbers were usually correct, the error rate at the SKU (stock keeping unit) level approached an unacceptable level of 5 percent.

The picking and packing process was re-engineered to solve these problems. In the new process, the picking ticket is a simple one-part form that includes a bar coded control number. The only change the person who fills the order needs to make is to note the change in the total number of units picked. When the order is picked it is transported to a scanning station connected to a personal computer on a local area network. The scanning stations consist of a control unit, a laser scanner, a thermal transfer label printer and a regular dot matrix printer.

When the order arrives at the scanning station, the control number on the picking ticket is scanned. This number is used to download the detailed information on the order from the network. The operator then scans the bar-coded label on the picking ticket, that signifies completion of special process, to verify that all processing for the order is complete. Next the garment tickets are scanned and checked against the order. Haggar prints a serial number bar-code in addition to the UPC on each ticket. By requiring that both bar codes be scanned on each garment ticket, the system accurately captures the contents of each carton and verifies that it matches the order.

Any errors detected by the system, such as missed special processing or picking errors, cause the order to be returned for correction to the area that made the mistake. When all errors are corrected, the shipping documentation, packing list and carton label, are printed at the scanning station. The garments are then packed in a carton and the documentation is applied. The system now has a listing of all the cartons in an order and the exact contents of each carton.

The shipping process has also been re-engineered to take advantage of the new carton-level information and to provide the necessary control measures to ensure that an accurate ASN is sent to every customer. This process involves quality control personnel equipped with handheld radio frequency scanners. The twenty-digit bar code identity of each carton is scanned using the radio frequency scanners as the cartons are loaded on pallets and as the pallets are loaded on trailers. The system verifies that the correct cartons are loaded on the correct trailer and that no carton is missing. The ASN is then produced by combining the information collected in the packing process with the information collected when the trailer is loaded. Due to the added labor needed for scanning of bar coded labels and for quality control checks, the cost of picking and packing process at Haggar has gone up slightly. However, the accuracy level is so high (SKU level error rate of less than 0.1 percent) that the retailer can trust ASN to automate the receiving process for reduced cost and time.

3.2.2 ASN Process at JCPenney—Automation of Stock Room Receiving. Automating the receiving process in the stock room is one of the most significant business process improvements made possible by use of the advanced shipment notice and carton bar code scanning. Figures 3 and 4 present the stock room receiving/check in process at JCPenney before and after the implementation of ASN. For retailers, verifying shipments and orders against the manifest and original purchase order has always been a slow, labor intensive, and error prone process. AT JCPenney these problems were compounded by the staggering number of cartons that were handled. In 1992, JCPenney processed over 65 million cartons through stock rooms and catalog distribution centers.

Prior to ASN implementation, it took JCPenney an average of 3.9 days to get merchandise checked in, to post the receipt to the open-order file, and to move it to the sales floor. If the receiving copy of the purchase order was misfiled, or if there were other problems with the shipment, the process took even longer than that. This time spent by the merchandise in stock room receiving meant that it was almost four days that the merchandise was *not* available for sale; almost four days during which the merchandise was in the stock room and yet the open-order file had not been adjusted to reflect the receipt; and almost four days that the replenishment merchandise, if needed, could not be reordered.

As recently as two or three years ago, four days was a commendable record, especially given the fact that larger JCPenney stores received two to three hundred cartons per day. However, in a Quick Response environment today, where shipments are smaller and frequent and most of the "safety-stock" is correctly viewed as excess inventory, four days is no longer acceptable. The ultimate goal here is to achieve a "just in time" system performance with a minimal level of inventory.

In the ASN scenario, a supplier generates the ASN and transmits it to JCPenney as soon as the doors close on the trailer. Valid ASNs are now available to JCPenney's retail stores and catalog distribution centers. When cartons arrive at the dock, the bar coded label on each carton is scanned. The serial number on the label is cross-referenced with the associated ASN to display detailed information about the merchandise contained in the carton. This is compared with the original purchase order. Assuming no discrep-

FIGURE 3

Manual receiving/check-in

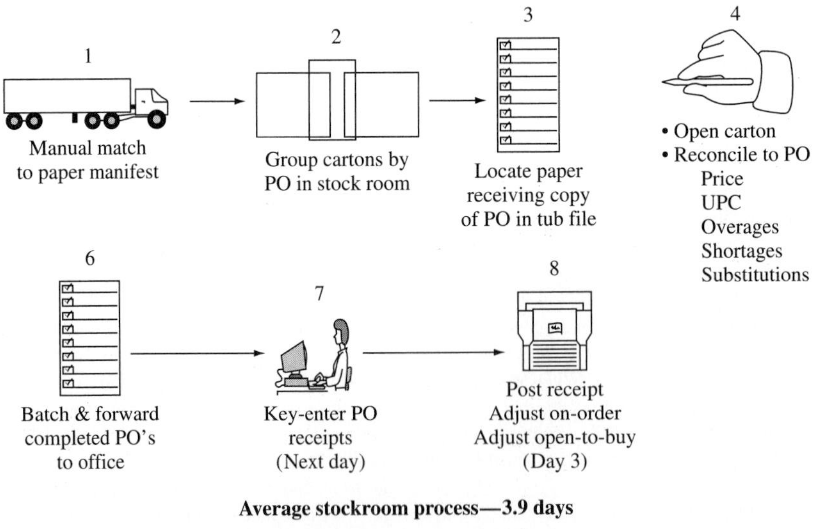

FIGURE 4

Automated receiving with ASN/Bar code carton labels

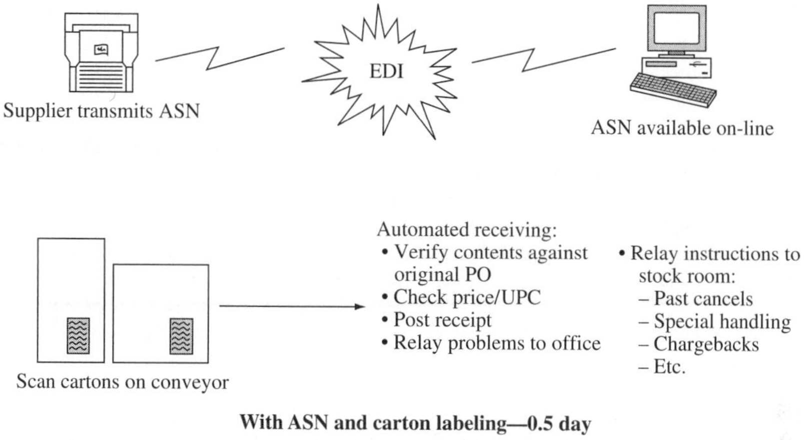

ancy, the merchandise receipt is posted to the open-order file, and the on-order quantities are suitably adjusted. A critical pre-requisite for the success of ASN is that the ASN sent by the supplier be *100 percent accurate.* In this case, there is no need for a retailer to open the carton to physically verify the contents. The carton can be moved directly to the floor and unpacked. Using the ASN and bar code carton labeling for the automation of stock room receiving process is estimated to have resulted in merchandise reaching the sales floor in 0.5 day or less. The financial benefits of ASN to JCPenney are estimated to be about $6 million per year.

Effective implementation of the advanced shipment notice, and the bar coding and scanning technologies require re-engineering and re-thinking of processes for both the trading partners. The retailer and the supplier must invest in scanning equipment, and in some cases, new or modified conveyor lines. In most cases both trading partners must also invest in new or enhanced software. For a supplier, this means new monitoring controls and validity checks in the packing and shipping operations.

3.2.3 ASN Process—Improved Information and Decision Making. The re-engineered ASN process also leads to improved decision making. Merchandisers now have the ability to know *exactly* what merchandise was shipped, the quantity that was shipped and the contents of each carton. This information is critical when merchandise is expected for advertised promotions and for placing back-orders for catalog shoppers. The store merchandiser can also "pre-check" in-transit merchandise against the original order and make decisions regarding shipments that are past the cancel date, or that involve shortages, overages, substitutions, or charge-backs. The merchandiser can relay these decisions, via the system, to a stock-room personnel *before* the shipment arrives. This eliminates the need to call a merchandiser away from customers on the sales floor, to the stock room to give instructions.

3.2.4 Point-of-Sale Process at JCPenney and Haggar. For the retailer, the consumer is the ultimate decision maker. Capturing sales data at the SKU level, and analyzing this data to uncover the underlying trends is the only way to know what is selling by style, color, and size. Knowing what is not *selling* is just as important. This information and analysis is critical to ensuring that the right product is available in the right quantity at the right store when a customer wants it.

In point-of-sale (POS) processing, the UPC bar code of each merchandise ticket is scanned at the check-out counter, and is cross-referenced with the pricing table using the UPC number. This scanning and price look-up process, in itself, saves considerable time for both the sales associate and the customer at the check-out counter while simultaneously improving the accuracy of sales transaction.

After POS scanning, the customer's sale is processed, and the data is distributed to the sales, inventory and financial management systems. Sales and inventory data is formatted into the EDI Product Activity Data transaction. This can be immediately made available to interested suppliers to identify sales trends, adjust and develop sales forecasts, determine manufacturing material requirements, plan production, and drive replenishment systems for basic items.

POS data received at Haggar is presently used to assist in the production planning process so that the inventory risk is minimized. Analysis of the styles, colors, and sizes that are sold at retail is used to detect trends and influence production plans.

3.2.5 Improvements in the Manufacturing Processes. In order to adequately support Quick Response Haggar has implemented a continuing stream of changes to the manufacturing process to reduce the throughput time for the production of a garment. Throughput time, 30 days only five years ago, is now routinely eight days. For a limited part of the production, it can be as low as two days. Even with this impressive manufacturing cycle time reduction, the time requirement of a "Quick Response order" is such that Haggar has been unable to respond to an order by manufacturing goods after the order has been received. Quick Response orders are still filled from inventory. Several other manufacturing process improvements, such as modular manufacturing and packing processes, are being studied which may permit some orders to be manufactured upon receipt.

4.0 Strategic Benefits

The literature on business strategy, till recently, has generally tended to focus on competitive strategy where firms are competing with one another and are trying to gain and sustain competitive advantage. The strategic benefits of quick response and EDI, in contrast, are derived mainly through cooperation among retailers, apparel manufacturers and textile mills, so that the maximum value at the least cost is delivered to consumers (Dearing 1990). Specific benefits of Quick Response and IT-based re-engineering of business processes at JCPenney and Haggar were identified as a part of the earlier discussion. We will briefly restate these benefits in general terms here.

Reduction in the cost of doing business is the most fundamental benefit of Quick Response and EDI. This results from reduction in human handling of business documents, savings in mailing expenses, reduced cost of re-keying of data, reduced cost of errors, and so forth. These benefits are accrued not only to a partnership of a retailer and an apparel manufacturer but also to a partnership involving the textile mill.

Implementation of Quick Response and EDI on both sides of the supply chain have resulted in significant cost reductions for Haggar. In order to gain competitive advantage with Haggar's customers (i.e., retailers), EDI implementations involving customers were given higher priority. Consequently, the implementations with suppliers (i.e., textile mills) have lagged behind. However, the process improvements derived from the implementations with suppliers have also been very significant. In partnership with a fabric supplier, who is technically advanced in implementing Quick Response and EDI, Haggar has been able to reduce the cycle time from order to the production of garments by as much as six weeks. Labor requirements have been simultaneously reduced by eliminating process steps that are no longer necessary.

Consider, for example, the fabric received from a supplier who is not EDI capable. This fabric must be measured by Haggar for width and length, the exact shade characteristics of each roll of fabric must be captured, the rolls must be put in inventory with the inventory locations properly noted, and all of this data must finally be manually entered into the system. Once this data is processed, the rolls of fabric are assigned to a specific production order, pulled at an appropriate moment from inventory, and suitably processed. With a technically advanced supplier, however, all the information about the rolls of fabric is transmitted via EDI transactions, and the rolls are assigned to a production order even prior to being picked up at the mill. The rolls are then transported directly to the production location bypassing the measurement steps and storage in a Haggar warehouse. The labor required to measure and process the rolls is completely eliminated and the inventory required to support manufacturing is dramatically reduced.

In addition to the reduction in cost, another important benefit is that of **reduced cycle time and improved ability to respond quickly to market needs.** These are very important strategic benefits to both trading partners. With the ability to better respond to changes in demand, the retailer can reduce the level of overstocked (and stockout) goods, and consequently the cost associated with mark-downs and promotional sales (and lost sales) is reduced. This important category of benefit is estimated to amount to more than ten million of dollars per year at JCPenney. The reduced cycle time also allows the apparel manufacturer to produce only the amount of goods the retailer has determined to be salable.

Improved planning and decision making capabilities through the sharing of information. EDI allows sharing and rapid receipt of accurate and more complete information on business transactions which can be used for improved forecasting and decision making by both parties. As examples, consider the potential improvements in decisions such as merchandise ordering at JCPenney and production planning of apparel at Haggar, and the resultant lowering of costs. Improved planning and decision making leads to increased responsiveness and lower costs which in turn can lead to increased market share.

Development of Quick Response partnerships is critically important since it is a pre-requisite to the realization of strategic benefits mentioned earlier. These partnerships will define new processes and inter-enterprise communications that "close the loop" on getting the right product to the customer at the right time. The Quick Response efforts at JCPenney have been coined "ASAP" which stands for Accelerated Supply And Purchasing. Although many ASAP programs are in their early stages, the results to date have been impressive. In addition to the partnership with Haggar Apparel, JCPenney's notable partnerships with suppliers include those with Ithaca Industries and Pacific Coast Feather. In all these cases, replenishment cycle times have reduced, sales have gone up, and the inventory turns and in-stock positions have improved. Ultimately these factors lead to improved customer service and profitability.

5.0 Managerial Implications

In implementing Quick Response and IT-based re-engineering of business processes over the past several years, JCPenney and Haggar have gained substantial experience that will prove to be very valuable as they move forward with these programs. The lessons learned in this experience are described below in two broad categories: those that relate more to managing information technology and those that have broader organizational and strategic implications.

Since EDI involves application-to-application communication, it is not surprising that **application enhancement and integration** represents a significant part of implementing EDI. In many cases, applications will need to be modified to ensure that they can generate and accept EDI data. In addition, some companies may need to buy application link software (i.e., interface software) that can map data from EDI software into internal databases of company applications such as purchasing, accounts payable, inventory control, etc.

In application integration, **a total process view should be kept in mind.** The main principle here is that the total cycle time of a process is ultimately dictated by the slowest element in the process. For example, an EDI transaction may be available to a trading partner in only a few hours, but if the business system that uses the EDI data is run on a daily (or a weekly) batch cycle, the cycle time for the complete business transaction will still be one day (or week) and not a few hours. Hence, to realize the full benefits of EDI, the associated business applications must also be suitably redesigned. As a specific example, consider the ASN transaction. For automated stock room receiving to work, it is essential that ASN be received and available for use before the arrival of merchandise. In some cases, due to once-a-day batch run, the merchandise arrives at the receiving dock before the system has acknowledged the receipt of ASN. In such cases, the stock room handling has to revert to the old manual process of verifying the receipt of merchandise. The real benefits of ASN are then lost. To avoid this loss, the batch system should be modified to work in real-time environment.

Some proven **tactics for using information technology in business process re-engineering** are to remove non value-added activities through automation, change a sequential task structure to a parallel one, use information itself as a powerful resource in decision making, and to use information-based trust in partnerships to eliminate (or simplify) transactions that serve only legal, and not operational, needs.

In understanding the implications of Quick Response movement, it is **important to separately consider two classes of goods** diametrically opposite to each other in sales characteristics. Sales of **basic goods,** such as undergarments, are fairly steady and are reasonably predictable. In contrast, the sales of **fashion goods** are highly unpredictable. Applying Quick Response to basic goods mainly involves use of EDI for speedy communications and replenishment systems that use on-hand inventory and sales data to automatically re-order basic goods from suppliers. In case of fashion goods, however, the key is to be able to commit to style, color, and quantity of garments as close to the selling season as possible. This is feasible only if manufacturing facilities are flexible and have short cycle production capabilities. Short cycle product line development capability of suppliers is equally important here.

For competitive reasons discussed earlier, it is unavoidable that Quick Response and EDI will become the norm in the retail, apparel and textile industries. Experience suggests that it is important to **aggressively implement Quick Response and EDI.** Suppliers that are left behind will find their customers forming partnerships with competitors, thus resulting in a lost market share. Retailers implementing EDI should consider providing incentives to suppliers for using EDI and dis-incentives to those not using it.

Use of Quick Response and EDI change the way business is done. Changes across the entire organization are typically needed to support these strategies. This means that **top management support to the Quick Response strategy is a must.** The impact of Quick Response and EDI on the people directly involved in the purchasing and logistics departments is obvious. But the impact on other functions is often overlooked. For example, the accounts payable staff will now have to deal with EDI invoice transactions, the legal department must deal with the lack of signatures on an electronic documents, and so forth. Given these sweeping changes, it is important that senior corporate and functional management be involved and committed to the success of Quick Response and EDI strategies.

How people are motivated, trained, and empowered is important to the success of EDI and Quick Response. For example, JCPenney has created a specific Quick Response training seminar for its buyers, and has planned a series of meetings with its suppliers involving all the right parties—IT staff, buyers, and principals of apparel manufacturers and textile mills—for promoting Quick Response and EDI. Changing performance incentives and measurements is also very important to ensure that the energy of all employees is properly focussed on the critically important strategies.

Given that the Quick Response and EDI strategies require crossing of organizational boundaries, close cooperation and coordination between companies engaged in a trading relationship is very important. In fact, **partnerships built on trust are a prerequisite for success in Quick Response,** since, only such partnerships allow sharing of information that is credible enough to be used in making important production and stocking decisions by both parties. Partnership means that support in the form of education, problem resolution, advice on technology selection, etc., be provided to trading partners. Long term commitment and relations with a relatively fewer trading partners are implicit in making these partnerships work.

Haggar has created "Partnership Teams" that work with customers in all affected functional areas including merchandising, transportation and receiving, and accounts payable. Relying on a single point of contact, as most old buyer-seller relationships did, will not work in this new environment. Establishing personal contacts in all functional areas of business will reduce the time and effort to resolve any problems that do occur.

It should be noted here that the partnership need not be based on a "blind" trust. Rather the information available through re-engineering of business processes be judiciously used by both the parties to ensure that the behavior and performance implicit in the partnership is adhered to by their counterparts. For example, JCPenney's ASN system produces reports that are used by stockroom associates to monitor and verify the accuracy of ASN sent by all suppliers through statistical sampling. In carrying out this task, stockroom associates adjust the sample frequency and the size of samples according to the accuracy of performance exhibited by individual suppliers.

According to Mr. James Oesterreicher,

> "The winners in the next generation of Retailing will be those who master the partnership. . . . A partnership based on communication, trust and commitment. . . . While partnerships are a key to Quick Response's success, information technology is also an enabler of the process."

Bibliography

Blackburn, J. D., "The Quick Response Movement in the Apparel Industry," in J. D. Blackburn, ed., *Time-Based Competition,* Business One Irwin, Burr Ridge, IL, 1991.

Brown, et.al., "Quick Response: Special Report," *Chain Store Age Executive,* March 1991.

Davenport, T. H., *Process Innovation,* Harvard Business School Press, Boston, MA, 1993.

Dearing, B., "The Strategic Benefits of EDI," *The Journal of Business Strategy,* Jan/Feb 1990, pp. 4–6.

Dolen, P., R. Grottke and J. Luker, *Quick Response,* Andersen Consulting, Arthur Andersen & Co., New York, 1989.

Evans, D. and U. M. Apte, "Catalog Forecasting System: A Graphics-Based Decision Support System," Working Paper # 92–061, Cox School of Business, Southern Methodist University, Dallas, TX, 1992.

Hammer, M., "Re-engineering Work: Don't Automate, Obliterate," *Harvard Business Review,* July–August 1990, pp. 104–112.

Hammond, J. A. and M. A. Kelly, "Quick Response in the Apparel Industry," *Harvard Business School Publication #9-690-038*, Boston, MA, 1991.

Oesterreicher, J. 1993. "Keynote Address," Quick Response '93 Conference, Dallas, TX.

Sokol, Phyllis K., *EDI: The Competitive Edge,* Intertext Publications, McGraw-Hill Book Company, New York, 1989.

C H A P T E R

8 MANAGEMENT OF NEW TECHNOLOGY

This chapter addresses four fundamental management challenges associated with the successful introduction of new technology in manufacturing environments. The first concerns accounting and performance measurement systems. We focus here on the changes being made to such systems to make them consistent with the new automated realities to be found on the plant floor and throughout the enterprise. The second is financial justification of the capital investment required to acquire, implement, and support automated technology. The debate on how to justify automation has been between proponents of quantitative and nonquantitative approaches. The third area to be considered is human resource management in automated process environments. While this issue has been implicit throughout our review of manufacturing technology, we will concentrate here on the impact of automation on factors such as job content, the organization of factory work practices, training requirements, and worker displacement. Finally, the chapter will consider some of the principal challenges of implementation. Automated technology represents a major innovation in many production systems. As we have seen throughout the book, this potential continues to be underused in many firms. Our discussion concerns the requirements (system, human resource, and infrastructure) for successful adoption.

8.1 Accounting and Measurement Systems in the Automated Environment

The impact of new technology on performance in manufacturing environments has been profound. It has redefined expectations for quality, precision, and the overall efficiency of the production process. It has also modified the economics of manufacturing by altering both fixed and variable costs, as well as by increasing capital investment requirements. It is not surprising, therefore, that traditional, unit-cost-based accounting and performance and measurement systems have proven to be inadequate in automated environments. These systems have also struggled to assess the overall costs and benefits of manufacturing automation. Indeed, traditional systems have been under attack for some time (see Kaplan [1990]). Issues concerning measurement only become more pressing as the capabilities and availability of automated production technology increases.

What, then, are the key challenges for effective performance measurement, and what are the implications of these challenges for management in an automated manufacturing environment? This section will address these questions by considering the key processes of *cost allocation* and *benefit measurement.*

We begin by considering a hypothetical example of a printer manufacturer who has improved its final assembly operations over a three-year period by investing in flexible manufacturing and automated testing equipment. Table 8.1 summarizes the impact of these programs on costs (as measured by the current accounting system) and on other measures of operational performance. As we can see, the unit cost of $156.30 in 1989, prior to implementation of the improvements, did not go down but increased to $179.49 by 1991. During this period, however, nonfinancial measures of performance showed dramatic improvement. How can this happen? Is the traditional accounting system accurate when it indicates that it costs more to produce the product in an operationally improved environment?

Clearly two principal challenges facing the traditional system are costing and benefits assessment. In response, improved methods for performance and cost measurement have been developed. As we shall see, *implementation* of these methods is a third management challenge. In the remainder of this section we will consider each of these challenges in more detail. We begin with the challenge of product costing.

- *Accurate Costing*—The principal function of an accounting system is to measure the "fully loaded" cost of producing each unit of every product. Traditional methods used to allocate indirect and overhead costs to individual products are based on measures directly related to the volume of production of each product, e.g., the direct labor hour content or the volume of materials consumed. Overhead cost categories are then divided up, in a proportionate manner, on the basis of such volume measures.

In flexible automated environments the level of product variety and customization increases. As a consequence the problem of *externality,* whereby the processing of one

TABLE 8.1 Cost/Performance Improvement Results for Printer Manufacturer

	1989	*1990*	*1991*
Cost/unit ($)	156.30	165.21	179.49
Defect rate	5%	2.4%	1.34%
Throughput (days)	35	9	3
Schedule performance	85%	89%	99%
Inventory (units)	47,520	15,488	4,895
Scrap ($000)	600	171	74

product affects the cost and/or performance of the system for processing another product, intensifies. For many of the support activities in manufacturing, the level of effort is not based on the volume of production, (e.g., the maintenance of the CAM software system, or the setting up of machines to accommodate a model changeover). As product mix diversity increases, volume-based allocations of indirect costs become less accurate.

This problem is more intense in automated systems where direct labor hours have been virtually eliminated. Consider again, the printer manufacturer. Table 8.2 provides additional information on the cost structure of the operation during the period of improvement. We note immediately that production volume has declined. The drop may be due to competition or the fact that according to the accounting system, it has become more costly to product printers in 1990 and 1991. The data also indicate that while production volume has declined by 50 percent, direct labor hours have been reduced by almost 80 percent. This follows since the new, automated processes require fewer workers. The situation for the allocated fixed cost pools is more complex; however, inspection costs have decreased and engineering/setup costs have increased. The engineering cost change may, in part, be due to the higher cost of maintaining the new equipment and the need to modify product and process designs (e.g., the number of engineering hours has increased in 1990, in spite of the drop in production volume). In addition, as direct labor hours decline throughout the plant, allocations to the printer line based on them can become distorted.

It is interesting to note that, historically, overall direct costs have decreased while indirect costs have increased significantly. Figure 8.1 illustrates this trend. Miller and Vollman indicate that these data, which concern direct labor and overhead as a percentage of value added (i.e., without other costs such as material), translate into overhead accounting for 35 percent of total manufacturing cost in 1975. A more recent data point puts the overhead fraction at 32 percent and direct labor costs at 34 percent. (see Howell et al. [1987]). Figure 8.2 illustrates how overhead, as a percentage of value-added, increases as the industry becomes more closely related to high volume production of standardized products, i.e., the primary users of automated manufacturing technology. Finally, Jaikumar (1993) reported that the ratio of indirect to direct labor cost varies significantly with the nature of the manufacturing process in use. In particular, this ratio is

TABLE 8.2 Cost Breakdown for Printer Production

	1989	*1990*	*1991*
Annual production volume (000)	120	101	60
Direct labor hours (000)	492	352	108
Engineering/setup hours (000)	5.280	8.467	5.040
Inspection hours (000)	30	25.2	15
Allocated engineering/setup cost ($000)	2,610	4,057	4,349
Allocated inspection cost ($000)	4,241	2,616	1,692
Variable cost ($000)	11,904	9,979	4,728

FIGURE 8.1

Percentage of value added (1855 to 1975)

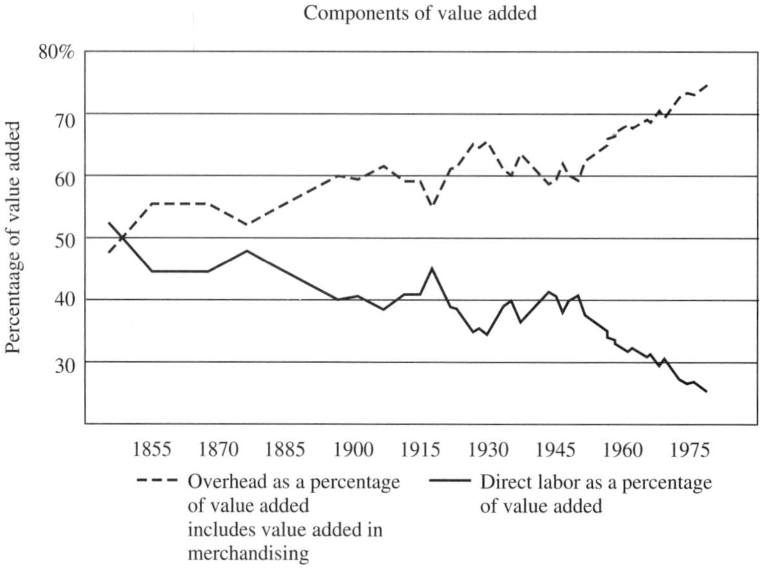

Source: Miller and Vollman (1985)

FIGURE 8.2

Overhead as a percentage of value added in the electronics industry

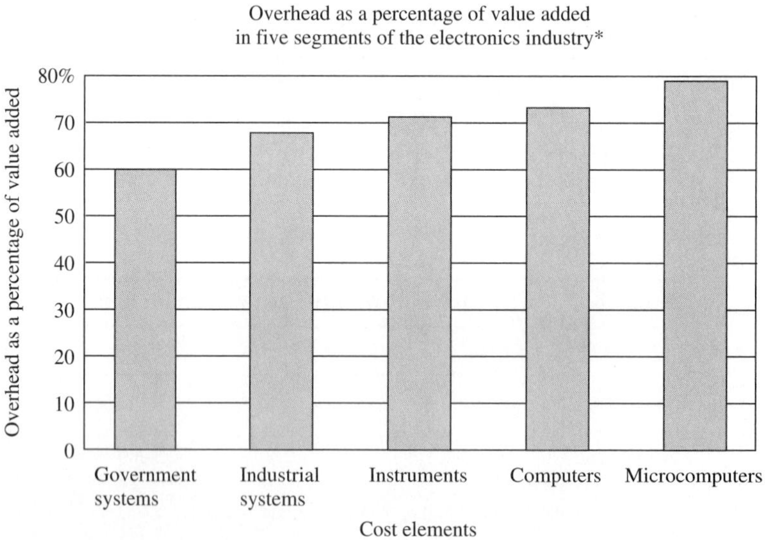

* Developed from data in the 1983 report of the American Electronics Association.
Source: Miller and Vollman (1985)

one to one for numerical control systems (consistent with the numbers noted above) and two to one for CIM.

The second challenge facing accounting systems is the assessment of benefits.

- *Benefits Assessment*—As indicated by the example and industry data (to be considered in Section 8.2), there is a wide range of operational benefits (pertaining to the performance of the production system and/or the perceived value of the product) that can be generated in automated manufacturing environments. Traditional accounting systems have typically ignored such measures altogether. Moreover, traditional systems are not able to link these improvements to cost measurement, e.g., how does improved quality or schedule performance affect cost?

A corollary to the proposition that benefits measurement is poor is the observation that the required justification of investment in new manufacturing technology is problematic. The so-called "soft" benefits of automation, which include quality, flexibility, and precision, are often based on synergies of various pieces of automated production equipment working together to improve overall factory performance. Such synergies, moreover, are based on enhanced coordination and integration at all levels of production planning and control and thus represent another example of externality (see Milgrom and Roberts [1990]).

The final challenge we consider is that of implementation.

- *Implementation Requirements*—There is widespread dissatisfaction with current product costing practices.[1] It is noteworthy that, many firms have experienced considerable difficulty in moving away from the status quo of accounting and performance measurement, (90 percent have failed according to Ness and Cucuzza [1995]).

A major stumbling block for adopting improved accounting systems has been organizational resistance to change. Such systems represent a specific innovation that should be viewed as part of the infrastructure required for the adoption of automated technology. The principal costs and organizational barriers that have blocked the adoption of accounting system improvements seem to be based on current management policies, and priorities as well as on overall organizational inertia. Indeed, as noted by Ness and Cucuzza (1995), the adoption of *activity based costing (ABC)* (to be discussed below) is viewed as a major change that requires significant investment in employee education, modification of operating procedures, development of extensive information systems, and adoption of new incentive and reward structures. Chrysler, which has been successful in rolling out ABC throughout the corporation, found that ultimately it had to train 18,000 employees, adopt a cross-functional team implementation approach, and

[1] The 1987 NAA/CAMI survey indicated that 54 percent of respondents felt the current system "needs improvement" or is "unsatisfactory." This disapproval rate was as high as 76 percent in high technology, 74 percent in electronics, and 62 percent in aerospace industries. The rate was only 37 percent in consumer products and 41 percent in metal, chemical, paper, and oil/gas industries.

involve senior management in a significant way. We will return to the issue of success requirements in our general discussion of technology implementation later on in this chapter.

Solutions to the performance/cost measurement problems noted above are generally well understood. They are based on the common sense idea that indirect costs should be allocated on the basis of the consumption of those resources that are utilized to support the processes engaged in the production of products. Consider, for example, the notion of the *hidden factory* (Miller and Vollman [1985]), which provided an early explanation of the origins of costs in a manufacturing system. Under this approach, *transactions* become the basis for assessing production overhead costs.

Miller and Vollman suggest that consideration of a wide range of transactions generated by production process activities can improve cost assessment. These transactions can be grouped into four classes: logistics, balancing, quality, and change. Logistics transactions relate to the movement of material (inputs, work-in-process, finished goods) within the factory and throughout the supply chain. Balancing transactions are required since it is necessary to match supply and demand of resources (material, labor, capital) throughout the manufacturing process. As a consequence, the firm must engage in a variety of scheduling, forecasting and related activities to ensure production of its mix of products. Quality transactions include activities associated with quality control (at the procurement, production, and distribution stages). It is also necessary to verify the completion of the other transactions related to production and to communicate product/order specifications, all of which gives rise to additional transactions. Finally, the operation of a production process requires constant updating of information to accommodate changes in design, schedules, standards, specifications, routings, bills of materials, and customer orders. Such changes in turn lead to all of the various transactions noted above.

These production process transactions are often not at all related to the level of direct labor activity. Hence, it is clear that their use as a basis for cost allocation will lead to more accurate estimates of production costs.

A more recent and precise approach to improving cost allocation, which is consistent with the transactions philosophy, is the activity-based, ABC method. This approach can be broken down into three steps. First, the complete set of activities associated with manufacturing must be identified. Accounting analysts have determined that these activities can be grouped hierarchically into four major categories: (1) unit level, (2) batch-related, (3) product-sustaining, and (4) facility-sustaining (see Figure 8.3). Activity in each of these areas occurs to support production, and thereby induces costs. At the second stage it is necessary to identify the "drivers" which link these activities to products, i.e., that relate the demand for activity resources that are generated by the production process to the level or production. Typical drivers include the number of machine setups, the number of engineering design changes, and the number of material procurement orders placed. Total activity costs are then allocated to individual products in the third step, on the basis of the volume of production (i.e., each product's requirement for a resource is a function of its volume and driver use rate). Total allocated indirect cost emerges as the weighted sum of component activity costs.

FIGURE 8.3

Activity hierarchy for the ABC method

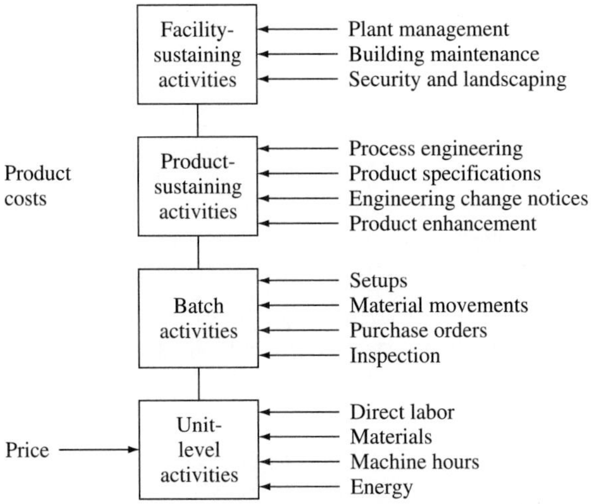

We return to our printer manufacturer to illustrate how the ABC method can be applied and how it can lead to significantly different conclusions on the cost of different products in a multiproduct manufacturing environment. Assume that the company has responded to the drop in sales and the increased flexibility potential of its new automated processes by increasing the number of printer models it produces at the plant (i.e., it moved from a single, generic printer to a set of customized designs for different regional markets). Table 8.3 contains data on three particular cost drivers: direct labor, engineering support, and inspection (under the assumption that total sales have returned to the 1989 level of 120,000 units). Direct labor is assumed here to be used at a rate that does not vary by product or with production volume (e.g., 4.1 hours/unit).[2] As a consequence, the total number of direct labor hours used for each product is proportional to the production volume. It then follows that all overhead costs which are allocated proportionately to the number of direct labor hours consumed, under traditional accounting, also will be proportional to volume. As a result the unit cost for each product is equal (i.e., $170.79 in this case).

Under the ABC method, the drivers of engineering hours and inspection hours are used to allocate the two fixed cost pools. It is assumed here that inspection effort also is proportional to volume and its use rate does not differ by product. Engineering hours, on the other hand, are required to support design changes and process setups. Consequently, a fixed amount (2,500 hours) is required for each customized product, independent of volume. This leads to an increase in the total engineering hours used when compared to the single product case (i.e., 10,000 hours vs. the 5,280 hours used in 1989 for

[2] A more realistic cost model could include economies of scale which lead to different labor use rates for each product. A primary factor leading to such scale based differences is local learning whereby plant floor productivity improves as the volume of production increases.

TABLE 8.3 Cost Comparisons: Traditional and Activity-Based Methods

	Multiproduct Plant (Country Localized Products)				
	Country 1	*Country 2*	*Country 3*	*Country 4*	*Total*
Volume	12,000	72,000	30,000	6,000	120,000
Direct labor hours	49,200	295,200	123,000	24,600	492,000
Engineering (hours)	2,500	2,500	2,500	2,500	10,000
Inspection time (hours)	3,000	18,000	7,500	1,500	30,000
Standard Cost Allocation:					
Engineering cost pool 1	$ 261,000	$ 1,566,000	$ 652,500	$ 130,500	$ 2,610,000
Inspection cost pool 2	$ 424,100	$ 2,544,600	$ 1,060,250	$ 212,050	$ 4,241,000
Variable cost 3	$ 1,190,400	$ 7,142,400	$ 2,976,000	$ 595,200	$ 11,904,000
Total cost	$ 1,875,500	$ 11,253,000	$ 4,688,750	$ 937,750	$ 18,755,000
Unit cost	$ 156.29	$ 156.29	$ 156.29	$ 156.29	
ABC Cost Allocation:					
Engineering cost pool 1	$ 1,235,000	$ 1,235,000	$ 1,235,000	$ 1,235,000	$ 4,940,000
Inspection cost pool 2	$ 424,138	$ 2,544,827	$ 1,060,345	$ 212,069	$ 4,241,379
Variable cost 3	1,190,400	$ 7,142,400	$ 2,976,000	$ 595,200	$ 11,904,000
Total cost	$ 2,849,538	$ 10,922,227	$ 5,271,345	$ 2,042,269	$ 21,085,379
Unit cost	$ 237.46	$ 151.70	$ 175.71	$ 340.38	$ 175.71

the generic product). As a result, there is an increase in the size of the engineering/setup cost pool from $2.61 million to $4.35 million. Consequently, unit costs differ significantly across products. Indeed, under the ABC method, printers for countries 1 and 4 are more expensive to produce, and the printers for country 2 are less expensive.

Figure 8.4 illustrates how the ABC and traditional methods differ in making cost allocations. The traditional method accurately captures costs associated with various overhead centers. The allocation of these overheads to products, however, is less accurate. Indeed, use of direct labor as the allocating variable implies that all overhead costs are proportional to production volume. The two step, ABC method starts with a definition of the activities performed by company resources. The total costs of these activities (i.e., the cost of running the department that sets up the machines) is easily obtained. The allocation of such costs to products is more accurate since it is based on product specific resource consumption rates (i.e., the number of setups performed for a particular product). This leads to an allocated cost that is proportional to the product's use of resources, (which may not be equivalent to a volume-based allocation).

The movement towards an improved, activity-based approach has been led by Kaplan (1990). He observed that the principal shortcomings of standard approaches that are especially relevant in an automated environment include: (1) distortions due to volume fluctuations where unit cost is computed on the basis of shipment volumes and not capacity, (2) the unmeasured cost of complexity generated by product mix (e.g., congestion and setup delays increase as more products are added to the plant's portfolio—see Banker, Datar, and Kekre [1988]), (3) overhead allocations based solely on the uti-

FIGURE 8.4

Two-stage process: traditional and ABC cost systems

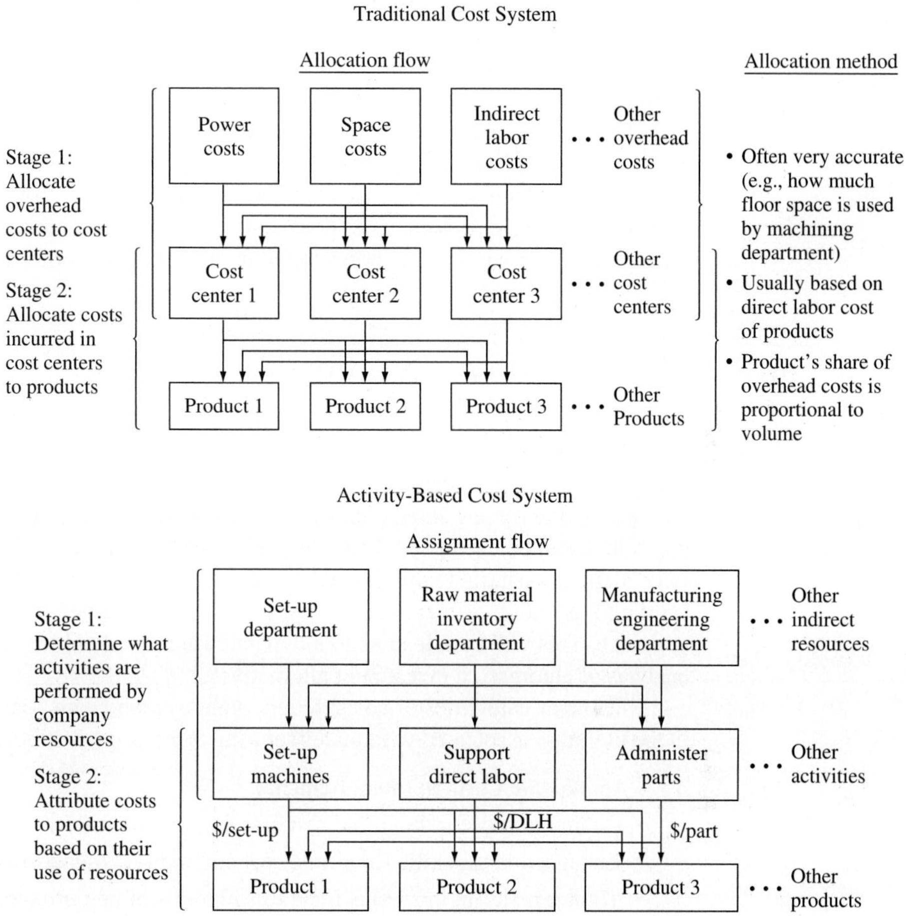

lization of direct labor hours, and (4) omitted cost savings through mechanisms such as work-in-process inventory reduction and flexible response to design changes.

Kaplan also noted that when the level of capital equipment investment increases there is added emphasis placed on machine utilization. This can lead to practices which add to costs (e.g., adding more products to the mix, increasing batch sizes, and launching a production run before firm orders are in place). Finally, Kaplan noted that there are many subtleties associated with value chain linkages that are ignored in standard costing systems. For example, as purchasing departments move to make large purchases from overseas suppliers in order to reduce costs, it is possible that the costs of delay and of defects borne by the manufacturer will go up. Similarly, the impact of delivery performance improvement on sales margins and market share is little understood and rarely quantified.

An important step that can address some of these issues involves the development of operational performance statements to supplement traditional financial performance statements. Examples of operational performance metrics include defect rates, obsolescence rates, throughput, lead times, fill rates, and customer satisfaction indices. Use of such performance metrics, however, requires a major paradigm shift in the minds of managers. Rather than using financial performance as the principal means to uncover operational problems (through cost variance analysis), managers should use operational performance to compute costs (via the activity-based approach noted above). They can then use these more accurate cost and operational metrics as the basis for measuring and improving strategic performance. The result of this change in mind set is to focus on operations problem solving and to use both internal and competitor benchmarking.

8.2 Technology Investment Justification

Any discussion of justification of the nontrivial capital investments required to implement manufacturing automation technology must begin with documentation of the costs and benefits of such systems. In this section we will identify the principal sources of cost and benefit and review some empirical evidence. We will then consider the methodologies which can be used to evaluate the investment. As noted earlier, the "finance versus faith" debate is central to this discussion. We will also introduce the notion that some insight into the value of a technology investment can be gained by using an option theory approach. This will then lead us to a dynamic framework which can be used to support analysis of automation investment alternatives.

The major categories of costs, requirements, and benefits that have been identified in the literature associated with manufacturing technology justification are listed below:

A. Automation Costs and Requirements

- Capital outlay for the initial and ongoing investments.
- Enhanced labor skills for both direct and indirect (supervisory) workers.
- Tighter material input specifications in terms of better material quality and increased availability.
- Enhanced system reliability and precision, i.e., less uncertainty in process equipment performance, supplier lead times, and market forecasts.
- A higher level of integration across functions and processes which leads to improved performance requirements for control systems and communication software.

B. Automation Benefits

- Productivity improvement through reduction in both direct and indirect labor.
- Flexibility achieved through reductions in design lead times, production cycle times, and equipment setup times.
- Quality improvement by reduction in rework, scrap, and improvement in product conformance.

- Cost reductions achieved through reduced material requirements, lower work-in-process inventory.
- Enhanced level of customer service by reductions in order delivery leadtimes and improvement in finished product availability.
- Improved product design principally through a higher level of customization and variety which is made economically feasible. Improved product reliability and performance can also be achieved.

Tables 8.4 and 8.5 are representative of early published accounts of such cost/benefit analyses. Table 8.4 indicates the measured values for metrics related to the benefits noted above for a sample of automation projects. Significant improvements have been

TABLE 8.4 Benefits of Flexible Manufacturing Systems

Comparison	*Prior Method*	*Flexible Manufacturing System*	*Average Improvement*	*Range of Improvement For Total Sample[1]*
Number of machines	29	9	70%	60–90%
Floor space	1500 m^2	500 m^2	66%	30–80%
Direct labor	70	16	77%	50–88%
Product cost	$2000	$1000	50%	25–75%
In-process time	18.6 days	4.2 days	77%	30–90%
Number of operations	15	8	47%	
Number of setups	13	5	62%	10–75%
Machine efficiency	20%	70%	50%	15–90%

Source: Frost & Sullivan, Inc. (New York).

[1]Based on a sample of 20 U.S. operating systems.

TABLE 8.5 Comparison of Manpower Requirements of Conventional Technology with Flexible Manufacturing Systems for Metal-Cutting Operations

Comparison	*Conventional Systems United States*	*Conventional Systems Japan*	*Flexible Manufacturing Systems Japan*
Engineering	34	18	16
Manufacturing overhead	64	22	5
Fabrication	52	28	6
Assembly	44	32	16
Total number of workers	194	100	43

Source: Jaikumar (1986)

achieved. Table 8.5, which is based on U.S. and Japanese FMS systems, indicates the enormous potential for reducing both direct and indirect labor.

Maffei and Meredith (1995) report on a framework for assessing the "level of success" of an advanced manufacturing technology, which was first introduced by Voss (1988). Four levels of success were proposed:

1. Technical—Does the system run without significant downtime?
2. Business/Direct—Has productivity increased (through reduced direct labor) and have costs gone down?
3. Business/Indirect—Have extra productivity gains been realized through reduced leadtimes, improved quality, and increased flexibility?
4. Competitive—Has success in the market place been achieved through an enhanced competitive advantage?

Maffei and Meredith analyzed six cases studies and developed conditions for success at each level (see discussion of CIM and human resources in Section 8.3 of the chapter).

Based on the data in Tables 8.4 and 8.5, it would seem that the case for investing in manufacturing technology is clear: Both variable and fixed costs can be reduced by making the capital investment. However, evidence suggests that many firms have been slow to adopt manufacturing automation innovations. Swamidass (1994), for example, discovered that the pattern of adoption is uneven, and has focused on some basic hard core technologies (such as CAD, CAM, CNC). These results were amplified in a field study conducted by Johansen et al. (1995), which found the level of integration in CIM to be low and the benefits realized to be less than expected.

There are a number of hypotheses which can explain this situation.

1. Manufacturing automation is priced too highly, i.e., its costs exceed its benefits.
2. It is impossible to quantify the costs and benefits with sufficient precision, and hence the level of risk associated with these investments is too high.
3. The preconditions, facilitating factors, infrastructures, and environments necessary to support the successful adoption and integration of manufacturing technology are not present, i.e., it will not work.
4. Managers are irrational and make bad decisions due to adoption of planning horizons which are too short or by being excessively risk averse.

Hypothesis 1 is plausible, and it suggests that in some cases automation technology for manufacturing requires further development to enhance its value. Hypothesis 2 is supported by our discussion of accounting and performance measurement systems. Hypothesis 3 is related to the thesis of this book, i.e., *successful adoption of automation requires the integration of multiple systems and technologies.* It is also supported by recent empirical evidence on the pattern of factory automation implementation (to be discussed in Section 8.4 of this chapter). Hypothesis 4 may be related to the methodology which managers use to evaluate automation investments. We will focus on this issue in the remainder of this section.

The management literature on manufacturing automation justification can be classified into two groups. One set of authors argues for a "strategic" approach. Because of the validity of Hypothesis 2, managers must abandon the effort to quantify costs and benefits in this case. Rather, they should focus on the strategic impact of making the investment and more importantly of not making the investment. Long-term competitive performance (and even survival) will be based on making the right investments now. Management cannot wait for all of the numbers. It is interesting to note that the survey by Johansen et al. (1995) discovered that the majority of companies surveyed did not adopt CIM for strategic purposes. Instead, they did so to make operational improvements (i.e., cost, quality, performance).

The alternative approach to strategic necessity is to utilize standard financial evaluation techniques based on the *discounted cash flow* method to estimate the *net present value* of the automation investment. To do so requires quantification of the costs and benefits as well as selection of an appropriate discount factor, planning horizon, and salvage value. In a seminal article (entitled "Can CIM Be Justified by Faith Alone," and revised in Kaplan and Atkinson [1989]), Kaplan points out that part of the criticism of the financial approach is, in fact, based on the incorrect application of the methodology. In particular, he points out some of the common mistakes that are made in this context (i.e., inconsistent treatment of inflation and excessive risk adjustment through the selection of an incorrect discount factor and/or planning horizon).

It has also been pointed out that there is an organizational bias that is introduced by using a financial return hurdle rate to screen all investment projects. Typically, the investment review process is bottom up, i.e., ideas are generated by lower level managers and are passed up to successively higher levels who are further removed from the relevant products and processes. In this process a proposal has to be rejected only once in order to exclude a project. As a result of capital rationing in many firms, these hurdle rates often are set at a very high level. The risk of automation investments is also perceived to be greater. As a consequence, it becomes easier to justify a short-term investment which does not require adoption of a new automation technology (e.g., invest in more marketing spending or upgrading of existing facilities). A related problem involves implicitly setting a zero value to the status quo ("do nothing") option. Indeed, as a result of competitor actions, this option may have a negative present value.

The difficulty in assigning values to many of the costs and benefits of automation has been documented above in our discussion of accounting practices. As a result, in many analyses that we have seen, the decision is made to ignore these "soft" factors when computing net present values. This approach is, of course, just as arbitrary as any other, i.e., the value of zero has been selected for those costs and benefits about which we are unsure.

In a recent survey, Slagmulder et al. (1995) evaluated the capital budgeting practices of 42 European companies for strategic investment in CIM technologies. They discovered that 70 percent of their sample used financial evaluation of some type, and 60 percent used a formal capital budgeting process to justify such investments. The average payback period and return (hurdle rate) required were 2.5 years and 23 percent respectively. These rather conservative numbers suggest that short-term, risk averse decision making is still prevalent. One of the most important nonfinancial metrics was "strategic fit." Projects that did not fit well with overall strategy vision were less likely to succeed. The primary motivation for these investments turned out to be productivity improvement and cost reduction. It is interesting to note that only 40 percent of the firms did a postimplementation audit.

We conclude this section by discussing an extension to the financial framework which addresses some of the difficulties noted above. This approach is based on viewing the investment in technology as a compound real option (see Mitchell and Hamilton [1988]). The value of an option is based on assessing both its downside and upside potential. The downside is associated with the cost of purchasing the option, i.e., the initial investment required to acquire and implement the automation technology (possibly on a pilot basis). Upon review of the performance of the technology, the firm can assess the likelihood that the technology will in fact deliver on its promised benefits. The upside return will accrue only if the firm exercises its option. This decision will be based on the contingency that the technology works and requires additional investments to support further development and implementation of the technology. Under the contingency that the technology does not work, these additional investments need not be made, and, hence, the loss is limited to the cost of the option (the initial investment).

The option framework suggests that increasing the time horizon and increasing the riskiness of an investment opportunity can serve to increase its value (as a real option). Implicit in this evaluation approach is a mechanism for looking at the complete range of decisions and contingencies that are associated with the future performance of the technology over a future, uncertain time horizon. Such dynamic decision frameworks have been modeled in a variety of contexts. See Cohen and Halperin (1986) for a dynamic programming model in which technology alternatives are characterized by their fixed and variable costs.

Figure 8.5 illustrates how this model could have been applied to the printer manufacturer example introduced earlier in the chapter. The figure is that of a decision tree, in which the square boxes denote decisions (concerning process and product), and the circles signify the realization of uncertain events that affect the company's fortunes (a competitor action to introduce customized niche products in this case). The sequence and timing of events and decisions is illustrated by the tree. The option involves modifying product mix and variety, in response to the competitor's action, by introducing country specific customized models. The correct decision here will, of course, be contingent on the earlier decision to invest in a flexible manufacturing process. The differential in profits at the end of the planning horizon (the right side of the tree) determines the option value of investing in the automated technology. The outcomes for each decision/competitor action combination are described at the end of each path. Clearly, the correct choice will depend on the probability of each outcome, the initial investment (the cost of the option), the potential market size, and the performance of the manufacturing system (with respect to cost and other operational dimensions).

Multistage, dynamic modeling frameworks have been extended in a number of ways. For example, Fine and Freund (1990) explicitly evaluate the value of selecting a more flexible (as opposed to dedicated) technology in a manufacturing system. This value is in fact maximized when demands (in a multiproduct setting) are perfectly negatively correlated. In general automated technology acts as a hedge against uncertainty

FIGURE 8.5

Option framework for technology investment analysis

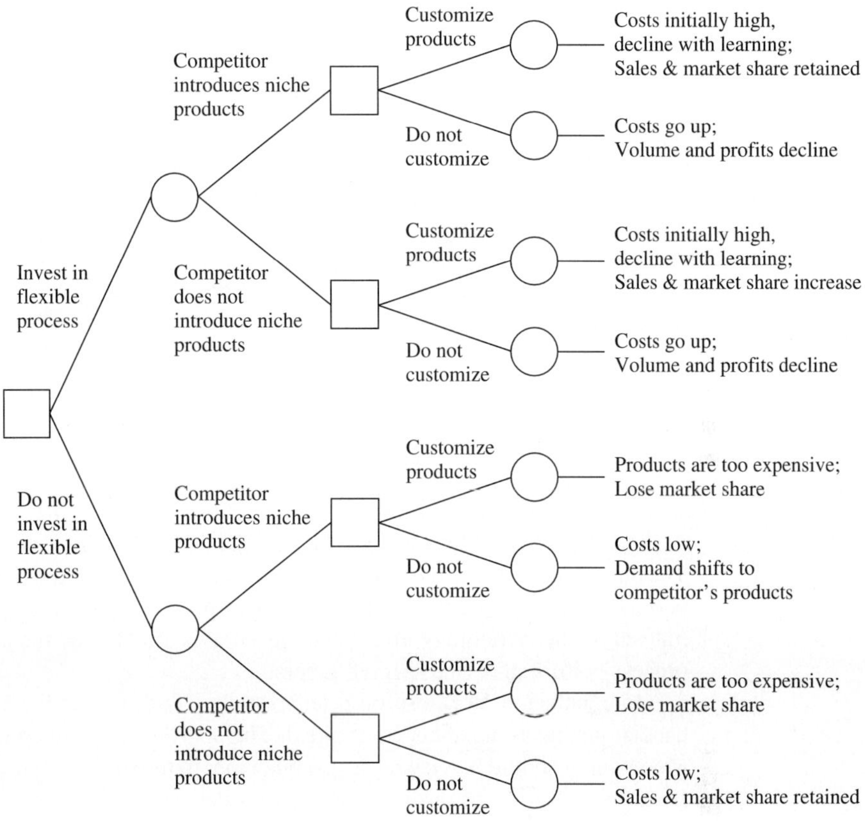

by enabling low cost adjustments of the plant's product mix in response to demand and supply contingencies.

Huchzermeier and Cohen (1994) apply the multiperiod, dynamic approach to the even broader problem of global supply chain strategy selection. In their model the investments relate to the opening and closing of factories in different parts of the world. A major source of uncertainty in this model is due to currency exchange rate fluctuations. The value of flexibility emerges here as a hedge that is based on limiting the cost impact of down-side risks and maximizing the value of up-side risks by shifting product mix throughout the global supply chain.

An interesting illustration of the dynamic nature of the process investment decision has been developed by McIvor et al. (1995) for semiconductor manufacturing. New fabrication facilities (fabs) cost in excess of one billion dollars, enjoy utilization rates of less than 50 percent, and have useful lives of less than five years. In this paper, the time paths of two key drivers of success in integrated circuit manufacturing are introduced: fab yield and market price for the device. Yields increase over time and prices generally

FIGURE 8.6

The Nutcracker in semiconductor fabrication

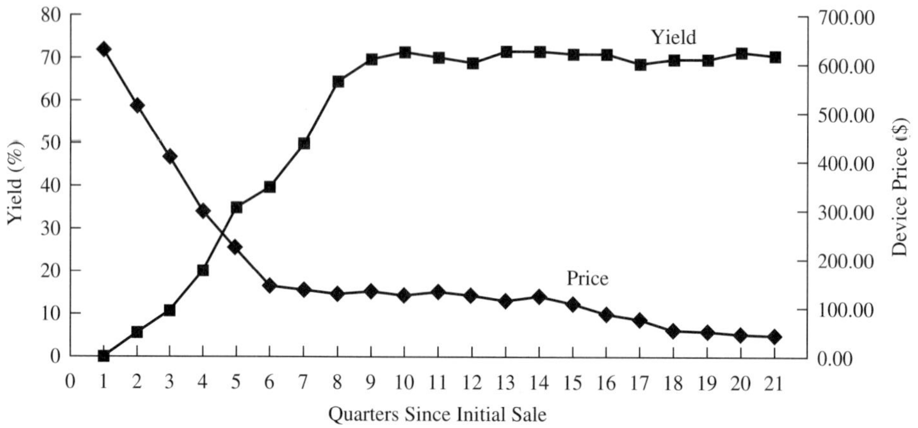

Source: R. McIvor, "Profiting from Process Improvements in the New Semiconductory Manufacturing Environment," Working Paper, Technology and Operations Review (TOR), 1995. Reprinted by permission from Technology and Operations Review (TOR).

decline. Figure 8.6 illustrates the "nutcracker" which faces semiconductor producers. The nutcracker clearly illustrates the value of improving yield during the *time to volume* period of the new product's life cycle when prices are declining rapidly. It also illustrates how the technology investment decision is affected by the dynamics interplay of process performance and market factors.

In conclusion, our position relative to the "faith versus finance" debate is that financial justification cannot be avoided. The analysis must be done correctly, however. Moreover, it should be viewed as just one input to the overall strategic planning process. In the next section of this chapter we consider one of the principal factors associated with successful implementation, i.e., managing human resources in an automated manufacturing environment.

8.3 Manufacturing Automation and Human Resources

The complexity and scope of change in the workplace, which is brought about by the introduction of manufacturing automation, necessitates careful planning. Ideally, human resource factors should be an integral part of the overall plan to introduce manufacturing automation. As noted by Adler (1992), firms often "muddle through" the management of the human resource impacts of technology implementation. Specifically, issues such as changing job content, training and retraining needs, worker compensation and incentives, organization structure and promotion policies, worker displacement, and health and safety must all be considered as automated technology options are evaluated and adopted. Indeed a comprehensive study of the impact of automation on the human resource function confirmed that those organizations that included human resources planning directly in their overall implementation plan were the most successful users of manufacturing automation (Helfgott [1988]).

Mafei and Meredith (1995) conducted six case studies to assess the impact of the operator's role on the performance of flexible manufacturing technologies. Based on the insights gained from the case studies, researchers conclude that human operators have a critical role to play in the successful implementation of manufacturing automation. Particularly, the functions of system monitoring, problem solving, and prove-out of new parts, all performed by human operators, can significantly improve productivity, quality, and responsiveness of flexible automation.

8.3.1 Changing Nature of Jobs and Organization

Computer integrated manufacturing has a profound impact on the nature of manufacturing jobs and the way in which production work is organized in the factory. As we discussed in the first chapter of the book, manufacturing technology has evolved from the craftsman stage through mass manufacturing to today's computer-based automation technologies. In the mass manufacturing age, management followed the dictum of Adam Smith and Frederick Taylor. Accordingly the emphasis was to implement the principle of division of labor, and to create narrowly defined, specialized jobs. This led to the notion that manufacturing automation results in the "deskilling" of the labor force. The debate around this issue has raged since the industrial revolution.

With reprogrammable machines and processing tools, we are returning to the concept of general-purpose machines. In spite of the fact that a certain amount of intelligence is now built into the computers that control such machines, production workers' jobs are becoming more complex as skills requirements have expanded in scope. The need for production workers to routinely handle preventive maintenance is a case in point. As Hirschhorn comments (1986), "The new cybernetic machines create new sources of error and failure with which only skilled workers, ready to learn and adapt to new production conditions, can contend." However, not all jobs benefit as a result of automation technologies. As programming of NC machines is increasingly done by engineers or NC programmers, the need for the traditional high-skill machinist is gradually decreasing.

The mass manufacturing era was characterized by an approach to work organization in which jobs were defined in a hierarchy of multiple job classes, each with a distinct wage rate. With the move towards multiskill and broader-scope jobs, this rigid system must give way to a system with fewer job types and a compensation plan where workers are paid based on their knowledge base. Such a compensation plan will then inherently provide incentive to all workers to seek and acquire the knowledge and skills necessary for high-performance automated factories. In this context, identifying and defining the knowledge base and skill set that should be considered as "ideal" will become a challenging and critical task for senior management. Automation reduces the direct influence that a worker may have on the output of a machine. Consequently, the traditional piece-rate incentive system must also be changed to account for significant factors other than output, such as the reduced downtime of machines through better maintenance.

Recently Adler (1992) has noted that there is evidence (based on the experiences of firms in the United States, Europe, and Japan) to support a generally optimistic view of the impact of technology on job content. He notes that workers in automated production environments need both upgraded skills and a new broadened role under the firm's division of labor in the manufacturing enterprise in order to be competent. (See Hirschhorn and Mokray [1992] for a case study based on Digital Equipment's experiences in this area.)

Finally we note that increased worker safety can be one of the main advantages of computer integrated manufacturing. As robots perform more and more of the jobs that are hazardous or that require arduous physical work, the work environment is becoming safer for human workers.

8.3.2 Training, Retraining, and Promotion Policies

The introduction of production automation invariably changes the nature of many jobs in a manufacturing company. In this context, the training and retraining of workers to achieve continuous upgrading of employee skills has become a critical element for the successful implementation of manufacturing automation technologies. As illustrated by the Toyota Production System (Ohno [1988]), the success of Japanese manufacturers in introducing automation can be attributed, in part, to Toyota's stress on employee training and retraining. By investing in training, the company secures a more versatile work force that is capable of efficiently operating new equipment in the changed manufacturing processes. Although training is expensive, it must be recognized that alternatives such as hiring new employees can be even more costly. These practices, of course, are consistent with and are aided by the policy of providing life time employment to production workers.

Programmable automation involves use of precise machines and processes, which in turn require more complicated work in performing proper maintenance of these machines. Increased use of electronics and computers also means that the skills of current maintenance workers often must be augmented, which requires both considerable retraining and a strengthening of selection criteria in hiring new workers. The need to maintain sophisticated computerized process equipment also requires that the jobs of production workers be enlarged to include preventive maintenance. These workers must also be trained accordingly. In fact, a policy that only includes worker retraining is inadequate, especially for "lead operator" jobs on the factory floor. The policy must also include worker promotion procedures that identify and promote the people with the right skills into the right jobs—much as firms do with management talent.

A mistake that must be avoided is to concentrate training efforts solely on the production and maintenance workers, while neglecting the training needs of salaried employees such as supervisors, engineers, and middle management. Given the increasing move to empower employees and to involve them in planning and implementing automation technologies, supervisors need training in not only how new technologies work but also in developing interpersonal and management skills. It is interesting to note that many engineers also need considerable retraining. For example, some engineers may have graduated a decade or so ago when they may have had little chance to work much with computers and much less with computer integrated manufacturing. Given the critical role that engineers need to play in the design, implementation, problem solving, and continuous improvement of computer-based automation, many engineers may require extensive retraining, perhaps in university sponsored educational programs. Similarly, middle management must also be retrained, both attitudinally and technically.

8.3.3 Worker Displacement

Automation technologies alter the production function and can result in higher productivity. Indeed, anticipated increase in productivity and associated lower costs are usually the primary motivations for the introduction of manufacturing automation. Unless a company is fortunate enough to be in a position of expanding its production volume, increased productivity also means a declining need for direct labor. Dealing effectively with worker displacement is therefore an important factor in the implementation of manufacturing automation. It should be noted, however, that increased productivity can make training relatively less expensive since the benefits of any skill improvement are leveraged by increased productivity through automation.

A decline in the need for labor does not always lead to layoffs. In fact, normal attrition and early retirement programs are two important strategies that can be used to deal with job displacement. Intraplant and interplant transfers of displaced workers, and the associated retraining, are other important actions available to prevent the loss of employment.

8.4 Planning and Implementation Considerations

As we discussed in Chapter 1, the evolution of manufacturing technology points to a consistent trend towards increased volume, increased variety, reduced cost, and increased flexibility. Given these developments in product and process technologies, we believe that the adoption of *computer integrated manufacturing (CIM)* over the long run by manufacturing firms of all sizes is unavoidable. However, as we noted in Section 8.2 of this chapter, the recent survey by Johansen et al. (1995) indicates that CIM is often not viewed as a competitive weapon or as a strategic necessity. Reasons for this include the previously noted difficulty in assessing benefits, as well as difficulty in understanding the potential for CIM on the part of nontechnical managers.

The Johansen et al. study (1995) also observed that integrated systems of manufacturing automation are being implemented predominantly in a "bottom-up" mode, with individual area or line managers undertaking technology projects based on their local needs. Ultimately, this practice will need to be changed, since automation technology is characterized by a high level of dependence on synergy. Accordingly, the benefits of implementing an integrated system are greater than the sum of the benefits of implementing the individual technologies.

Individual automation technologies, such as NC milling machines, are available "off-the-shelf" from vendors. It would be a mistake, however, to consider CIM as a "generic" system that can be purchased and plugged in. As discussed in Chapters 4 and 5, implementation of manufacturing automation technology invariably involves customization to meet the unique needs of the firm and to support a specific production process. For example, a software program that controls the NC machine and the automated material handling system in an FMS is, of necessity, unique to the product line and market needs of the firm. In general, the larger the firm, the more customization is needed. The observation that larger firms are more successful in the adoption of automation may in part be based on the superior ability of the larger firms to support such customization.

As we discussed in Chapter 5, the infrastructure technologies, such as the software control system noted above, should be viewed as enabling technologies that enhance the value of production automation technologies. Interestingly, the idiosyncratic nature of a software system can also give a certain degree of competitive advantage to the firm since, generally, it is very hard for a competitor to simply copy such a system. A case in point is the experience of the National Bicycle Company, a subsidiary of Matushita, in developing a CIM system using in-house capability that allowed the company to manufacture customized bikes in accordance to the unique specifications of individual customers. The company's competition, Bridgestone Company, tried to imitate the mass customization strategy by hiring a software vendor, Nintendo, to develop the needed CIM software. The difficulties in developing CIM software were such, however, that the project was scrapped by Bridgestone after two years and a multimillion dollar expenditure. Thus, in-house software development capability enabled the National Bicycle Company to enjoy a competitive advantage in the market for this mature product (Fisher [1993]).

Hayes and Jaikumar (1991) point out the importance of learning in successful implementations of automation technologies. They conclude that the companies that are best able to exploit the capabilities of a new technology are usually those that adopt it early, continually experiment with it (encountering and overcoming setbacks in the process), and keep updating their skills and equipment as the technology evolves.

Manufacturing automation can have a dramatic impact on multiple functions (operations, marketing, human resource management). Hence, its successful implementation requires commitment and consistent, ongoing support from the senior management of the company. Given the breadth of these technologies, the project team entrusted with the planning and implementation effort must be cross-functional involving participation of personnel from all functions at various levels in the organization. Moreover, as experienced in most successful product design projects, the technology team should ideally be led, especially in the case of major CIM project, by a "heavy-weight" manager who can wield the clout necessary to make implementation a reality.

Manufacturing automation technology is a comprehensive innovation and must be linked to the overall business objectives and strategy of the firm. The business, technical, and human resource aspects of a project must be carefully coordinated. A one-shot implementation of a complete automation system is neither practical nor advisable. An argument also can be made that automation should be implemented in a gradual incremental fashion. This allows the firm to start with those elements that are easier to implement and that have more concrete, visible, and immediate benefits. Having established credibility and having gained crucial implementation experience, other more difficult elements can then be tackled at a later stage. The cost reductions gained from the implementation of the earlier elements can also be used to justify the investment required for implementing the later elements.

Although it is impossible to dictate a list of standard steps to follow in implementation, a generally accepted set of steps includes (Czernik and Quint [1992])

1. Define objectives.
2. Form cross-function project team and initiate project.
3. Identify and prioritize manufacturing system needs.
4. Estimate costs and benefits and evaluate project justification.
5. Prepare system design (general architecture and detailed design).
6. Implement and integrate automation technologies.
7. Provide the necessary training and retraining.
8. Begin operation.
9. Monitor and analyze progress.
10. Develop ideas for system extension and repeat the above steps.

In most companies, these steps do not always get executed in a sequential fashion. Rather, iteration over time and among alternatives is inevitable. As we noted in Chapter 2, in the context of new product development, implementation of manufacturing automation technology requires a structured, cross-functional, consistent, and rational management approach. The ideas and concepts introduced in this book are directed toward providing input to support this approach.

8.5 Summary and Conclusions

This chapter has introduced two management challenges which are fundamental to the successful implementation of automated technology: *performance measurement* and *investment justification.* In both cases, methodologies for addressing these challenges have been reviewed. The ABC method in conjunction with operations performance measurement can provide accurate estimates of the costs and benefits of introducing process innovation. The investment/option approach to justification can provide a more complete picture of how these costs and benefits will accrue over time and thus support analysis of the decision to make the initial technology investment. The chapter has also briefly reviewed issues concerning human resource management in automated environments and the overarching challenge of implementation.

The issues and methods discussed in this chapter can only provide input to what is ultimately a strategic policy decision: adoption of automated process technology. Successful use of manufacturing automation will require in-depth knowledge of current and emerging technology trends and an understanding of the management challenges that have been discussed throughout this book. There is no doubt, however, that future competitiveness will depend on the ability of companies to come to terms with changing technology and to find ways to use it effectively.

Study Questions

1. How has product costing been affected by the implementation of manufacturing automation?
2. What is a cost driver in the *activity-based costing (ABC)* method? Give an example how it is affected by the introduction of automation?
3. What are the major noncost performance attributes that should be measured in a manufacturing environment? How does automation affect them?
4. What are the major costs and resource requirements for the implementation of a CIM system?
5. The level of implementation of manufacturing automation has consistently lagged behind predictions. Explain why this is so.
6. "The *discounted cash flow (DCF)* method is inappropriate for the justification of investment in factory automation." Do you agree or disagree? Why?
7. How has automatiom changed the character of work in a factory? What are the long-term implications of these changes?
8. The "lights out" factory is the ultimate goal of factory automation. Do you agree or disagree? Why?
9. What is the role of learning in the successful implementation of factory automation?

References

Adler, P. S. (1992), *Technology and the Future of Work,* Oxford University Press, Oxford.

Banker, R. D., S. M. Datar, and S. Kekre (1988), "Relevant Costs, Congestion and Stochasticity in Production Environments," *Journal of Accounting & Economics,* 10, pp. 1–27.

Cohen, M. A. and R. Halperin (1986), "Optimal Technology Choice in a Dynamic-Stochastic Environment," *Journal of Operations Management,* Vol. 6, no. 3, pp. 317–331.

—— (1984), "Computerized Manufacturing Automation: Employment, Education and the Workplace," Report of the Office of Technology Assessment, Government Printing Office, Washington, DC.

Fine, C. H. (1989), "Developments in Manufacturing Technology and Economic Evaluation Models," in *Logistics of Production and Inventory,* eds. S. C. Graves, A. Rinnooy Kan, and P. Zipkin, North Holland.

Fine, C. H. and R. M. Freund (1988), "Optimal Investment in Product-Flexible Manufacturing Capacity," Working Paper, MIT Sloan School of Management, Cambridge, MA.

Fisher, M. (1993), "National Bicycle Company," Case material of the Wharton School, University of Pennsylvania, Philadelphia.

Hayes R. and R. Jaikumar (1991), "Requirements for Successful Implementation of New Manufacturing Technologies," *Journal of Engineering and Technology Management,* Vol. 7, pp. 169–175.

Helfgott. R. B. (1988), *Computerized Manufacturing and Human Resources,* Lexington Books, Lexington, MA.

Hirschhorn, L. (1986), *Beyond Mechanization: Work and Technology in a Post-Industrial Age,* MIT Press, Cambridge, MA.

Hirschhorn, L. and J. Mokray (1992), "Automation and Competency Requirements in Manufacturing: A Case Study," Chapter 2 in *Technology and the Future of Work,* ed. P. S. Adler, Oxford University Press, Oxford.

Huchzermeier, A. and M. A. Cohen (1996), "Valuing Operational Flexibility under Exchange Rate Risk," *Operations Research,* Vol. 44, no. 1, pp. 100–113.

Jaikumar, R. (1993), "200 Years to CIM," IEEE Spectrum, September, pp. 26–27.

Johansen, J., U. S. Karmarkar, N. Nanda, and A. Seidmann (1995), "Computer Integrated Manufacturing: Empirical Implications for Industrial Information Systems," Working Paper, Simon School of Business Administration, University of Rochester, NY.

Kaplan, R. S. (1990), "Limitations of Cost Accounting in Advanced Manufacturing Environments," Chapter 1 in *Measures for Manufacturing Excellence,* Harvard Business School Press, Boston, MA.

Kaplan, R. S. and A. A. Atkinson (1989), "Justifying Investments in New Technology," Chapter 12 in *Advanced Managerial Accounting.*

Maffei, M. J. and J. Meredith (1995), "Infrastructure and Flexible Manufacturing Technology: Theory Development," *Journal of Operations Management,* 13, pp. 273–298.

McIvor, R., J. D. Martin, H. Matsuo, and J. Ng (1995), "Profiting from Process Improvements in the New Semiconductor Manufacturing Environment," Working Paper, University of Texas, Austin.

McLoughlin, I. and J. Clark (1994), *Technological Change at Work,* 2nd ed., Open University Press, Philadelphia, PA.

Milgrom, P. and J. Roberts (1990), "The Economies of Modern Manufacturing: Technology, Strategy and Organization," *The American Economic Review,* Vol. 80, no. 3, pp. 511–514.

Miller, J. G. and T. E. Vollman (1985), "The Hidden Factory," *Harvard Business Review,* pp. 142–150.

Mitchell, G. R. and W. F. Hamilton (1986), "Managing R&D as a Strategic Option," *Research Technology Management,* May–June, pp. 15–22.

Ness, J. A. and T. G. Cucuzza (1995), "Tapping the Full Potential of ABC," *Harvard Business Review,* pp. 1–11.

Ohno, T. (1988), *Toyota Production System: Beyond Large-Scale Production,* Productivity Press, Cambridge, MA.

Slagmulder, R., W. Bruggeman, and L. Van Wassenhove (1995), "An Empirical Study of Capital Budgeting Practices for Strategic Investments in CIM Technologies," *International Journal of Production Economics,* 40, pp. 121–152

Swamidass, P. M. (1994), "Technology on the Factory Floor II: Benchmarking Manufacturing Technology Use in the United States," Working Paper, Technology Management Center, Auburn University, Auburn, AL.

Wall, T. D., C. W. Clegg, and N. J. Kemp, eds., (1987), *The Human Side of Advanced Manufacturing Technology,* John Wiley & Sons, NY.

GLOSSARY

This glossary defines or explains the key terms associated with computer integrated manufacturing.

AGV (automated guided vehicle) AGVs are used on plant floors to move materials, tools, and other items. They are self-controlled or under the control of a computer and follow a specified path.

AI (artificial intelligence) Programmed intelligence in computers that imitates the capabilities of human beings, such as self-learning, adaptation, and reasoning. Expert system is a type of AI system that is particularly useful in manufacturing.

APT (automatically programmed tools) One of the main languages for programming numerically controlled machine tools and other NC/CNC equipment.

Application program or software A computer program that carries out a task specific to the users' needs.

ASCII (American Standard Code for Information Interchange) An industry-standard character code widely used for information interchange among data processing, communication, and associated manufacturing systems. It is a seven-bit code system prevalent in microcomputer hardware.

AS/RS (automated storage and retrieval system) A high-density rack storage system for rapid storage and retrieval of parts and tools; it may be computer-controlled.

ATC (automatic tool changer) A mechanism, usually under the command of part programs, that automatically changes the tool on NC machines.

Attribute A characteristic of an entity or object that identifies or describes it. For example, a part's attributes may be part number, material, weight, and so on.

Batch production The type of manufacture in which products are made in batches. Batch production of discrete parts and their assembly into end products benefits the most from CIM.

Bits The two binary digits 0 and 1. A bit is the smallest unit of digital data.
BOM (bill of materials) A list of parts and material data referring to the end product.
Buffer A temporary area for holding something. On the shop floor, a buffer is an area or space for storing items (blanks, semifinished or finished products, tools, etc.) until the next machine or material-handling device such as a robot is ready to fetch it.
Byte A set of eight bits, processed together as a unit; it can be a subset of a computer word. A byte represents one alphanumeric character in ASCII. It is also used to express a computer's memory size.

CAD (computer-aided design) This term is broadly used to denote the use of computer systems in designing components and end products. It includes all activities associated with product design, such as modeling, drafting, and analysis. In the beginning, it meant drafting only, but nowadays the letter D stands for design, which includes drafting as well. Modem CAD systems also help the designer carry out analysis for stress, vibration and noise, thermal deformation, and more to optimize the design.
CADD (computer-aided drafting and design) In the early stages of CAD development, this term was coined to emphasize the design capabilities of software packages beyond drafting. Now, since most packages include both design and drafting, the term CADD has become superfluous. CAD is a better terminology and is preferred to CADD. However, CADD is used sometimes to emphasize the drafting part of design.
CAD/CAM (computer-aided design and computer-aided manufacturing) This term was coined to highlight the link between CAD and CAM. In the 1970s, computer use in design and in NC programming developed separately. When it was realized that the data generated to represent a part in CAD could also be used for NC programming (i.e., in CAM), the term CAD/CAM was coined to emphasize this link. Later versions of both CAD and CAM systems began to offer the capability of transporting CAD data into the CAM arena. Such a capability is a standard feature of today's CAD/CAM products.
CAE (computer-aided engineering) The use of computer systems in functions that are essential to engineering a product. Examples of such functions are DFM considerations, modal analysis, material selection, and finite element analysis. CAE ensures that the product will carry out its intended function during normal operation.
CAM (computer-aided manufacturing) This term was coined to denote the second stage of computers' use in manufacturing, the first being CAD. In the early days of CAM, the term was limited to mean computer use in NC programming tasks only. Later its scope broadened to include other areas of production as well. Today, CAM represents all uses of computers in management, control, and operation of a manufacturing plant, except those in design and business functions. It includes tasks such as process planning, part programming, robotics, material handling, tool management, fixturing, inspection using CMM, and more. In fact, CAM is as broad as CIM, except that the latter emphasizes integration. CAM connotes piecemeal applications that may lead to *islands of automation*—an undesirable result for CIM.
CAPP (computer-aided process planning) Use of computer systems in process planning, which involves selecting processes and their parameters for embodying the design into a tangible product. CAPP bridges the gap between CAD and CAM.

Cell A production unit consisting of two or more workstations with the material handling units, storage buffers, and other associated devices necessary for their logical connection. The cell is an independent facility that can process most effectively a family of parts or models. When it is flexible, the term FMC (flexible machining or manufacturing cell) is used to denote the facility.

Cellular manufacturing The concept and practice of cell-based production.

CIM (computer-integrated manufacturing) Integrated use of computer systems in manufacturing. In its broadest sense, as in this book, CIM includes all the functions of operating a manufacturing business. It covers CAD, CAE, CAM, and all other uses of computers in managing the entire enterprise.

CIM wheel A conceptual wheel model developed by CASA/SME to portray the meaning of CIM.

CNC (computer numerical control) The technology and devices in which a computer numerically controls the process or equipment such as a machine tool. The use of computers makes NC systems more flexible since changes are made more easily.

Communication hardware Hardware especially designed or tailored for the needs of communication.

COMPACT II A source language used in programming NC machines.

Compiler A program, also called a language processor, that translates the statements of a high-level language into machine codes.

Computer language Language used in programming computers.

Computer program A set of commands a computer can understand; also called software.

Concurrent engineering The concept and practicc of various functions or departments working together, from the beginning, to engineer a product.

COPICS (communications-oriented production information and control system) A computer system developed by IBM for managing the various production functions in an integrated way.

CPU (central processing unit) The hardware responsible for interpretation and execution of instructions given to a computer.

Database A collection of interrelated data or information stored on a data storage device.

Data dictionary The database's table of contents. It defines the database and identifies the data items used by application and program modules. A necessary tool for the database administrator.

Data model A conceptual method of structuring data. The three common models are hierarchical, network, and relational.

DBA (database administrator or administration) The person or department in charge of data administration in an organization. Duties include the purchase and development of appropriate DBMS, design and control of the database or databases, responsibility for backup and recovery, and so on.

DBMS (database management system) A set of system software that facilitates the management and control of the database.

Dedicated automation Automation that is dedicated to a specific product, process, or task, and is difficult to modify for processing a different product. Also called hard automation or fixed automation.

Detroit-type automation Detroit is the hub of the U.S. automotive industry. The type of automation practiced in such facilities before the advent of programmable or soft automation is referred to as Detroit-type automation. Because of its inflexibility, this is also called hard automation.

DFA (design for assembly) The concept and practice of taking all elements into account at the product design stage to ensure ease of assembling.

DFM (design for manufacturability or manufacturing or manufacture) This term means that the design must be adapted to match the capabilities of the available production resources. With the use of computers as CAD/CAM systems, design's shortcomings in terms of manufacturability become apparent. The practice of DFM minimizes these shortcomings; it leads to a set of rules which, if followed at the product design stage, facilitate production.

DFMA (design for manufacturing and assembly) Consideration of all the relevant factors at the product design stage itself to achieve ease of both manufacture and assembly.

Discrete manufacturing The type of manufacture in which discrete components, subassemblies, and/or assemblies are the end products. The term is used to differentiate this type of operation from process-type manufacturing, in which production is continuous.

DNC (distributed numerical control) A system of interconnected NC equipment that allows downloading of NC programs from a central computer to computer control units dedicated to NC machines. DNC once was an acronym for direct numerical control, where it meant the control of a machine tool by a mainframe.

End effector A gripper or actuator at the end of a manipulator or robot arm.

Entity An object or thing about which data is stored.

Factory of the future A conceptual, hypothetical automated factory likely to be developed in the near future. Being futuristic, such a factory is like a mirage that moves further away as you reach it.

Field A specified area in a record to hold data.

File The storage space for a set of data.

Flat file A file with rows and columns for displaying the data.

Flexible automation The type of automation that can be modified easily, primarily through software, to accommodate design changes in the product within the constraints of a production facility. Also called soft automation. (See Detroit-type or hard automation.)

Flexible fixturing The process of designing, developing, or using a fixture that can be modified easily for other parts.

FMC (flexible machining or manufacturing cell) A flexible production facility to process a "family" of parts.

FMS (flexible machining or manufacturing system) A highly sophisticated production facility, costing millions of dollars, that can manufacture any product within a family. It comprises several FMCs with appropriate material-handling systems serving them. FMS can easily accommodate product variations since the system is primarily software-driven.

Forming Manufacturing processes in which part shape and size are formed in a cavity or otherwise, rather than machined off a blank.

GT (group technology) A concept in which parts' similarities are identified, and used to group the parts for benefit of design and manufacture.

Hard automation Same as Detroit-type automation or dedicated automation.

Hardware The physical items of a computer system.

High-level language A computer language, such as BASIC, C, or FORTRAN, that has been developed for convenience of human beings.

Host computer A networked computer through which communications among the rest of the computers and other devices take place.

IGES (initial graphics exchange specification) A standard developed by private companies and the U.S. National Bureau of Standards; initially released in 1981 and revised several times. IGES aids in the exchange of design data between CAD systems of different makes.

Inference engine Expert system module responsible for drawing conclusions based on rules and data.

Input device A device through which a user can communicate with computers.

I/O (input or output) device A device that can be used to send input or receive output from a computer.

Interface Physical or logical link between two devices or systems.

ISO (International Standards Organization) A United Nations agency charged with the task of developing and implementing standards worldwide.

Islands of automation Flexible automation trend of the 1970s in which individual operations or stations were automated, mostly computer-based, one at a time with little or no regard to similar efforts elsewhere in the company. It was thought that the individually automated workstations or processes, called islands of automation, could easily be interconnected later when needed. But this proved to be difficult, forcing early thinking about CIM.

JIT (just-in-time) A new approach to processing or producing items just before their use. JIT reduces the investment in inventory.

Knowledge based systems Computer systems, especially software, that allow development, storage, and execution of inference rules in addition to the usual data storage and manipulation. Also called expert systems.

Labor productivity An indicator of how well available human resources are being used. Measured as output per employee hour.

Ladder diagram A diagram useful in programming PLCs.

LAN (local area network) A facility, usually a combination of wiring, transducers, adaptor boards, and software protocols, that interconnects workstations, computers, machine tools, robots, and other digital devices in a CIM environment within a plant or several plants in close vicinity. LAN is a data communication network covering limited distance of a mile or two. It is proprietary and provides high-bandwidth communication over inexpensive media (usually coaxial cable or twisted pairs) as well as switching capability.

Language processor Software that translates or interprets computer languages.

Lean production The concept and practice of improving productivity by reducing or eliminating unnecessary resources.

Low-level language Computer language in a format that is easily understood by microprocessors, but difficult for humans.

Machine control unit The microprocessor-based control unit of a modem machine tool.

Machine language The language of digital computers based on 0 and 1.

Machine tool A machine such as a lathe that can be used to produce another of its own kind.

Machining center A modem machine tool with milling as the primary operation.

Mainframe Shortened form of mainframe computer.

MAP (manufacturing automation protocol) A communication standard specifically tailored for the needs of automation in manufacturing.

Mass storage External memory for storing large amounts of data and/or programs.

MCU (machine control unit) See machine control unit.

Microprocessor A miniature CPU on a single integrated-circuit chip.

Minis Clipped form for microcomputers.

MIPS (million instructions per second) A unit for measuring the processing speeds of CPUs.

MIS (management information system) Software programs especially designed to produce information in formats suitable for managers.

MODEM An acronym combining the terms modulator and demodulator. Modems use digital data to alter a signal so the signal can be transmitted or received over an analog medium such as a telephone line.

MRP (materials requirement planning) Planning for the materials required to manufacture a product. MRP is mostly computer-aided.

MRP II (manufacturing resources planning) Denotes the planning of all the resources, such as materials, machines, and operators, required for the manufacture of a product.

MTBF (mean time between failures) A measure of product reliability.

NC (numerical control) See numerical control.

Network topology The pattern of connection between points in a network. The common types of network are mesh (each point connected to all others), star, bus, and ring.

Numerical control A technique for effecting control through numbers and symbols. Initially, NC technology developed on its own, but later merged with computer technology. This resulted in rapid growth of computer use in manufacturing.

Object-oriented DBMS A database management software that organizes information using models that represent real-world entities as objects rather than records.

Object-oriented programming A programming technique in which sections of the program code and data are represented, edited, and used in the form of objects, such as graphic elements and window components, rather than as strict computer codes. Such an approach allows development of toolkits that facilitate programming.

Off-line programming The process of developing a program on a system that physically simulates the actual system on which the program will be run. This approach frees the actual system to continue to operate normally, thereby maintaining high utilization.

Operating system The group of software that handles a computer's housekeeping chores while using the hardware for applications.

Optical character recognition A computer-based, optics-oriented system that can recognize characters.

Output device A device such as a printer that receives the output of a computer's calculations, analyses, and the like.

Packaged-software Ready-made software.

Parity The number of 1s in an eight-bit character or other groups of bits, which may be odd or even. Parity check is a technique of ensuring error-free transmission. It can verify against one (or odd number of) error only.

Part A component that is either an end product on its own or is assembled into an end product. Also called workpiece.

Part programming The process of writing appropriate codes or programs for processing or machining the part on NC machinery.

Payload Maximum weight that can be carried at normal speed of operation; used as one of the specifications for robots and other machines.

Peripheral devices Devices such as a printer or mouse that are connected to the computer to improve its effectiveness.

Personal computer A desktop-sized computer for individual use. Also called microcomputer.

Point-to-point The direct connection between two communicating devices. Tool movement between two points with no consideration of the path locus.

Process capability A measurement of the spread in data collected during normal operation.

Process planning Planning for the steps necessary to manufacutre a part or product. The results of process planning are summarized on a route sheet that lists the sequence of operations, machines, and other pertinent information.

Product life cycle An expression to describe the length of time a product lasts in the market. It comprises the periods of conception, design, production, use, and, finally, termination when not needed by the market. Flexible technologies allow shorter product life cycles, which means more frequent design and model changes.

Production control Actions taken to remove the bottlenecks when actual production fails to follow the planned schedule. Production control also encompasses supervisory tasks necessary to ensure that the products are ready for shipment on due dates.

Productivity A measure of the effectiveness of various resources. Productivity is often compared with a baseline (past data) to assess how (relatively) well the resources are used. Alternatively, it can be quantified by the ratio between the value of outputs to that of the inputs.

Programmable logic controller A modem controller based on PC technology for controlling a process; its logic is programmable. It monitors parameter levels and initiates programmed actions whenever the levels rise or fall beyond the set limits.

Protocol A set of rules for communication.

Record Data about an entity or object. Files contain records of individual data items.

Relation Table of data about an entity type; the *join* operation connects data about different entity types.

Relational database management system A system based on a relational model in which data are stored in tables, called relations, that consist of rows (tuples) and columns (domains). Relationships among data items are explicitly specified as equally accessible.

R&D (research and development) Efforts involved in conceiving, researching, and developing a new product or major modifications in an existing one. It also includes efforts to keep the facilities up to date.

ROI (return on investment) A percentage value as the basis of selecting the best from possible alternatives. ROI is commonly practiced in industry for making economic decisions.

Route sheet A list of the processes, machines, and other pertinent information necessary to manufacture a part or product.

Rule-based system Another term for knowledge-based system.

Run (a program) Command or function that makes the computer process a given program.

Schema A definition of the database. These are of three types: *conceptual schema* defines the logical structure of the database; *internal schema* specifies how data are physically stored and accessed; and *external schema* describes how applications or users would see the data. Communication among the three schema is the task of the DBMS.

Shop-floor control Control of the various shop floor activities such as materials handling, resource utilization, and efficient routing by keeping track of the containers, tools, work-in-process inventory, and so on.

SIC An acronym for Standard Industrial Classification. SIC codes are used to classify industrial activities and products; manufacturing falls in the major group 20 through 39. Every type of manufactured product is allocated a four-digit code, of which the first two are in the 20–39 range.

Software Computer programs to make the hardware work or to carry out an application task.

Solid model The geometry of a part of end product, based on solid primitives, as defined in the CAD system.

Source program An original program or code that is translated in machine language by an interpreter or compiler prior to running it. The compiled version is usually saved as an object program code to avoid the need of retranslation.

SPC (statistical process control) The use of statistical techniques in measuring, analyzing, and controlling a process. SPC helps improve the process to its ultimate capability, thereby improving product quality.

SQC (statistical quality control) The use of probability and statistics in achieving and maintaining product quality.

Syntax A set of rules that describe the structure of statements allowed in computer communications. The commands and routines must be in the same syntax.

System An assemblage of interacting elements to carry out a specific task. Manufacturing is in essence a system. A system approach to manufacturing improves its overall performance.

System commands A set of commands to activate a CAD/CAM or any other system. As an example, typing RUN runs the current program.

System integrator A person or device whose role is to integrate the various modules of the system.

System software A set of software consisting of operating systems, language processors, and utility programs.

Teach pendant The control box operators use to program and guide the robot motions, which can be recorded by the control memory and replayed to repeat the cycle.

Terminal Any input/output device for entering information into a computer or receiving information from it.

Tool (or tooling) management The efforts of managing tool inventory to keep the tooling cost low. In modem plants consisting of FMCs and FMSs, tool inventory may become high, especially if the tools are dedicated to individual machines, and not shared.

TQM (total quality management) The practice whereby the entire organization, from the president down, is committed to, and involved in, continuous improvement of product quality.

Transfer line A production facility, usually for automobiles, in which the product is assembled while being moved on a conveyor system from the start end of the line to the finish end. Transfer line is an example of fixed automation.
Turning center A machining center whose primary operation is turning.

UNIX A multiuser, multitasking interactive operating system developed by AT&T Bell Laboratories. It has been widely used and significantly improved by user universities. UNIX is becoming increasingly popular in a wide range of commercial applications including manufacturing.

WAN (wide area network) A network that allows communication over long distances of large amounts of data. In CIM, WANs would connect two or more plants physically remote from each other.
WIP (work in process) Products in various stages of processing; a plant's work inventory awaiting processing.
Workstation A generic term for a place, equipment, process, or device where work is accomplished.

I N D E X

NOTE: Page numbers in **bold** refer to terms listed in the glossary